Lecture Notes in Computer Science 14181

The series Lecture Notes in Computer Science (LNCS), including its subseries Lecture Notes in Artificial Intelligence (LNAI) and Lecture Notes in Bioinformatics (LNBI), has established itself as a medium for the publication of new developments in computer science and information technology research, teaching, and education.

LNCS enjoys close cooperation with the computer science R & D community, the series counts many renowned academics among its volume editors and paper authors, and collaborates with prestigious societies. Its mission is to serve this international community by providing an invaluable service, mainly focused on the publication of conference and workshop proceedings and postproceedings. LNCS commenced publication in 1973.

Jérémie Guiochet · Stefano Tonetta ·
Friedemann Bitsch
Editors

Computer Safety, Reliability, and Security

42nd International Conference, SAFECOMP 2023
Toulouse, France, September 20–22, 2023
Proceedings

Editors
Jérémie Guiochet
University of Toulouse/LAAS-CNRS
Toulouse, France

Stefano Tonetta
Fondazione Bruno Kessler
Trento, Italy

Friedemann Bitsch
GTS Deutschland GmbH
Ditzingen, Germany

ISSN 0302-9743 ISSN 1611-3349 (electronic)
Lecture Notes in Computer Science
ISBN 978-3-031-40922-6 ISBN 978-3-031-40923-3 (eBook)
https://doi.org/10.1007/978-3-031-40923-3

This Springer imprint is published by the registered company Springer Nature Switzerland AG
The registered company address is: Gewerbestrasse 11, 6330 Cham, Switzerland

Preface

This volume contains the papers presented at SAFECOMP 2023, the 42nd International Conference on Computer Safety, Reliability and Security, held in Toulouse, France, in September 2023. The "European Workshop on Industrial Computer Systems, Technical Committee 7 on Reliability, Safety and Security" (EWICS TC7), established the SAFECOMP Conference series in 1979. Since then it has contributed considerably to the progress of the state of the art of dependable computer systems and their application in safety-related and safety-critical systems, for the benefit of industry, transport, space systems, health, energy production and distribution, communications, smart environments, buildings and living.

This year, we received 100 high-quality submissions, and the international Program Committee (more than 50 members from 16 countries), selected 20 of them for presentation and for publication in the SAFECOMP 2023 Proceedings in the Springer LNCS Series (LNCS 14181). Each submitted article was single-blind reviewed by at least three independent reviewers; the decision on the conference program was jointly taken during the International Program Committee meeting in April 2023.

The program was enriched by three keynotes given by renowned speakers: "Designing Software for Certification" by Shankar Natarajan; "Securing the (Open Source) Software Supply Chain: Challenges and Opportunities" by Roberto Di Cosmo; and "Certification Use Reliance when Integrating Machine Learning Solutions in Safety Related and/or Critical Systems", by Emmanuelle Escorihuela and Gary Brown. As in previous years, the conference was organized as a single-track conference, allowing intensive networking during breaks and social events, and participation in all presentations and discussions.

For this edition, we had 6 high-quality workshops in parallel the day before the main conference: ASSURE, DECSoS, SASSUR, SENSEI, SRToITS and WAISE) These workshops differed according to the topic, goals and organizing group(s), and are published in separate SAFECOMP Workshop Proceedings (LNCS 14182).

We would like to express our gratitude and thanks to all those who contributed to make this conference possible: the authors of submitted papers and the invited speakers; the Program Committee members and the external reviewers; EWICS and the supporting organizations; last but not least, the Local Organization Committee who took care of the local arrangements.

We hope that the reader will find these proceedings interesting and stimulating.

Jérémie Guiochet
Stefano Tonetta

Organization

EWICS TC7 Chair

Francesca Saglietti — University of Erlangen-Nuremberg, Germany

General Chair

Jérémie Guiochet — LAAS-CNRS, Université de Toulouse, France

Conference Program Co-chairs

Jérémie Guiochet — LAAS-CNRS, Université de Toulouse, France
Stefano Tonetta — Fondazione Bruno Kessler, Italy

General Workshop Co-chairs

Erwin Schoitsch — AIT Austrian Institute of Technology, Austria
Matthieu Roy — LAAS-CNRS, France

Position Papers Chair

António Casimiro — Universidade de Lisboa, Portugal

Publication Chair

Friedemann Bitsch — GTS Deutschland GmbH, Germany

Publicity Chair

Barbara Gallina — Mälardalen University, Sweden

Industry Chair

Magnus Albert	SICK AG, Germany

Web Chair

Joris Guerin	Espace-Dev, IRD, Univ. Montpellier, France

Local Organization Committee

Marie Laure Pierucci	LAAS-CNRS, France
Isabelle Lefebvre	LAAS-CNRS, France
Karama Kanoun	LAAS-CNRS, France

International Programme Committee

Eric Alata	LAAS-CNRS, France
Magnus Albert	SICK AG, Germany
Uwe Becker	Drägerwerk AG & Co KGaA, Germany
Peter G. Bishop	Adelard, UK
Friedemann Bitsch	GTS Deutschland GmbH, Germany
Andrea Bondavalli	University of Florence, Italy
Jeroen Boydens	Katholieke Universiteit Leuven, Belgium
Jens Braband	Siemens Mobility GmbH, Germany
Simon Burton	Fraunhofer Institute for Cognitive Systems, Germany
António Casimiro	Universidade de Lisboa, Portugal
Marsha Chechik	University of Toronto, Canada
Peter Daniel	EWICS TC7, UK
Ewen Denney	SGT/NASA Ames Research Center, USA
Felicita Di Giandomenico	ISTI-CNR, Italy
Wolfgang Ehrenberger	Fulda University of Applied Sciences, Germany
John Favaro	Intecs, Italy
Francesco Flammini	Linnaeus University, Sweden
Jesús Friginal	SCASSI/JUNE, Spain
Barbara Gallina	Mälardalen University, Sweden
Georgios Giantamidis	United Technologies Research Center, Ireland
Janusz Górski	Gdańsk University of Technology, Poland
Lars Grunske	Humboldt University Berlin, Germany

Jérémie Guiochet	LAAS-CNRS, Université de Toulouse, France
Ibrahim Habli	University of York, UK
Maritta Heisel	University of Duisburg-Essen, Germany
Andreas Heyl	Robert Bosch GmbH, Germany
Yan Jia	University of York, UK
Phil Koopman	Carnegie Mellon University, USA
Youssef Laarouchi	Électricité de France SA, France
John McDermid	University of York, UK
Ganesh Pai	KBR/NASA Ames Research Center, USA
Philippe Palanque	ICS-IRIT, University of Toulouse, France
Yiannis Papadopoulos	University of Hull, UK
Michael Paulitsch	Intel, Austria
Peter Popov	City, University of London, UK
Andrew Rae	Griffith University, Australia
Alexander Romanovsky	Newcastle University, UK
Matteo Rossi	Politecnico di Milano, Italy
Juan Carlos Ruiz Garcia	Universitat Politècnica de València, Spain
Francesca Saglietti	University of Erlangen-Nuremberg, Germany
Behrooz Sangchoolie	RISE Research Institutes of Sweden, Sweden
Christel Seguin	Office National d'Etudes et Recherches Aérospatiales, France
Erwin Schoitsch	AIT Austrian Institute of Technology, Austria
Oleg Sokolsky	University of Pennsylvania, USA
Wilfried Steiner	TTTech Computertechnik AG, Austria
Mark-Alexander Sujan	University of Warwick, UK
Kenji Taguchi	UL Solutions Japan
Stefano Tonetta	Fondazione Bruno Kessler, Italy
Mario Trapp	Technical University of Munich, Germany
Elena Troubitsyna	KTH Royal Institute of Technology, Sweden
Martin Törngren	KTH Royal Institute of Technology, Sweden
Meine van der Meulen	DNV, Norway
Hélène Waeselynck	LAAS-CNRS, France

Sub-reviewers

Koorosh Aslansefat	University of Hull, UK
Gabriel Ballot	Électricité de France SA, France
Stylianos Basagiannis	United Technologies Research Center, Ireland
Sylvain Bertrand	Office National d'Etudes et Recherches Aérospatiales, France
Marc Carwehl	Humboldt University Berlin, Germany

Gold Sponsor

SICK AG

IMAGINARY

Silver Sponsor

Critical Systems Labs Inc.

Supporting Institutions

European Workshop on Industrial Computer Systems
Technical Committee 7 on Reliability, Safety and Security

Laboratory for Analysis and Architecture of Systems, Carnot Institute

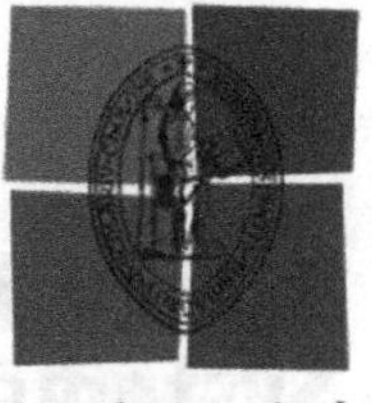

Université de Toulouse

Université Toulouse III Paul Sabatier

Fondazione Bruno Kessler

Austrian Institute of Technology

THALES

GTS Deutschland GmbH

Lecture Notes in Computer Science (LNCS),
Springer Science + Business Media

European
Network of
Clubs for
REliability and
Safety of
Software

Technical Group ENCRESS in GI and ITG

Gesellschaft für Informatik (GI)

Inside Industry Association

Informationstechnische Gesellschaft (ITG) im VDE

Austrian Computer Society

Electronics and Software Based Systems (ESBS) Austria

European Research Consortium for Informatics and Mathematics

Austrian Software Innovation Association

Contents

Model-Based Security and Threat Analysis

Safety of Autonomous Driving

Security Engineering

AI Safety

Neural Networks and Testing

Safety Assurance

Assurance Case Arguments in the Large: The CERN LHC Machine Protection System

Laure Millet[1], Simon Diemert[1(✉)], Chris Rees[1], Torin Viger[2], Marsha Chechik[2], Claudio Menghi[3], and Jeffrey Joyce[1]

[1] Critical Systems Labs, Inc., Vancouver, Canada
{laure.millet,simon.diemert,chris.rees,jeff.joyce}@cslabs.com
[2] University of Toronto, Toronto, Canada
{torinviger,chechik}@cs.toronto.edu
[3] University of Bergamo and McMaster University, Hamilton, Canada
claudio.menghi@unibg.it

Abstract. Most public assurance arguments are used to introduce, discuss, and present novel concepts and techniques related to structured argumentation. These examples often rely on generic claims such as "All hazards have been identified" and generic patterns of reasoning and are quite different from their fully developed industrial counterparts. This practical experience report describes a medium-size assurance case argument for the CERN Large Hadron Collider Machine Protection System expressed using Eliminative Argumentation. This assurance case with 509 nodes was created in approximately three months, validated in collaboration with CERN experts, and is now publicly available. We also report on our practical experience in creating this argument and reflect on the support provided by the features of the collaborative assurance case editor we used called *Socrates*.

Keywords: Assurance Case · Large Hadron Collider · Nuclear · Goal Structuring Notation (GSN) · Tools

1 Introduction

Producing high-quality Assurance Case (AC) arguments for industrial systems is complex and time-consuming, especially when the system design is highly innovative and experts cannot benefit from a previously established structure for the argument. Examples of such complex industrial arguments are not generally publicly available: they are typically proprietary and protected by non-disclosure agreements. The absence of representative examples of industrial arguments is a severe limitation to the scientific and industrial communities. Scientists and researchers need representative examples to evaluate new methods and techniques. Instead, they can only rely on small-scale showcase examples that often

J. Guiochet et al. (Eds.): SAFECOMP 2023, LNCS 14181, pp. 3–10, 2023.
https://doi.org/10.1007/978-3-031-40923-3_1

do not represent the challenges faced by practitioners. On the other side, industry cannot often assess the maturity and applicability of the different research results since they lack validation on industrial-scale examples.

This problem motivated a collaborative effort involving Critical Systems Labs (CSL), the University of Toronto, McMaster University, and the European Organization for Nuclear Research (CERN) to produce a representative assurance case argument to be publicly shared. We chose the CERN Large Hadron Collider (LHC) Machine Protection System (MPS) for this case study for four reasons. First, the MPS is a large safety-critical system that was the result of more than 10 years of effort involving highly knowledgeable and skilled personnel. As such, it likely shares some similarities with other safety-critical systems. Second, there is a substantial amount of openly available technical documentation about the MPS design [11]. Third, some of the authors from CSL were previously involved in technical reviews of the MPS during its commissioning and were already familiar with the system [3]. Finally, we could rely on the feedback from CERN experts to validate our argument.

We prepared this assurance case using the Eliminative Argumentation (EA) method and notation [4]. We selected EA since it explicitly supports the expression (and resolution) of doubt about the validity of the argument through the inclusion of *defeaters*. We have also used EA to prepare ACs for other safety-critical systems [1]. Tool support for EA-based arguments is also provided by *Socrates* [12], a collaborative assurance case tool developed by CSL.

This paper reports on the following contributions. First, we present our medium-size EA argument for the MPS and make it freely accessible by members of the scientific community and industry [2]. This argument consists of 509 nodes of different types, including 146 claim nodes and 105 defeaters. Second, we reflect on features the AC tool we used and their impact collaborative development of the argument by a geographically distributed team.

The rest of this paper is organized as follows. Following a brief overview of the CERN Large Hadron Collider MPS (Sect. 2) and an introduction to EA (Sect. 3), this paper briefly describes the structure of the assurance case argument for the MPS (Sect. 4). Then we discuss features of the assurance case editor that facilitated the collaborative development of the argument (Sect. 5). Finally, we present our conclusions (Sect. 6).

2 The Machine Protection System

Built by the European Organization for Nuclear Research (CERN), the Large Hadron Collider is the world's largest particle accelerator and collider. It is a 27-kilometer ring that contains thousands of superconducting magnets that increase the energy of two particle beams until they reach nearly the speed of light, and then make them collide. Collision experiments are performed to analyze phenomena related to the collision: testing theories and investigating unanswered questions in particle physics.

The hyper-accelerated particle beams generated during the experiments release a significant amount of energy which could damage to the system if their trajectories become unstable; this phenomenon is called "beam loss". The *Machine Protection System (MPS)* is a monitoring system composed of several interdependent sub-systems designed to ensure that the Large Hadron Collider does not become damaged during operation due to unstable particles [7]. When a beam loss is detected, the MPS is responsible for performing a "beam dump" (i.e., extracting all particles before hazardous scenarios occur) within 400µs of the occurrence of a beam loss event. A beam permit signal is used by components of the MPS to communicate that the LHC can continue to operate safely.

The MPS is comprised of four main sub-systems: the Beam Loss Monitoring System (BLMS), the Beam Interlock System (BIS), the Beam Dumping System (BDS) and the Safe Machine Parameter Controller (SMPC).

- The *Beam Loss Monitoring System* consists of approximately 4000 monitors distributed within the LHC to measure the beam loss. It is designed to detect and communicate beam losses to the Beam Interlock System within 80µs and uses the Safe Machine Parameter Controller to track beam losses events accurately. When intolerable beam losses are detected, it signals the Beam Interlock System to initiate a beam dump by withdrawing the beam permit.
- The *Beam Interlock System* takes between 20 µs and 120 µs to receive, process, and transmit a beam permit signal to the Beam Dumping System and transmits loss of the beam permit in at most 100 µs. It also determines whether the Beam Dumping System should initiate a beam dump due to other conditions, such as loss of redundancy of critical components.
- The *Beam Dumping System* is responsible for extracting the beams from the LHC's rings without damaging the system. The BDS directs the beams towards a large graphite sink designed to absorb the beam's energy. Components called "kicker magnets" are used to divert the beams from the main LHC ring towards the sink. The magnets also disperse the beams into smaller clusters to reduce the energy density when beam reaches the sink.
- The *Safe Machine Parameter Controller* computes Safe Machine Parameters used by the other systems to identify dangerous or spurious beam losses within the LHC.

Considering the consequence of a failure of the MPS, engineers need to ensure that risk is properly mitigated. The Eliminative Argumentation (EA) method and notation for preparing ACs can support engineers in this activity.

3 Eliminative Argumentation for Assurance Cases

Eliminative Argumentation (EA) [4,5] is a graphical notation to express assurance cases. EA allows engineers to reason about properties such as safety or security and and their confidence in the argument. It is similar to *Goal Structuring Notation* (GSN) [6,8]. Like GSN, EA uses a directed acyclic graph to organize the argument structure. EA and GSN arguments both approximate a tree data structure that starts with (at the "root") a high-level claim about the system that is decomposed into sub-claims until it can be directly supported by evidence, thus providing traceability between evidence and the claims. Unlike GSN, EA enables analysts to express "doubts", a.k.a., *defeaters*, they might have in the validity of claims, evidence, or inferences in an AC. EA is founded on the notion of 'defeasible reasoning' where confidence in the validity of a claim (or evidence or inference) increases as reasons to doubt that claim are resolved. Thus, EA defeaters can express doubt about claims, evidence, if it does not accurately support its parent's claim, and inference rules to identify doubts one may have about the the soundness or completeness of the argument.

EA enables analysts to use different types of nodes [1] to build an argument: *claim nodes* express statements that must be supported by further argumentation to demonstrate their validity; *evidence nodes* express observations, data, or artifacts that support the argument; *strategy nodes* describe the approach used to organize a collection of nodes; *inference rules* describe how to logically combine a collection of nodes; *context nodes* provide background information or missing details that may be necessary to understand the argument; *assumption nodes* list conditions related to the system or its operational environment assumed to be true in the argument, and *defeater nodes* express doubts about the validity of an assurance argument.

In EA, confidence in the top-level claim is established by showing that reasons to doubt the case have been resolved. If a doubt not resolved, it is marked as "residual" and contributes to the overall doubt in the argument. In our experience, residual doubts are a helpful when communicating sources of risk with stakeholders, such as management, regulators, or customers [1].

4 The Machine Protection System Assurance Case

This section describes the AC for the Machine Protection System (MPS) created using the EA. The argument consists of 509 nodes and is publicly available [2] as both a PDF report and a machine-readable archive (JSON). Among the 105 defeaters that can be found in the argument, 10 expose unmitigated risks in the MPS that were confirmed to be valid by CERN experts.

Figure 1 presents the high-level structure of the EA argument for the MPS. The argument starts with a top-level claim (C0001) that asserts that "*The MPS protects the LHC against damage from potential beam losses*". This claim is decomposed into four sub-arguments ("branches"), one for each MPS sub-system (see Sect. 2).

For each sub-system, the corresponding sub-argument argues that design is acceptably safe. The argument relies on various decomposition strategies. For example, Fig. 2 presents a portion of the argument that decomposes the claim C0030, that states that "*The BIS will transmit loss of the beam permit to the BDS in less than 100 microseconds*", via strategy S0654, by arguing over the foreseeable failure modes that may block, delay or otherwise interfere with the transmission of a beam dump request from the BIS to the BDS. These failure modes are represented by the defeater nodes D0031, D0036, D0438, and D0512.

Each defeater branch is followed by a sub-argument that addresses how the associated risk is mitigated (not shown in Fig. 2). For example, the sub-argument

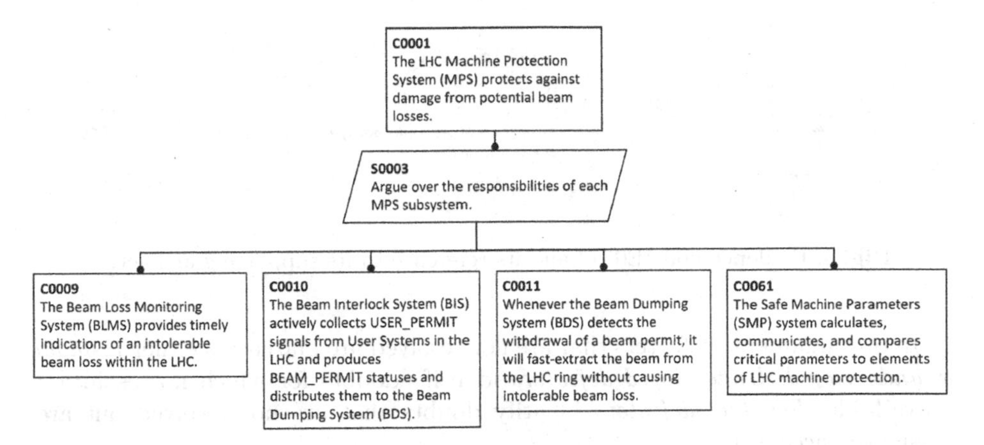

Fig. 1. High-level argument structure for the Machine Protection System.

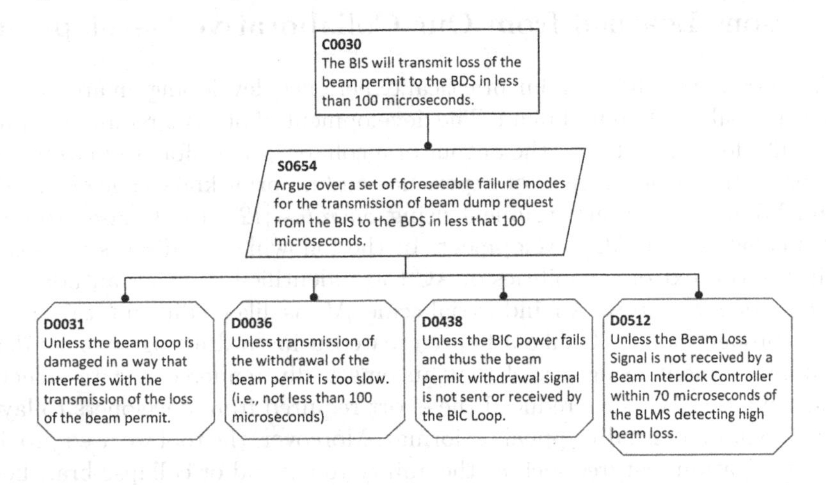

Fig. 2. A portion of the argument for the Beam Interlock System.

rooted at D0031 takes into account the fail-safe design of the mechanism responsible for the transmission of the beam permit including redundancy of critical components.

Through the argument, each claim is either refined into a set of evidence nodes or defeaters. Each evidence node is associated with an artifact that directly references specific details of the MPS design that justifies why the risk is sufficiently mitigated. For example, Fig. 3 shows that the evidence E0543 ("*In the event of one or all transmission lines being damaged, the bean permit loop will have no 10MHz signal or noise and subsequently result in the request for a beam dump*") is contained in the document "Architecture of Beam Interlock System". In the AC tool *Socrates*, the user can click to open the corresponding document on CERN's file server.

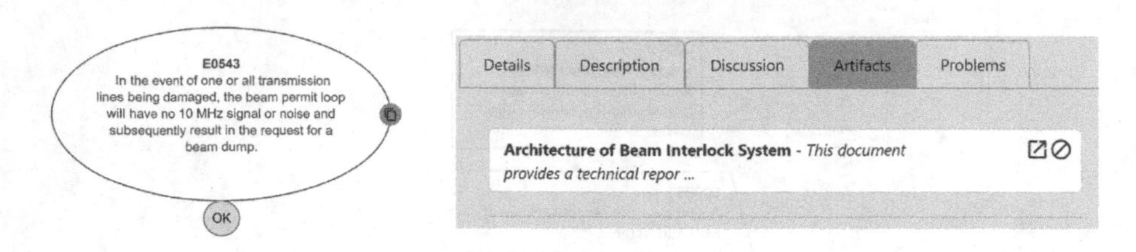

Fig. 3. Evidence node E0543 and its reference to its supporting artifact.

In our argument, defeaters that are not resolved are marked as either *residual* or *undeveloped*. *Undeveloped defeaters* identify doubts for which no resolution is available. *Residual defeaters* identify doubts that are not resolved but are considered acceptable.

5 Lessons Learned from Our Collaborative Development

In this section, we reflect on our practical experience developing an argument by a geographically distributed team. The development of our AC required approximately 92 effort-days. It was the output of a collaborative effort from four safety engineers with various degrees of experience and no prior knowledge of the MPS system. We developed our argument using *Socrates* [12], a web-based tool that enables collaborative AC development. In the following, we discuss our experience in the context of capabilities of AC tools identified by other authors [9].

Navigation Features. An industrial-scale AC is likely difficult to navigate without proper tooling. Unlike a figure created using a drawing tool, we developed our argument using a tool that automatically rendered a data structure representing the AC. This reduced the effort required from developers to layout the argument in a visually appealing format. Moreover, the tool we used provides several navigation features such as the ability to expand or collapse branches of the argument or the search function to navigate to specific sub-trees. These navigation features were increasingly useful as the size of the AC increased.

Collaborative Development. Collaborative tools, such as *Socrates*, enable developers to work on the AC in parallel, by editing the argument simultaneously. This feature was effective in preventing conflicts as this project's team was distributed over two continents and four different time zones. Our team extensively used a feature that enables discussion between developers by attaching comments to nodes in the argument.

Natural Language Processing. From our experience, we learned that tools with a grammar checker or other basic language processing might be helpful during the development of an AC. For example, to avoid ambiguous phrasing. The tool we used offered no language processing support. Like many real-world ACs, specialized terminology is used in the LHC MPS argument. Inconsistent use might be avoided by the presence of a built-in glossary.

Linking Artifacts. This feature allows developers to link nodes in the argument to external artifacts. As with many large ACs for complex systems, one of the main challenges in the development of the MPS argument was mapping aspects of the argument to details in the documentation. We linked AC nodes to the documentation that defines or acts as evidence for these nodes.

Version Control. This feature provides the ability for developers to save stable versions of the AC. When preparing and maintaining an AC, developers often restructure the argument to improve its understandability or to reflect changes to the system. In our project, we found this version control useful, especially to consult previous versions of the AC.

Rule-based Static Analysis. This feature provides rule-based checks on the argument structure that address both the syntactic correctness of the argument and detects common "anti-patterns" in the argument. For most of the project team, this was their first experience with the development of a large AC using EA. Static analysis does not ensure that an argument is logically sound and complete, but it helped developers minimize the number of basic errors. The real-time feedback provided by this feature also helped team members learn EA.

Impact Analysis. Modifying, adding or deleting a node may affect the structure of the entire sub-argument in which it is contained. In such cases, an analyst may find it necessary to apply the same modification to other unrelated sub-arguments that share a similar structure, in order to ensure the overall cohesion of the argument. Beyond basic search and linking functions, the tool we used did not offer sophisticated impact analysis functions. However, had these been available our experience suggests they would have reduced the effort associated with argument maintenance, particularly in the later stages of development.

Metrics. The tool we used reports metrics about the argument, such as a timeseries showing how the number of nodes in the argument (stratified by node type) changed over time. For example, we used the rate at which new defeater nodes were created to understand when the developers had reached a level of technical understanding sufficient to pose probing questions to the CERN experts participating in this project. Overall, this feature helped us to gauge our understanding of the LHC's MPS and the maturity of the AC.

6 Conclusion

This paper presented a medium-size assurance case argument for the CERN Large Hadron Collider Machine Protection System that is expressed using Eliminative Argumentation. We reported on our practical experience in creating this argument and reflected on the support provided by the capabilities of the collaborative assurance case tool that we used.

In future work, we plan to develop a framework that can monitor the activity of the analysts and provide practical recommendations as mandated by our manifesto, which proposes using assurance cases as data [10]. We are also empirically studying assurance case development.

Acknowledgements. We acknowledge the support of the Natural Sciences and Engineering Research Council of Canada (NSERC) [funding reference numbers RGPIN-2022-04622, DGECR-2022-0040, RGPIN-2015-06366].

We thank Mateo Delgado and Rolf Lippelt for their contribution, and CERN experts Jan Uythoven Markus Zerlauth, and Lukas Felsberger for their review and feedback on the AC.

References

1. Diemert, S., Joyce, J.: Eliminative argumentation for arguing system safety - A practitioner's experience. In: International Systems Conference (SysCon), pp. 1–7. IEEE (2020)
2. LHC MPS argument. www.cds.cern.ch/record/2854725/files/ (02 2023)
3. Ghafari, N., Kumar, R., Joyce, J., Dehning, B., Zamantzas, C.: Formal verification of real-time data processing of the lhc beam loss monitoring system: a case study. In: Salaün, G., Schätz, B. (eds.) FMICS 2011. LNCS, vol. 6959, pp. 212–227. Springer, Heidelberg (2011). https://doi.org/10.1007/978-3-642-24431-5_16
4. Goodenough, J., Weinstock, C., Klein, A.: Eliminative argumentation: A basis for arguing confidence in system properties. Tech. Rep. CMU/SEI-2015-TR-005, Software Engineering Institute, Carnegie Mellon University, Pittsburgh, PA (2015). www.resources.sei.cmu.edu/library/asset-view.cfm?AssetID=434805
5. Goodenough, J.B., Weinstock, C.B., Klein, A.Z.: Eliminative Induction: A Basis for Arguing System Confidence. In: International Conference on Software Engineering (ICSE), pp. 1161–1164. IEEE (2013)
6. Group, A.C.W.: Goal structuring notation community standard - version 3. Tech. rep., Safety Critical Systems Club (2021). www.scsc.uk/r141C:1?t=1
7. Holzer, E.B., et al.: Beam loss monitoring for LHC machine protection. Phys. Procedia **37**, 2055–2062 (2012)
8. Kelly, T.P.: Arguing safety-a systematic approach to safety case management. DPhil Thesis York University, Department of Computer Science (1999)
9. Koopman, P.: The UL 4600 Guidebook: What to Include in an Autonomous Vehicle Safety Case. Carnegie Mellon University (2022)
10. Menghi, C., Viger, T., Di Sandro, A., Rees, C., Joyce, J., Chechik, M.: Assurance Case Development as Data: A Manifesto. In: International Conference on Software Engineering: New Ideas and Emerging Results (ICSE-NIER), pp. 135–139. IEEE/ACM (2023)
11. CERN Website. www.cern.ch (04 2022)
12. Socrates. www.safetycasepro.com/welcome (04 2022)

Identifying Run-Time Monitoring Requirements for Autonomous Systems Through the Analysis of Safety Arguments

Richard Hawkins(✉) and Philippa Ryan Conmy

Assuring Autonomy International Programme, Department of Computer Science, University of York, Deramore Lane, York, England YO10 5GH,
{richard.hawkins,philippa.ryan}@york.ac.uk

Abstract. It is crucial that safety assurance continues to be managed for autonomous systems (AS) throughout their operation. This can be particularly challenging where AS operate in complex and dynamic environments. The importance of effective safety monitoring in ensuring the safety of AS through-life is already well documented. These current approaches often rely on utilising monitored information that happens to be available, or are reliant solely on engineering judgement to determine the requirements. Instead, we propose to use a systematic analysis of the safety case as the basis for determining the run-time monitoring requirements.

Safety cases are created for AS prior to deployment in order to demonstrate why they are believed to be sufficiently safe to go into operation. The safety case is therefore inevitably based upon predictions and assumptions about the system and its operation which may become untrue due to changes post-deployment. Our approach identifies specific run-time monitoring requirements for AS based upon a dialectic analysis of the safety case developed for the system. The advantage of the approach described is that it is systematic (through explicit consideration of elements of the safety case for the AS) and provides a way to justify the sufficiency of the resulting monitoring requirements (through creating explicit links the safety claims made about the AS).

Keywords: Monitors · Safety arguments · Run-time

1 Introduction

It is crucial for the assurance of safety-related and safety-critical systems that the safety of the system can be demonstrated throughout its entire operational

This work is funded by the Assuring Autonomy International Programme https://www.york.ac.uk/assuring-autonomy. Parts of this work were undertaken as part of the "LOPAAS" project of the Fraunhofer-Gesellschaft "ICON" programme.

J. Guiochet et al. (Eds.): SAFECOMP 2023, LNCS 14181, pp. 11–24, 2023.
https://doi.org/10.1007/978-3-031-40923-3_2

life. Many safety assurance activities (such as design analysis and testing) can be undertaken during development prior to the deployment of the system to operation. However, it is also important that safety continues to be managed and assessed post-deployment, particularly to identify and respond to unanticipated changes in the system or the operating environment. As part of this it is important to ensure that effective monitoring is in place during operation that can identify when a response is required to ensure acceptable safety is maintained. Effective monitoring is important for all safety-related systems, however it is of particular importance for autonomous systems (AS) [3,7], since it is expected that AS will experience more change during operation. This may be in the form of changes to the AS itself (updates to machine learning models or unanticipated failure modes of system components), or changes in the complex, dynamic operating environment in which AS are required to operate.

Other work has previously discussed the need for safety monitors for AS. For example, in [6] the use of safety performance indicators (SPIs) is suggested as a way of defining safety metrics for an AS that can be monitored during operation. In [2], the authors propose the use of runtime monitors to assess assurance properties of AS by measuring confidence/uncertainty in those properties at runtime. And in [9], the use of probabilistic runtime risk monitors is proposed as a way of supporting dynamic risk assessment of AS during operation. A lot of literature relating to run-time monitoring is focused on highly situational monitoring to mitigate specific hazards. For example, [8] provides a method for ensuring safe distance between platooning trucks. In [5], the authors propose a safety concept, which monitors internal health and plausibility checking for an autonomous driving control system. However, underpinning the successful use of any safety monitors is the need to be able to identify and justify the selection of what is required to be monitored, and how the use of the information from that monitor can be shown to be effective in maintaining safety during operation. Whereas existing approaches rely largely on engineering judgement to define monitoring requirements, this paper describes an approach for identifying specific run-time monitoring requirements for AS based upon an analysis of the safety case developed for the system.

The advantage of the approach described in this paper is that it is systematic (through explicit consideration of elements of the safety case for the AS) and it provides a way to justify the sufficiency of the resulting monitoring requirements (through creating explicit links to the safety claims made about the AS). The paper is structured as follows; Sect. 2 introduces our approach based on the use of dialectic arguments; Sect. 3 discusses the activities required pre-deployment of the AS in order to identify run-time monitoring requirements; Sect. 4 then discusses how the monitors are used post-deployment to ensure the validity of the AS safety case is maintained; We draw conclusions and describe further work in Sect. 5. We use an illustrative example throughout the paper to illustrate our approach.

2 Background

Safety cases can be created for systems prior to deployment in order to demonstrate why they are believed to be sufficiently safe to go into operation [10]. The safety case for any system is therefore inevitably based upon predictions and assumptions about the system and its operation. Despite the best efforts of engineers when developing the safety case to ensure all predictions and assumptions are valid, some of these may turn out to be incorrect during operation, or may be correct at the point of deployment, but later become incorrect due to changes during operation. For example, during operation, upredicted emergent properties of the system may become evident, unanticipated changes may occur in the external environment, or the operational performance of system components may start to diverge from predictions. Safety assurance requires that confidence can be provided in the continued validity of the safety case post-deployment once the AS is in operation. In order to do this we need to understand what are the correct things to monitor that will provide the required information about safety case validity. We also need to know at what point the information obtained from the monitors indicates that the validity of the safety case may be undermined. This requires the definition of triggers associated with each monitor. The triggers are used to indicate that corrective actions are required to restore the validity of the safety case. The analysis that we propose in order to define the safety monitoring requirements for an AS is based around the use of dialectic argument techniques. Below we introduce dialectic arguments and explain how we use them in defining safety monitors for AS.

2.1 Dialectic Safety Arguments

Dialectic arguments provide a way of explicitly representing not just the argument and evidence that support the truth of the claims that are made (as is done in a conventional safety argument), but also a way of representing argument and evidence that could undermine the truth of those claims. Dialectic arguments are created by identifying challenges to elements of the safety argument (claims evidence, context, assumptions etc.). The challenges can take the form of either claims that, if true, undermine the argument (challenge claims) or evidence that undermines the truth of the argument (counter-evidence). The Goal Structuring Notation (GSN) standard [1] defines a way of documenting counter evidence as part of a safety argument. Figure 1 shows a simple example of argument challenges and defeat (a key to the GSN symbology is provided in Fig. 2). The diagram on the left hand side (a), shows a small GSN argument fragment regarding a claim that an algorithm is correctly implemented. The diagram on the right hand side (b), shows how various elements of that argument may be challenged. Challenges are represented by dotted arrows that link from the challenge claim to the element being challenged. In Fig. 1 challenges are shown to three elements of the argument: strategy S12, evidence E121 and claim C12. Challenges are represented as claims (CC12.1, CC121.1 and CC2.1 respectively). For challenges to be compelling they should be supported by some evidence (referred to as counter-evidence). In Fig. 1, an example of counter-evidence is shown as EC2.1. If a challenge presented to the safety case is determined to be valid, then the challenged element in the safety argument is said to be defeated. Figure 1 shows an example of how a defeated claim, C12, can be represented.

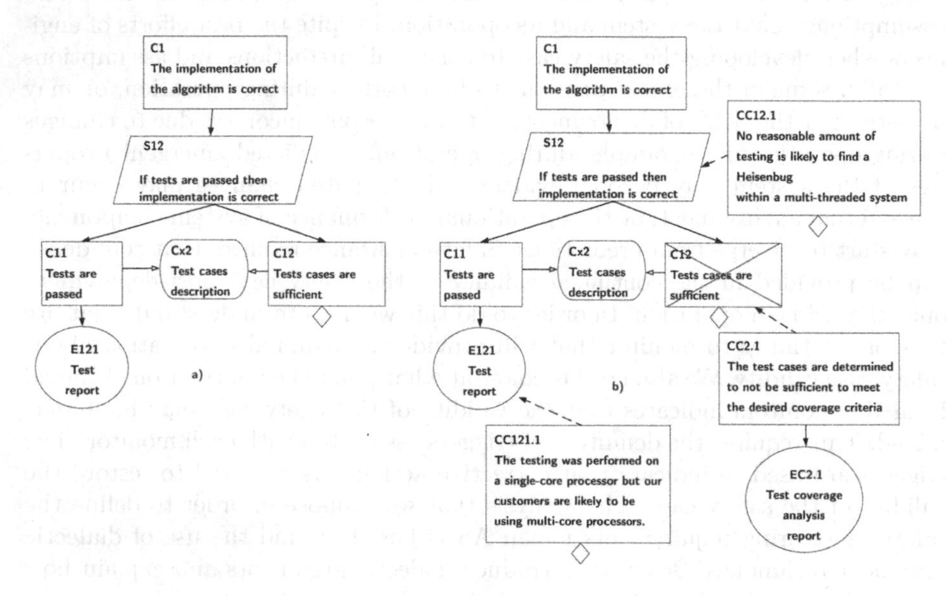

Fig. 1. A simple example GSN safety argument (a) and a dialectic argument showing how challenges and defeat of elements of the GSN structure can be represented (b).

In order to identify how the validity of an AS safety case may be undermined during operation, we propose that dialectic argumentation can be used to explicitly identify, prior to deployment of the AS, how the safety argument or evidence may be challenged at run-time. This enables detailed challenge claims and counter-evidence specific to the operation of the AS under consideration to be identified. As this dialectic argument focuses on potential run-time challenges, we refer to it as an operational dialectic argument. Our approach says that if, at any point in the lifecycle of the AS, the counter-evidence or challenge-claim we have identified in the operational dialectic argument exists, then the safety case is undermined, since elements of that argument become defeated. Therefore by monitoring explicitly at run-time for occurrence of this counter-evidence, we can have confidence that the safety case will remain valid during operation.

Once the operational challenges have been identified using the operational dialectic argument, the mechanisms for successfully monitoring and responding to the occurrence of that counter-evidence must be determined and implemented. We discuss this in Sect. 4. Firstly in the next section we look at how an operational dialectic argument can be created for an AS.

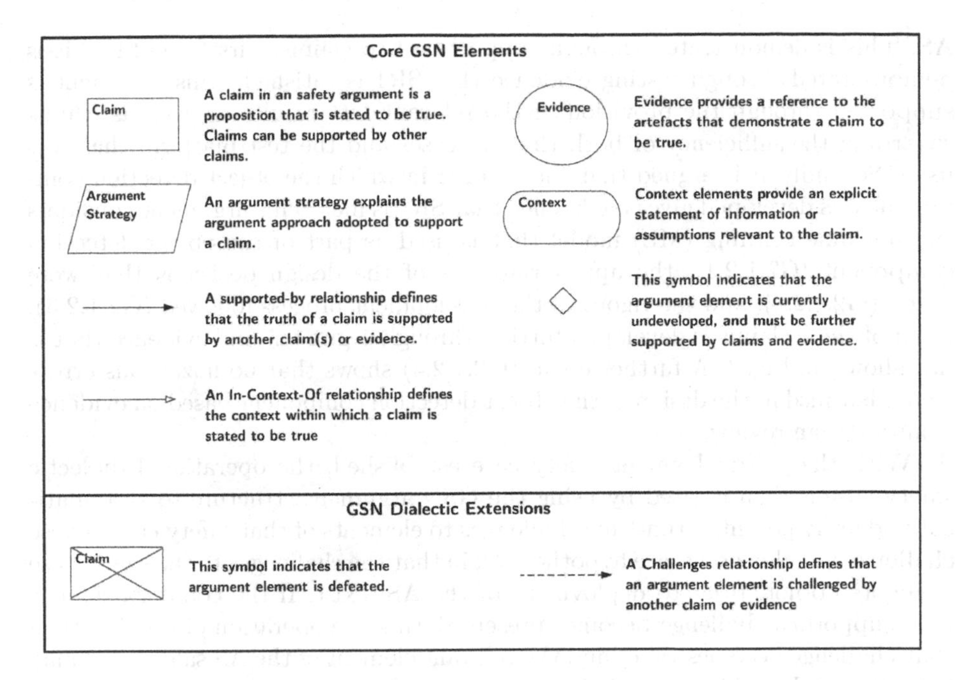

Fig. 2. A key to the GSN symbols used this paper

3 Pre-deployment

In this section we describe the activities required in our approach prior to deployment of the system in order to identify a sufficient set of run-time monitors and to be able to justify their sufficiency. This firstly requires the development of an operational dialectic argument, which is discussed in Sect. 3.1.

3.1 Operational Dialectic Arguments

The starting point for creating a dialectic argument for an AS is the AS safety case itself. This safety case will have been created prior to deployment of the AS during the development of the system to demonstrate, through a safety argument supported by evidence, that the AS is sufficiently safe to operate in its defined context. There is existing guidance that can be used to help in the creation of AS safety cases [4]. Figure 3 shows a simplified example of part of the safety case for an autonomous vehicle that will operate on public roads. The safety case is represented using GSN.

This part of the safety case shows the argument that a particular safety requirement (SR1) has been addressed through the design and implementation of a particular system component (in this case the object detection component of the vehicle). The object detection component in this case is responsible for identifying objects that are present in the operating environment of the

AS. This is demonstrated through supporting two claims. Firstly (G2.1.1) it is demonstrated through testing evidence that SR1 is satisfied. This argument is supported through the provision of the relevant test results, as well as claims regarding the sufficiency of both the test cases and the test platform that was used. Secondly, it is argued that the manner in which the object detection component was developed gives confidence that SR1 is met. This argument considers the machine learning (ML) model that is used as part of the object detection component (G2.1.2.1), the appropriateness of the design decisions that were taken (G2.1.2.2), and the rigour of the development process followed (G2.1.2.3). Each of these claims is developed further through argument and evidence that is not shown in Fig. 3. A further claim (G2.1.2.4) shows that no hazardous errors were identified in the design of the object detection component based on evidence from a design review.

With the pre-deployment safety case established, the operational dialectic safety case is then created by using the GSN argument structure to systematically identify potential run-time challenges to elements of that safety case. These challenges, at this point, are hypothetical, in that the challenges to the safety case elements do not, prior to deployment of the AS, exist. If the counter-evidence that supports a challenge becomes present during the operation of the AS, then that challenge becomes valid, and the relevant element of the AS safety case may be defeated. For this reason it is important to be able to know, at run-time, if any of this counter-evidence exists. Unless sufficient monitoring is put in place prior to deployment of the AS, then the counter-evidence that can defeat elements of the safety case may exist without the knowledge of the system operator. By identifying the potential run-time challenges prior to deployment, sufficient monitors can be put in place to identify the presence of counter evidence.

Figure 4 shows a simplified version of the argument fragment from Fig. 3 with the potential run-time challenges and counter-evidence identified (for clarity in the diagram only those argument elements with challenges are fully represented). This dialectic argument was created by systematically considering each element of the safety argument and considering whether any events that could foreseeably occur during operation of the AS could defeat that element. These challenges are captured as challenge claims, which are stated as propositions that would become true in the presence of particular counter-evidence arising during operation. For example, the claim G2.1 that the object detection component satisfies the requirement SR1 would be directly challenged at run-time if the performance that is actually being observed by the system in operation is seen not to meet that specified by the requirement. This has been captured by the challenge claim CC1 in Fig. 4. CC1 would be become a valid challenge during operation of the AS if there was evidence to show that this challenge claim was true. OpEv1 states the nature of the operation counter evidence that would support this. OpEv1 will therefore be used as the basis for a monitoring requirement in Sect. 3.2 to ensure that the presence of the counter evidence is known.

Another example provided in Fig. 4 relates to claim G2.1.1.3, that the test platform represents the actual AV platform on which the component is operating.

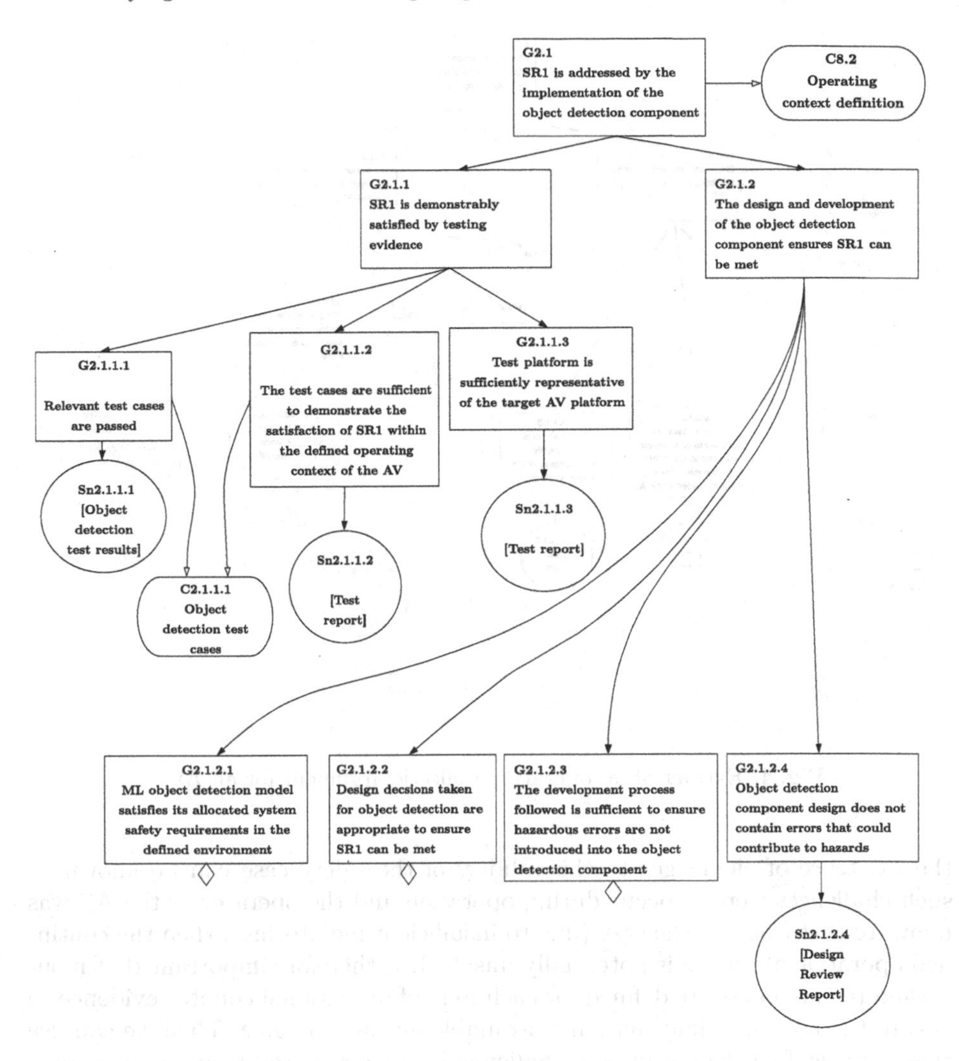

Fig. 3. GSN safety argument extract from a safety case for an autonomous vehicle.

This claim could be challenged if the AV platform is changed during operation such that the test platform is no longer indicative of the AV. Such changes could mean that the testing results are no longer valid. This has been captured by the challenge claim CC3 in Fig. 4. CC3 would be become a valid challenge during operation of the AS if there was evidence to show that such a change had indeed occurred to the vehicle in operation. OpEv3 identifies change reports as the mechanism by which evidence of this could be identified. Once again OpEv3 can be used as the basis for a monitoring requirement in Sect. 3.2.

Each of the items of operational counter-evidence from Fig. 4, along with other examples from related parts of the wider safety case are listed in Table 1. It is only by identifying the presence of this counter-evidence at run-time that

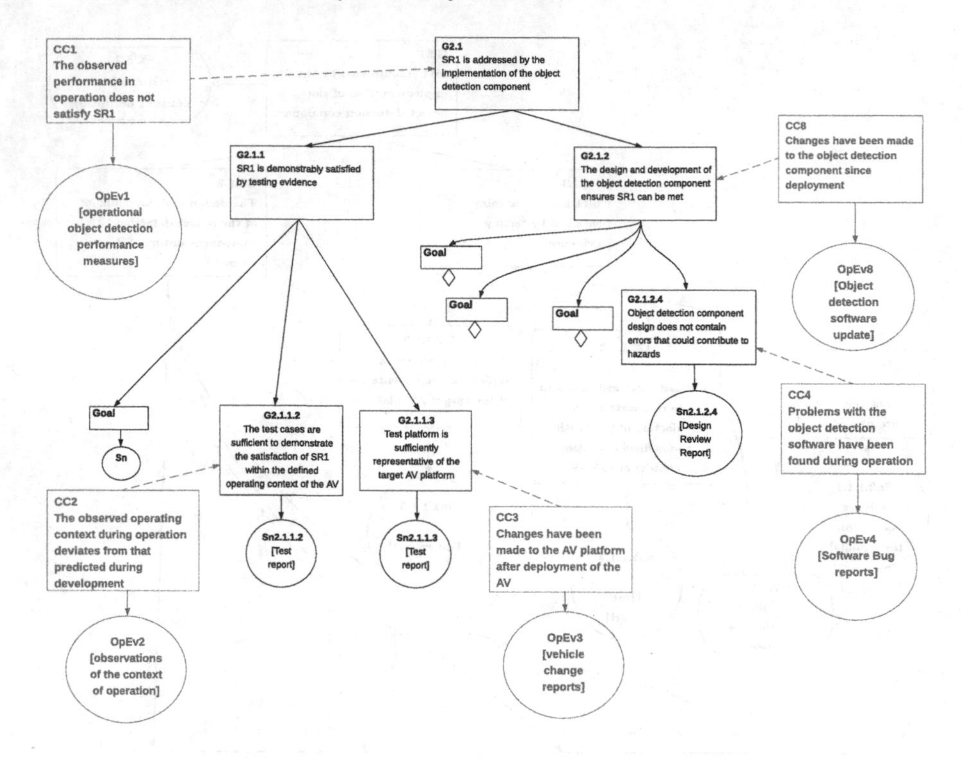

Fig. 4. Extract of an operation dialectic argument for an AV.

the existence of challenges to the validity of the safety case can be known. If such challenges were to occur during operation and the operator of the AS was unaware that this was the case (due to insufficient monitoring), then the continued operation of the AS is potentially unsafe. It is therefore important that monitoring requirements are defined for each item of operational counter-evidence to ensure the necessary information is available during run-time. Then we can use this to know if a safety argument challenge has occurred and hence an appropriate response can be enacted to ensure the safety of the AS is maintained. The next section discusses these monitoring requirements.

3.2 Identifying Run-Time Monitoring Requirements

For each item of operational counter-evidence identified, it is necessary to define:

- what needs to be monitored at run-time in order to be aware of the presence of the counter-evidence?
- what the criteria are that will be used for judging whether what has been monitored represents counter-evidence?
- what will be used as the trigger for determining the presence of counter-evidence?

Table 1, shows this information captured in the form of a table for the counter-evidence identified for the example system. This is captured as 'Monitor', 'Criteria' and 'Trigger' respectively. It can be seen in these examples that a diverse set of information must be monitored for the AV to ensure that all the relevant counter evidence will be identified. This includes things such as the number of missed detections across the fleet of AVs, information about changes to the vehicle platform, and detection of software errors. Each of these requires some mechanism to be put in place to obtain that information at run-time. This may include, for some of the information, the installation of physical sensors on the AS, but may also include more procedural mechanisms to check and record events. In this paper we consider all such mechanisms to be run-time monitors.

Table 1. Identifying requirements for run-time monitoring from counter-evidence

Op. Evidence	Monitor	Criteria	Trigger
OpEv1 - [operational object detection performance measures]	Number of missed pedestrian detections across the vehicle fleet	Missed detections observed per 1000 mi of operation	#misses/1000 mi exceeds rate reporting in test results by 10%
OpEv2 - [observations of the context of operation]	Input images arising from the camera for operation within defined ODD	Measurement of key parameters within images (e.g., light levels, surfaces, colours etc.)	Operational images outside of test distribution
OpEv3 - [vehicle change reports]	Physical changes to vehicle platform (such as updates to sensors, processors etc.)	Changes that may impact software performance	Notification of AV platform modification
OpEv4 - [Software bug report]	Software errors discovered during operation	Errors identified in object detection during operation	Notification of error found in object detection
OpEv5 - [AV incident reports]	Reports raised by operators of the vehicle	Incidents that relate to object detection	Notification of object detection incidents that may be hazardous
OpEv6 - [Camera maintenance records]	Calibration of camera	Time since last calibration	Greater than 6 months since last calibration
OpEv7 - [Camera drift measurements]	Drift measurement of camera images	Rate of drift in operation	Rate of drifting exceeds design assumption
OpEv8 - [Object detection software update]	Software version	Change to object detection software	Non-approved version of software running
OpEv9 - [Lidar error status]	Lidar health monitoring	Lidar availability	Lidar fails to provide output to object detection component

We can illustrate the monitoring requirements by considering examples from Table 1. The operational counter-evidence OpEv1 has been identified as operational object detection performance measures. It can be seen in Fig. 4 that this was identified as counter-evidence because it would support a claim that the observed performance of the component in operation does not satisfy the defined safety requirement. In order to know if this evidence has arisen, we determined that it would be necessary to monitor the number of missed detections that are seen to occur across the fleet of vehicles that are operating. The criteria that would be used to determine if counter-evidence is observed is the number of missed detection that are observed in a set period (in this case 1000 mi of operation). This can then be compared against the defined trigger to determine if this represents counter-evidence. In setting the trigger we must consider the context of the safety argument, and the specific element to which the counter-evidence would relate. In this case we must consider the number of misses per 1000 mi that would represent a deviation from the performance claimed in the safety case. Here, it would be when the number of misses exceeds what was demonstrated in the pre-deployment testing of the vehicle. Since in this case the performance seen in testing comfortably exceeded that required to satisfy the safety requirement (SR1), we do not need to consider every small deviation from this as a trigger for counter-evidence. Instead we have set a 10% threshold, which is still within the safety requirement, but indicates a threat to the validity of that aspect of the safety case that warrants assessment and potential action (see Sect. 4).

As a second example we can consider OpEv8, which has been identified as an update to the object detection software. It can be seen in Fig. 4 that this was identified as counter-evidence because it would support a claim that the object detection software that is running on the operational vehicle is not the software for which the claim about the soundness of the development was made. In order to know if this has occurred we need to monitor the version of the object detection software that is running on the operational system. The criteria is if the version number of the executing software indicates that there has been a change to the software, and this will trigger counter-evidence if the version that is running is not approved (since this could mean that we are no longer able to claim that the development of the object detection software is sufficient). Further investigation would again be required at this point to determine the nature and impact of the changes to the software.

Having identified the run-time monitoring requirements prior to deployment of the AS, those requirements then have to be managed post-deployment to ensure their effectiveness. This is discussed in Sect. 4.

4 Post-deployment

In this section we discuss in more detail the post-deployment monitoring mechanisms and how they might be implemented in practice. First we need to identify responsible organisations and create processes to keep track of the specified monitors and triggers (the effectiveness of these also needs to be justified in the safety case). Second we need to perform those processes when required.

4.1 Organisation and Continual Monitoring Processes

Once the AS is in-service, safety management processes will need to be put in place to review each of the identified criteria. Whilst the triggers provide clear thresholds beyond which the safety case can be considered no-longer valid, and some may be considered higher priority than others, we would expect all of the criteria to be reviewed regularly as part of planned safety case review. Additionally, we would expect a periodic review of the safety case in case additional run-time monitors have been identified, or could be improved over time, e.g., if there is new technology for monitoring or a changed legal requirement. Further, there may be other changes in the operating environment which impact on the safety case. We emphasise that the run-time safety monitoring is a continual process which evolves during the life of the project.

To be effective, a responsible organisation must be identified, who have the role of collating the specified run-time monitored information. The precise organisation, or organisations, required to support this may be dependent on the nature of the run-time monitoring requirements, the type of AS, and the regulatory regime. For our autonomous vehicle example this will include fleet owners, independent servicing centres and/or original manufacturers as well as individuals who own the vehicle. A further complicating factor would be the need to consider international boundaries, and national requirements.

4.2 Impact Assessment Process

Assuming that data indicates the trigger threshold has been reached, or a planned review shows concern about a particular trigger, there will need to be an impact assessment. This will first need to identify areas of potential impact on the safety case. This will differ depending on the type of issue. At a minimum, all branches which refer to a particular trigger would need to be reviewed, but the impact on the case could be more wide ranging, e.g., if a new hazard was discovered.

For example, if the trigger is *Op Ev1 - #misses/1000 mi exceeds rate reporting in test results by 10%*, the potential impact of this could include one or more of the following:

- the validity of the goals below G2.1 in Fig. 4 are challenges, as if the observed performance relating to SR1 is not as expected which implies a problem with the test and design evidence
- more specifically, the deployment environment may differ to that anticipated - implying a further issue for *Op Ev2 - observations of the operating context*
- the risk of collision with a pedestrian was higher than that anticipated due to one or more causal subsystems having lower performance than anticipated and each of these subsystems should be investigated for shortfall
- the *Op Ev1* trigger criteria was defined incorrectly, essentially a false positive problem
- the validity of one or more higher branches of the safety argument above G2.1 are undermined
- the scope of the impact could be an entire fleet of vehicles, or a subset of the fleet (for example operating in a particular region)

Alternatively, if we consider a low level trigger such as *Op Ev 7 - Rate of drifting exceeds design assumption*, the potential safety impact could extend to the performance of any subsystems using the camera. We should investigate any claims in the safety argument about the performance of the camera, which might be in different branches of the case.

Having assessed the impact of the operational counter-evidence on the safety case and the AS itself, it is necessary to either provide a rebuttal to this, or propose actions to address the impact. We should do this considering the impacted claims and their context, rather than considering the trigger in isolation. In practice a shortfall in measured performance may still be within tolerable safety limits and be localised to different parts of the case and this should be investigated. To continue our examples, if the trigger for *Op Ev1 - #misses/1000 mi exceeds rate reporting in test results by 10%* is breached but there have been no significant incidents or other subsystems are still compensating for shortfall in one systems performance (e.g. where there are diverse means of detecting pedestrians) then it may be possible to argue that the AS is still acting well within tolerable levels of risk. Alternatively, it could be that the trigger has only been observed in an isolated case with unusual environmental conditions which it could be argued are unlikely to be seen again. Such rebuttals should be explicitly document and could be added to the dialectic argument during operation to document the resolution of the challenge claim.

If the trigger *Op Ev 7 - Rate of drifting exceeds design assumption* is breached, it may be that this results in no significant performance alteration in any of the systems which are using the camera and/or that a more regular recalibration process can be automatically performed without impacting on the AS user. Again, such rebuttals could be added explicitly to the dialectic argument.

Where the monitors indicate that there is impact on the validity of the safety case but no rebuttals for the counter-evidence are identified, action must be taken at run-time to address this. Typical actions could include limitations on use, for example, avoiding using the AS in a particular environment until an issue is fully investigated and fixed, or limiting the operating speed. The safety case would need to be updated for this interim period, with the limitations made explicit at the top level of the case, and monitoring processes would continue. As an extreme example, action may require grounding of a fleet of AS. Once the safety issue was resolved, the safety case would again be updated to reflect this.

5 Conclusions and Further Work

The importance of effective monitoring in ensuring systems remain safe throughout their operational life is well understood. Due to the nature of AS, it becomes particularly important that the sufficiency of the monitoring that is in place at run-time can be justified. It is not sufficient therefore that the monitoring requirements for an AS be defined in an ad-hoc manner, or relying solely on engineering judgement. Instead it is important that a systematic and defensible method for deriving the run-time monitoring requirements is established. In this paper we have discussed how the continued validity of the AV safety case should be the focus of the run-time monitoring, with monitors used to identify if counter-evidence to the safety case occurs during operation, such that mitigations can be enacted. The approach we have described uses the concept of creating dialectic arguments as a way to systematically anticipate and identify operational counter-evidence, and thus to derive effective run-time monitoring requirements.

The work presented in this paper will lead to further related work. We have demonstrated how our approach can be applied in practice using an example from a self-driving vehicle and have provided simplified examples in this paper. Further evaluation of this approach will be undertaken through additional case studies undertaken by independent engineers. We will consider the application of the approach to different types of AS in other domains in order to show the generalisabilty of the approach. We will also seek to evaluate, through observation during operation, the effectiveness of the monitoring requirements arising from following our approach. The approach we have described in this paper can be used to enhance the safety case for the operation of the AV, by enabling a compelling argument to be made about the sufficiency of the run-time monitoring. Further work will develop and demonstrate the structure for such arguments.

We will use the results obtained from case studies in order to investigate further the nature of the run-time monitoring requirements that are derived from applying our approach. It was seen in this paper that the nature of the monitoring requirements can be diverse in nature, and consequently the monitoring and mitigation mechanisms are also diverse. We will look to characterise the run-time requirements for AVs as the basis for providing further guidance on their effective management.

References

1. ACWG: Goal Structuring Notation Community Standard. Tech. Rep. SCSC-141C v3.0, Safety Critical Systems Club (2021). www.scsc.uk/scsc-141C
2. Asaadi, E., Denney, E., Menzies, J., Pai, G.J., Petroff, D.: Dynamic assurance cases: a pathway to trusted autonomy. Computer **53**(12), 35–46 (2020)
3. Haupt, N.B., Liggesmeyer, P.: A runtime safety monitoring approach for adaptable autonomous systems. In: Romanovsky, A., Troubitsyna, E., Gashi, I., Schoitsch, E., Bitsch, F. (eds.) SAFECOMP 2019. LNCS, vol. 11699, pp. 166–177. Springer, Cham (2019). https://doi.org/10.1007/978-3-030-26250-1_13
4. Hawkins, R., Osborne, M., Parsons, M., Nicholson, M., McDermid, J., Habli, I.: Guidance on the Safety Assurance of Autonomous Systems in Complex Environments (SACE). arXiv preprint arXiv:2208.00853 (2022)
5. Hörwick, M., Siedersberger, K.H.: Strategy and architecture of a safety concept for fully automatic and autonomous driving assistance systems. In: 2010 IEEE Intelligent Vehicles Symposium, pp. 955–960 (2010)
6. Laboratories, U.: UL 4600: Standard for Evaluation of Autonomous Products (2020). Standard for safety, Underwriters Laboratories (2020)
7. Machin, M., Guiochet, J., Waeselynck, H., Blanquart, J.P., Roy, M., Masson, L.: SMOF: a safety monitoring framework for autonomous systems. IEEE Trans. Syst. Man, Cybern.: Syst. **48**(5), 702–715 (2016)

8. Reich, J., Sorokos, I., Papadopoulos, Y., Kelly, T., Wei, R., Armengaud, E.: Engineering of runtime safety monitors for cyber-physical systems with digital dependability identities. In: Computer Safety, Reliability, and Security, pp. 3–17 (2020)
9. Reich, J., Trapp, M.: Sinadra: towards a framework for assurable situation-aware dynamic risk assessment of autonomous vehicles. In: 2020 16th European Dependable Computing Conference (EDCC), pp. 47–50. IEEE (2020)
10. Sujan, M.A., Habli, I., Kelly, T.P., Pozzi, S., Johnson, C.W.: Should healthcare providers do safety cases? lessons from a cross-industry review of safety case practices. Saf. Sci. **84**, 181–189 (2016)

Redesigning Medical Device Assurance: Separating Technological and Clinical Assurance Cases

Spencer Deevy[1](✉), Tiago de Moraes Machado[1], Amen Modhafar[2], Wesley O'Beirne[2], Richard F. Paige[1], and Alan Wassyng[1]

[1] McMaster Centre for Software Certification, McMaster University, Hamilton, Canada
{deevys,machadotiagom,paigeri,wassyng}@mcmaster.ca

[2] Arrayus Technologies Inc., Burlington, Canada
{amodhafar,wobeirne}@arrayus.ca

Abstract. The safety and clinical effectiveness of medical devices depend on their use in specific clinical treatments. Due to the variability in physiology and genetics, different people react differently to exactly the same treatment. High-intensity focused ultrasound systems and radiation therapy machines are examples of systems where this needs to be taken into account. If we use a conventional monolithic assurance case for such systems, the inherent complexity affects our ability to construct an argument so that manufacturers and regulators are sufficiently confident that the device is safe and effective for a given treatment. We propose separating the assurance of these types of systems into two linked assurance cases. The first assurance case demonstrates the safety of the medical system independent of its clinical effect. The second demonstrates the safety and clinical effectiveness of the system when it is used within specific clinical treatments. Based on our experience in the ongoing development of a high-intensity focused ultrasound system, we introduce these separate assurance cases, and show how to structure them. We present definitions that are useful in determining boundaries, interfaces and dependencies between the two assurance cases, and include observations related to the effectiveness of this approach.

Keywords: Assurance Case · Separation of Concerns · Medical Devices · Safety-Critical · Software-Intensive · Safety · Certification · Focused Ultrasound

1 Introduction

Over the past two years, the McMaster Centre for Software Certification (McSCert) and Arrayus Technologies Inc., have been working together on constructing an effective assurance case (AC) for Arrayus's new therapeutic focused

Partially supported by the Natural Sciences and Engineering Research Council of Canada.

J. Guiochet et al. (Eds.): SAFECOMP 2023, LNCS 14181, pp. 25–38, 2023.
https://doi.org/10.1007/978-3-031-40923-3_3

ultrasound (FUS) system. This system emits FUS energy waves to deliver precise treatment for several medical conditions, including uterine fibroids and pancreatic cancer. It uses an external magnetic resonance imaging (MRI) scanner for the guidance of the treatment and monitoring of the patient. The combination of such non-invasive technologies forms a software-intensive system of systems commonly known as a Magnetic Resonance-guided Focused Ultrasound (MRgFUS) medical device [8].

Safety cases, a precursor of assurance cases, were introduced more than 50 years ago to help manufacturers document a structured, explicit argument that the system of interest is acceptably safe [7]. Modern assurance cases have the same intent, but can be used to guide development as well as document that a system possesses properties of concern, including but not limited to safety. They are thus well suited for use in developing and assuring safe and effective medical systems.

McSCert has been researching and developing assurance cases for more than 10 years. The McSCert researchers were confident that they could bring this expertise to aid Arrayus in assuring the safety and effectiveness of their new system. After a relatively easy start, progress slowed as the team grappled with structuring arguments that were both convincing and understandable by the different readers who need to use the assurance case. Making an argument that the medical system produces the intended outputs and would not deliver more ultrasound energy than requested, or outside of the focus area, was similar to many systems that McSCert researchers had worked on. However, dealing with the clinical effects of those outputs raised the complexity of the overall argument to a level that was not useful. The difficulty with systems such as MRgFUS devices is that there are so many variations in how different people react to exactly the same system outputs. Dealing with the system and its clinical effects in a single, monolithic AC seemed the wrong way to handle the assurance.

Out of that realization grew the idea that we now propose – separate the assurance into two linked assurance cases. The Technological Assurance Case (TAC) argues the safety and effectiveness of a medical system viewed solely as a machine that produces deterministic outputs given specific inputs. This is independent of the effect these outputs have on a patient during clinical treatment. The Clinical Assurance Case (CAC) argues the overall safety and effectiveness of the medical system, and specifically deals with that part of the assurance as to how the outputs from the medical system affect patients during clinical treatment. The CAC makes reference to the TAC in that part of the assurance that depends on the system producing its intended functionality "safely".

2 Preliminaries

2.1 Background

In discussion related to assurance cases and in the structure of the assurance case figures, we have used Goal Structuring Notation (GSN) [3] with some minor changes in terminology. For example, we prefer to talk about *claims* and *evidence* rather than *goals* and *solutions*.

2.2 Basic Concepts

Prior to delving into our proposed separation of assurance, we need to make an important distinction between two concepts relating to medical devices. This distinction provides a basis for our rationale for the separation of assurance discussed in Sect. 3.

Technological Effects: When considering the *technological effects* of the medical device, we consider the device solely as a machine that produces deterministic outputs given specific inputs. For example, a MRgFUS machine can be viewed simply as a machine that delivers high-intensity ultrasound energy focused at a certain location in space as required by its user. Similarly, a radiation therapy machine can be viewed simply as a machine that delivers radiation energy to a certain confined location in space as required by its user. In both examples, it does not include how the outputs of the medical device would affect patients.

Clinical Effects: We are concerned with the use of the medical device for a particular clinical treatment, in which patient variability leads to variability in response to the outputs of the medical device. As such, *clinical effects* of the device refer to the physiological responses of a human patient to the use of the medical device and its operating procedures during a specific clinical treatment. This is meant to cope specifically with the fact that different people can react differently to the exact same treatment. In our example of the MRgFUS, the intended clinical effects from using the device refer to how the device can be used to produce outputs, technological effects, that result in the intended biological/physiological response for the specific treatment. This includes how the amount, focus and duration of ultrasound waveforms produced by the machine may affect a human patient. Operating procedures and treatment plans also influence the clinical effects. Pre-treatment dietary changes or medication are actions that can be taken that do not change the behaviour of the medical device and its technological effects, but may alter how the patient's physiology responds to the technological effects. Thus, operating procedures, pre-treatment and treatment plans are included in how the intended clinical effects are produced.

The actual definitions of these terms and derivative terminology are shown in the relevant assurance cases in the following section on the proposed separation of assurance cases.

3 Separation

We typically create a single, comprehensive AC for engineered systems in non-medical domains. In these assurance cases, safety is considered in relation to the overall behaviour of the system and the respective effects produced in a given environment. This strategy has been applied to medical devices as well, and has been effective for many of them, but is problematic for those that have to reckon with the fact that people's shapes and sizes vary extensively, and their bodies can respond differently to the same clinical treatment protocol. This variability, however, is not limited to patient physiology. It may also extend to the types of treatments a single medical device can perform and the different regions of the body that can be treated with that medical device. Therapeutic MRgFUS and radiation therapy machines, as mentioned previously, are examples of how a complex system may be used to treat a wide variety of medical conditions, over several regions of the body. All of this makes it difficult to document a compelling assurance argument. The argument must demonstrate that the machine works as intended, delivers the correct outputs within a safe range and will not harm the environment or users of the system. It must also demonstrate that it is unlikely that any output will be delivered outside of the intended treatment location, but if it does, it will not cause unacceptable harm to the patient. Finally, it must show that the system will achieve the desired physiological response for a particular patient.

While grappling with this, we realized an analogy between this situation and the complexity inherent in very sophisticated control systems that are also safety-critical. A number of industries have used separation of concerns to deal with such problems. For example, many countries have mandated the separation of control and safety in nuclear power plants. This results in much simpler safety systems that can then be built and certified to be safe, independent of what the control system does. This separation is enforced in the system itself. It occurred to us that for some medical systems, the separation of concerns to control complexity could be applied to the assurance case.

Thus, instead of constructing a monolithic assurance case, we propose splitting the argument of safety and effectiveness for certain medical devices into two linked assurance cases: one based on the *"Technological effects"* as defined in Sect. 2.2, and another based on the *"Clinical effects"*, also as defined in Sect. 2.2. The former presents the argument pertaining to the safety and the effectiveness of the medical device and the therapy-agnostic operating procedures in relation to the medical device's ability to deliver its promised behaviour independent of any clinical context. The latter presents the argument pertaining to the safety and the effectiveness of the medical device and therapy-specific operating procedures/treatment plans in relation to achieving the intended biological/physiological response required to treat a particular medical indication. Overall assurance of the medical device used for a specific therapy is obtained by the combination of two linked ACs. The final assurance is documented in the AC that is focused on the *clinical effects.* This is the AC that we have called the "Clinical Assurance Case". The CAC is dependent on assurance provided by the AC that focuses on the *technological effects.* This is the AC that we have called the "Technological Assurance Case".

3.1 The Monolithic Assurance Case

To discuss this separation in more detail we consider the top level of a monolithic assurance case for a MRgFUS device that provides clinical treatment for uterine fibroids, achieved by thermally ablating problematic tissue. This is shown in Fig. 1, in which the top claim is shown along with its top-level GSN decomposition. The GSN components are labelled as follows: C indicates a *Claim*; S indicates a *Strategy*; and X indicates *conteXt*. We have removed Assumptions and Justifications in the interest of saving space. The claims with the tabs on the top left edge are *modules*. The lower levels of the argument are contained within those modules. The evidence that supports terminal claims in the argument are visible only in the content of those modules, and are not described in this paper.

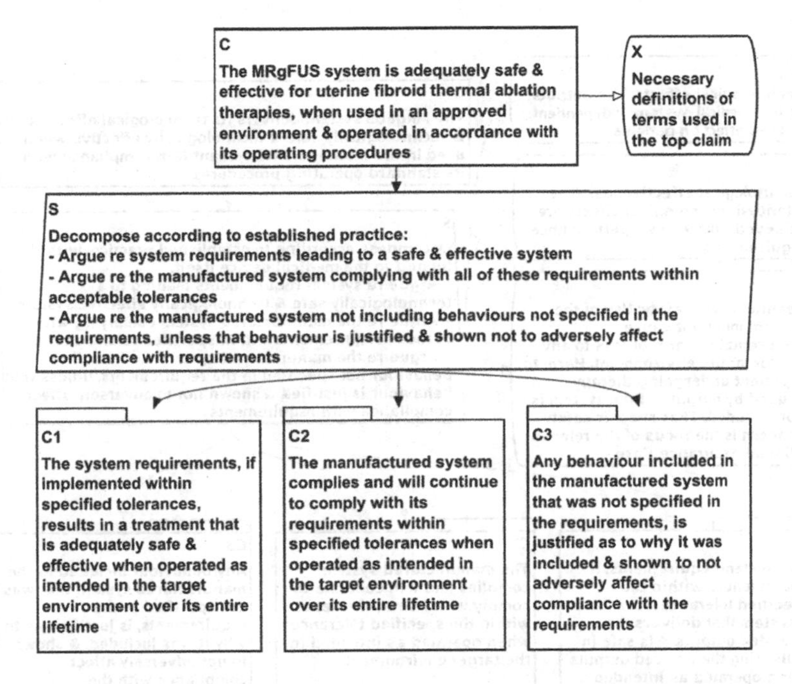

Fig. 1. Top-level of a monolithic assurance case for a MRgFUS system

We now compare this monolithic AC example with the proposed separation of assurance cases in the following sections.

3.2 Technological Assurance Case

The top claim and top-level decomposition of the *TAC* is shown in Fig. 2. This is slightly different from the monolithic version in that it disregards the clinical application and treats the safety of the system according to its capability in delivering the technological effects, or outputs, independent of any clinical effect. The device must be able to focus and deliver ultrasonic energy to a particular location in space from its ultrasonic transducers within the required specifications. Such performance-based properties of the technological effects are referred to as technological effectiveness. The medical device must also handle safety concerns that affect everyone using the system, for example, not having exposed wiring, not having sharp edges on the chassis/housing, having power limits that are enforced, focusing the energy at a designated location within the specified tolerance, etc. These safety concerns are identified through hazard analyses and deal with the intended behaviour of the system as well as how the machine interfaces with its environment and interacts with other medical devices. These safety concerns are also independent of any specific clinical effects. The safety-based properties of the technological effects are referred to as *technological safety*.

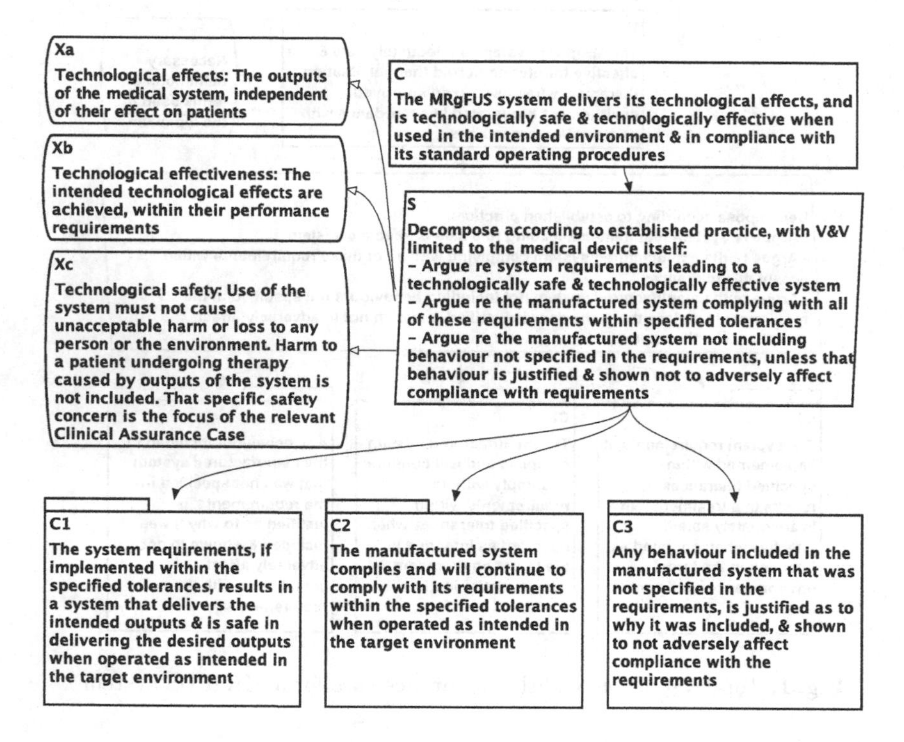

Fig. 2. Top-level of the TAC for a MRgFUS system

Our standard practice in structuring the argument is described in the Strategy, S. This directly leads to the supporting claims shown in C1, C2 and C3. The context component Xa defines the technological effects, or outputs, provided by the device. In the TAC for a particular system, this would be complemented by including references to the manufacturer's documents that specify the system's outputs. Context Xb and Xc define the effectiveness and safety properties associated with the technological effects, and all three of these terms reinforce that the effect of the outputs of the medical device on the patient is not considered within this argument. Those safety and effectiveness concerns are to be separated out and dealt with in the CAC. It is important to note that C1 and C2 in the TAC assure the original manufacture of the system as well as the maintenance plans in place so that the system remains safe and effective over its lifetime.

3.3 Clinical Assurance Case

The second assurance case we have called the *CAC* and its top-level is shown in Fig. 3. We use the same notation in the CAC as we did in the TAC.

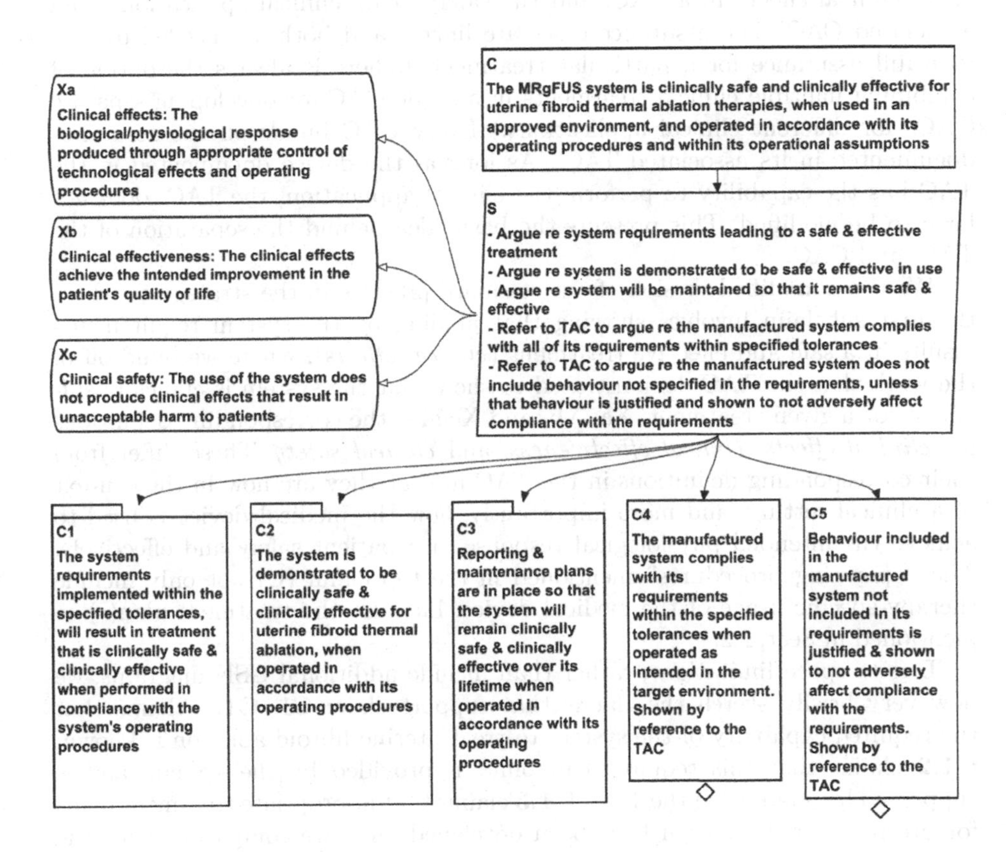

Fig. 3. Top-level of the CAC for a MRgFUS system

As one can see, the top claim in the CAC is different from the top claim in the TAC, since the concept and meaning of safety and effectiveness for the CAC addresses the intended clinical effects relevant to the clinical application of the medical device. The purpose of the MRgFUS device is to perform medical procedures, so we must cope with therapy-applicable hazards related to using the system to treat patients. Additionally, clinical safety also takes into account the safety concerns addressed already in the TAC, but those are expanded within the context of the planned clinical application. Hence, a clinical safety example for this particular machine is when it is used to perform thermal ablation of uterine fibroids in the body. Depending on where the fibroid tissue is, treatment may or may not result in a hazard caused by the amount of energy deposited in other critical biological structures within the beam path of the ultrasound wave, since different biological structures absorb energy differently. If such hazards are not detected and mitigated in advance, the system can potentially damage the tissue that was not the intended target.

In practice, we can thus assure the safety of the medical system independent of its clinical effects in a TAC, and the safety of its clinical application in an associated CAC. The assurance cases are linked, and both are needed to provide full assurance for a particular treatment. (There is always the option of combining multiple clinical treatments in a single CAC, or developing separate CACs for different clinical applications.) Every CAC builds on the assurance documented in its associated TAC.. As long as the device documented in the TAC has the capability to perform the clinical application, the TAC. does not have to be modified. This presents the basic idea behind the separation of the TAC. and CAC.

We use a modified version of our standard practice in the strategy S, where the first subclaim involves showing that abiding by the system requirements results in a safe and effective treatment (*clinical effects*), where we build off of the work that the TAC demonstrated to show that the system is clinically sufficient for a given treatment. Xa, Xb, and Xc give the corresponding definitions for *clinical effects*, *clinical effectiveness*, and *clinical safety*. These differ from their corresponding definitions in the TAC in that they are now in the context of a clinical setting, and more importantly, how the medical device is used to achieve the intended physiological response in a patient safely and effectively. The "operating procedures" mentioned in the top claim (C) not only include therapy-specific usage of the medical device, but also the treatment plan(s) as mentioned in Sect. 2.2.

Due to space limitations, rather than include additional GSN diagrams, we now very briefly sketch the claims that support claim C1. **C1.1** claims that the required capability of the system to treat uterine fibroid ablation is known. **C1.2** claims that this required capability is provided by the system, and is supported by referencing the TAC. **C1.3** claims that appropriate treatment plans for uterine fibroid ablation have been developed, and are compatible with the system and its operating procedures.

We see in Fig. 3 that claims C4 and C5, that are similar to claims C2 and C3 in the monolithic assurance case, are dealt with by reference to the associated TAC. They do not have to be re-argued in the CAC! Clearly, the CAC is dependent on its associated TAC in that the safety of the machine itself in delivering its functional outputs is documented in the TAC. This implies that the outputs of the medical system required for a clinical treatment must be documented explicitly in the CAC, and then verified as provided by the system as documented in the TAC. In general, the CAC may reference any items in the TAC. However, it is crucial that there are no references from the TAC to any dependent CAC. The diamonds below the claims C4 and C5 are GSN symbols to indicate that the claims are not further developed in the CAC. The required information is documented in context nodes that support claims C4 and C5.

3.4 Maintenance and Dependencies in the TAC/CAC Combination

ACs are always related to a snapshot in time, and that time is chosen with respect to what we want to achieve with the AC. A very common snapshot that is useful for medical systems is when the manufacturer applies to a regulatory body for approval to market the system. With this example in mind, we can explain how the TAC/CAC combination is designed to work together in practice.

At this point in time, the manufacturer has iterated on the development and manufacture of the system so that the TAC describes a technologically safe and effective system, that can be used for the planned treatment therapy. The system has been used in practice successfully and the CAC documents that this is so by showing that the system is both technologically and clinically safe and clinically effective. In order to proceed to market, the manufacturer has to demonstrate that there are reporting and maintenance plans that should ensure that the system remains safe and effective over the lifetime of the system. This is one aspect of maintenance. Another aspect of maintenance is that "gaps" that arise in the TAC and/or CAC over time will also lead to maintenance analyses and possible changes to the system. All of this is included in the monolithic AC shown in claim C2 in Fig. 1. It is also included in claim C2 in the TAC (Fig. 2), and in claim C3 of the CAC (Fig. 3). The dependencies between the CAC and TAC are shown in Fig. 4.

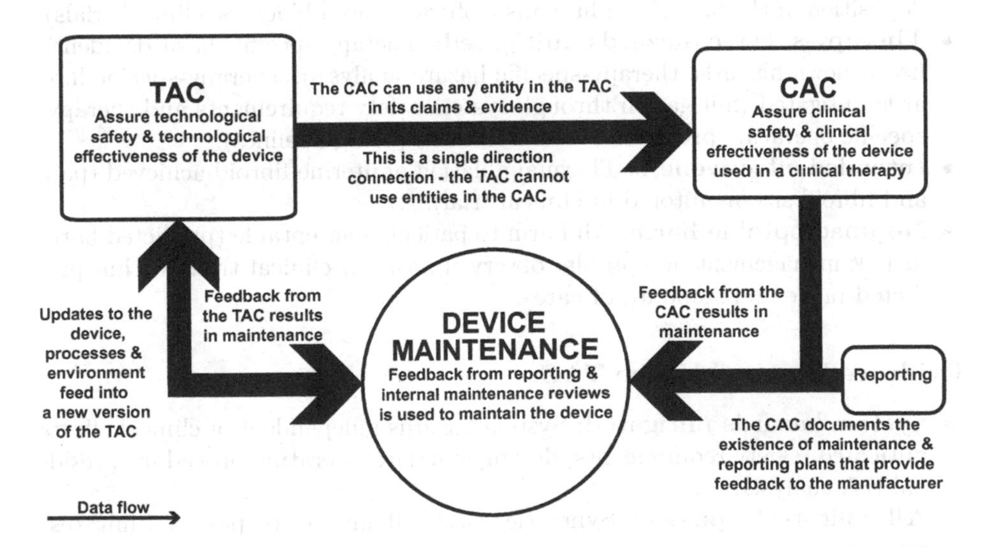

Fig. 4. Dependencies between the TAC and CAC

In summary, the system itself may need to be updated because of maintenance feedback from evidence generated in either the TAC or the CAC, or by reporting mechanisms detailed in the CAC. Changes to the system may affect the TAC, and may flow through into the CAC. There is clearly a dependence of the CAC on its associated TAC. The arguments in the CAC can refer to elements (claims/evidence/context/assumptions) in the TAC. The TAC cannot use elements from the arguments in the CAC.

3.5 Evidence in the TAC and CAC

Traditionally, one of the most important reasons for creating an AC very early in the development process is that it can direct us to tailor development to produce the evidence the AC argument requires. In the case of the TAC and CAC this guidance is especially useful since it separates what needs to be provided about the system itself, versus what needs to be provided about clinical results. These simple examples concerning the MRgFUS system used for uterine fibroid ablation show the differences in the type of evidence in the two ACs. The format for each example is:
Keywords: Required demo 1 (evidence 1). Required demo 2 (evidence 2). ...

Evidence Required by the CAC:

- **Thermally ablate uterine fibroids:** Successful thermal ablation of human uterine tissue (Pre-existing literature, pre-clinical trials). Successful thermal ablation of uterine fibroids (Clinical trials).
- **Far field effects at proposed energy levels are safe:** Acceptable energy deposition in the far field in humans (software-based blockers, clinical trials).
- **Therapy-specific hazards mitigated:** Therapy-specific hazards identified (known hazards, therapy-specific hazard analysis). Therapy-specific hazards mitigated (mitigation through system safety requirements and therapy-specific operating procedures and treatment plan guidelines).
- **Intended effectiveness:** Thermal ablation of uterine fibroid achieved (pain and fibroid size monitored in clinical trial).
- **No unacceptable harm:** All harm to patient is acceptable (predicted harm in risk management acceptable, observed harm in clinical trials within predicted range and standard of care).

Evidence Required by the TAC:

- **System hazards mitigated:** System hazards independent of clinical effects mitigated (safety requirements, design, standard operating procedures, residual risk analysis).
- **All unit tests passed:** Syntactic check all unit tests passed (unit test report).
- **Unit test coverage is sufficient:** Unit test coverage meets required standards (unit test report).
- **Unit tests correct:** Review of unit tests determines correctness (unit test report review).

- **Energy level and beam forming within tolerance:** Requirements specify intended output and acceptable tolerances (validation of requirements). Manufactured system complies with requirements within tolerance (verification report).

4 Observations

When we started this work we thought it would be straightforward to define the boundaries of the TAC and CAC. It was not. We encountered consistent problems in this regard until we arrived at the current definitions of technological and clinical effects, safety, and effectiveness. These definitions enabled us to determine the rules for separating the TAC and CAC that we now use.

For example, the analysis that evaluates whether the benefits of delivering the treatment outweigh the treatment risks for the patient, has argument components split/shared between the TAC and CAC. These definitions helped us determine more precisely where to document how the company should address these concerns.

The separation of concerns had a positive side effect. The Arrayus/McSCert team has held weekly meetings throughout this project. The discussions during those meetings became much more focused once we separated the TAC and CAC, as the teams were able to more accurately focus their feedback and revisions on more concise and specific parts of the overall argument. In effect, this is what is often seen when reasoning about or analyzing software architectures: the decomposition into different views lends itself to more focused reasoning, deeper understanding, and faster improvements during revision.

The separation of a and its associated has been beneficial in detailed discussions on the arguments necessary to support different clinical therapies such as for uterine fibroids and pancreatic cancer. Those discussions also helped motivate and shape our approach to defining the technological and clinical properties we need.

We have not had any reason to abandon this approach while constructing the TAC and CAC for the use of the MRgFUS system for the treatment(s) of interest. Particularly, as our experience with this style of decomposition for assurance cases has increased, our effectiveness in addressing assurance questions and concerns has, in our opinion, improved.

5 Related Work

Increasingly complex software-intensive safety-critical systems have become the norm, introducing a multitude of challenges in the medical industry [11,12,16]. Previous work in the railway domain has leveraged the modularity of particular systems to split assurance cases across application-specific subsystems [6], in effect using the system architecture to define views on the assurance case structure. Our method is distinct from this strategy in that we are not splitting across the system itself, but rather, between the outputs of the system and the effects of those respective outputs on a patient. The assurance of system-level properties, such as safety of medical devices, has likewise adopted new strategies to address this complexity. One such strategy involves applying software architectural principles to safety case development, whereby interrelated modular safety cases may be "composed" to produce a complete safety case [9,15]. In our case, we are not suggesting that safety, or any other system-level property, is compositional. We have arranged the TAC and CAC such that they are *linked* rather than *composed*, recognizing that portions of a typical monolithic assurance case are independent from the clinical effects produced by the outputs of the medical device.

Assurance cases have also been investigated for their role in medical device certification [21], where they may be used to improve and expedite the approval process. The Food and Drug Administration (FDA) has conducted reviews on assurance cases for insulin pumps [17] and has recognized guidelines for medical device assurance cases [4], specifically in its applicability to insulin pumps. The adoption of assurance case templates approved by regulatory bodies would promote their use in the medical industry. Currently, compliance with relevant standards [1,2] is used to evaluate the safety and readiness of a medical device to enter the market. These standards, however, often describe development processes that should be followed rather than describing acceptance criteria on the artefacts of the system. A shift in focus towards this product/evidence-based approach has been suggested, using assurance case templates as standards [19,20].

Dealing with the complexity of software-intensive medical devices and the need for rapid and thorough certification of such systems necessitates the reuse of proven assurance where possible. In assurance cases, reuse has been investigated in the form of assurance case patterns [22] and their extension in the form of assurance case templates [19,20]. One framework for assurance case patterns involves *independent co-assurance* [14], where at least two system-level properties are both assured in parallel, including the influence each property may have on other properties. This approach has been used to co-assure system safety and security properties in infusion pumps [14], collaborative industrial robot applications [10], and in the automotive industry [18]; the ideas are also motivating novel schemes for assurance case structuring in Industry 4.0 [13].

The FDA has also published guidance for medical devices [5], where they indicate that validation involving the outputs of the medical device are different from the validation of the clinical outcomes of using the medical device. This distinction is central to our proposed separation of assurance, that was designed in a manner that would focus on the issues specifically associated with medical device assurance.

6 Conclusion

Monolithic assurance cases for medical devices typically mix technological factors with clinical factors. The resulting assurance case quickly becomes unwieldy, especially when new clinical therapies are added to an existing assurance case.

We have shown that separation of concerns can be used in assurance cases to reduce the complexity of demonstrating safety and effectiveness for software-intensive medical systems, such as the MRgFUS. By separating the demonstration of system safety in producing the intended deterministic machine output independent of clinical safety, we believe we can significantly reduce the overall complexity of the safety assurance argument.

It is common to find that a particular medical device is used for different clinical procedures as is the case for the MRgFUS that is capable of offering treatment for several medical conditions. This is similar to the design of product lines. If we can assure the technological safety and effectiveness of the device independent of its clinical effects, this raises the possibility that a single TAC could be used. The TAC can be linked with multiple CACs, to assure the safety and effectiveness of the device used in multiple clinical treatments.

The TAC and CAC can be developed and maintained separately. To reap the full benefit of this approach we must ensure that the TAC is not dependent on the claims and argument in the CAC. However, the CAC *will be* dependent on many entities in its associated TAC.

We already noted that an assurance case represents an argument for the properties of interest at a snapshot in time. We also believe that the assurance case(s) should be developed as early as possible, should guide development and be updated as development proceeds. The separation between TAC and CAC can provide extremely useful guidance on what evidence has to be produced to provide a solid foundation for the different arguments.

References

1. Medical Devices - Application of Risk Management to Medical Devices (ISO 14971:2019). Standard, International Organization for Standardization, Geneva, CH, December 2019
2. Medical Device Software - Software Life Cycle Processes (IEC 62304:2006). Standard, International Electrotechnical Commission, Geneva, CH, May 2006
3. GSN Community Standard (Version 1). Standard, Origin Consulting (York) Limited, York, GB, November 2011
4. Medical Device Safety Assurance Case Guidance (AAMI TIR38:2019). Standard, Association for the Advancement of Medical Instrumentation, Arlington, VA, USA, January 2019
5. Design control guidance for medical device manufacturers: Guidance document. Food and Drug Administration, Silver Spring, MD, USA, March 1997
6. Althammer, E., Schoitsch, E., Sonneck, G., Eriksson, H., Vinter, J.: Modular certification support - the DECOS concept of generic safety cases. In: 6th IEEE International Conference on Industrial Informatics, pp. 258–263, August 2008
7. Bloomfield, R., Bishop, P.: Safety and assurance cases: past, present and possible future - an Adelard perspective. In: Dale, C., Anderson, T. (eds.) Making Systems Safer, SSS 2010, pp. 51–67. Springer, London (2010). https://doi.org/10.1007/978-1-84996-086-1_4
8. Bradley, W.G., Jr.: MR-guided focused ultrasound: a potentially disruptive technology. J. Am. Coll. Radiol. **6**(7), 510–513 (2009)

9. Despotou, G., Kelly, T.: Investigating the use of argument modularity to optimise through-life system safety assurance. In: Proceedings of the 3rd IET International Conference on System Safety, November 2008
10. Gleirscher, M., Johnson, N., Karachristou, P., Calinescu, R., Law, J., Clark, J.: Challenges in the safety-security co-assurance of collaborative industrial robots. In: Aldinhas Ferreira, M.I., Fletcher, S.R. (eds.) The 21st Century Industrial Robot: When Tools Become Collaborators. ISCASE, vol. 81, pp. 191–214. Springer, Cham (2022). https://doi.org/10.1007/978-3-030-78513-0_11
11. Gordon, W.J., Stern, A.D.: Challenges and opportunities in software-driven medical devices. Nat. Biomed. Eng. **3**(7), 493–497 (2019)
12. Hatcliff, J., Wassyng, A., Kelly, T., Comar, C., Jones, P.: Certifiably safe software-dependent systems: challenges and directions. In: Future of Software Engineering Proceedings. FOSE 2014, pp. 182–200. Association for Computing Machinery, New York, NY, USA, May 2014
13. Jaradat, O., Sljivo, I., Hawkins, R., Habli, I.: Modular safety cases for the assurance of industry 4.0. In: Proceedings of Safety-Critical Systems Symposium, pp. 105–124 (2020)
14. Johnson, N., Kelly, T.: Devil's in the detail: through-life safety and security co-assurance using SSAF. In: Romanovsky, A., Troubitsyna, E., Bitsch, F. (eds.) SAFECOMP 2019. LNCS, vol. 11698, pp. 299–314. Springer, Cham (2019). https://doi.org/10.1007/978-3-030-26601-1_21
15. Kelly, T.: Using software architecture techniques to support the modular certification of safety-critical systems. In: Proceedings of the 11th Australian Workshop on Safety Critical Systems and Software. SCS 2006, vol. 69, pp. 53–65. Australian Computer Society Inc, Sydney, AU, May 2007
16. Lee, I., et al.: High-confidence medical device software and systems. Computer **39**(4), 33–38 (2006)
17. McGowan, R., Stevens, A., Chapman, R.: Food and drug administration review of safety assurance cases for medical devices. J. Clin. Eng. **39**(2), 96–98 (2014)
18. Warg, F., Skoglund, M.: Argument patterns for multi-concern assurance of connected automated driving systems. In: Asplund, M., Paulitsch, M. (eds.) 4th International Workshop on Security and Dependability of Critical Embedded Real-Time Systems (CERTS 2019). OpenAccess Series in Informatics (OASIcs), vol. 73, pp. 3:1–3:13. Schloss Dagstuhl-Leibniz-Zentrum fuer Informatik, Dagstuhl, Germany, July 2019
19. Wassyng, A., Joannou, P., Lawford, M., Maibaum, T., Singh, N.K.: New standards for trustworthy cyber-physical systems. In: Trustworthy Cyber-Physical Systems Engineering, pp. 337–368 (2016)
20. Wassyng, A., et al.: Can product-specific assurance case templates be used as medical device standards? IEEE Des. Test **32**(5), 45–55 (2015)
21. Weinstock, C., Goodenough, J.: Towards an assurance case practice for medical devices. Technical report CMU/SEI-2009-TN-018, Software Engineering Institute, Carnegie Mellon University, Pittsburgh, PA, USA, October 2009
22. Yamamoto, S., Matsuno, Y.: An evaluation of argument patterns to reduce pitfalls of applying assurance case. In: 2013 1st International Workshop on Assurance Cases for Software-Intensive Systems (ASSURE), pp. 12–17. IEEE, May 2013

Software Testing and Reliability

A Cognitive Framework for Modeling Coincident Software Faults: An Experimental Study

Bo Zhao[2], You Song[2], Wenhao Xu[2], and Fuqun Huang[1](✉)

[1] Department of Computer Science, Western Washington University, Bellingham, WA 98225, USA
huangf2@wwu.edu
[2] School of Software, Beihang University, Beijing 100191, China
{Sy2021122,songyou,xuwenhao}@buaa.edu.cn

Abstract. The question of when different programmers tend to commit the same errors is a critical issue for achieving fault diversity in fault tolerance. This problem is interdisciplinary and related to theories of human error in cognitive psychology. This paper proposes a psychological framework that combines Rasmussen's performance levels with cross-level errors, represented by post-completion error, to model situations in which different programmers are prone to making the same errors. To validate the framework, we conducted an experiment where 200 student programmers independently solved the same problem, with the same tool and language. The results indicate that programmers unlikely commit the same errors in skill-based performances, most likely make the same errors in rule-based performances. These findings suggest that natural independent development may be less effective in preventing common errors in functions involving rule-based performance and post-completion scenarios, whereas it could be effective in preventing common errors in skill-based and knowledge-based performances. The results provided new insights into the strategies for avoiding coincident faults in N-version programming, from a human factor perspective.

Keywords: Software Reliability · Fault Tolerance · Coincident fault · Software Diversity · Cognitive Model

1 Introduction

Fault tolerance is critical for ensuring operational reliability, where a safety-critical software system remains functional even in the presence of residual faults. N-version programming, as one of the two primary approaches for achieving fault tolerance, is especially effective in tolerating design faults [1]. N-version programming [2, 3] is a software development practice in which multiple programmers independently create multiple versions of a program based on the same requirements. It has been widely adopted as a strategy to achieve fault tolerance and improve reliability in safety-critical software systems [1]. Recently, N-version programming utilized to enhance the fault tolerance of deep neural network systems [4] and improve the security of cloud systems [5].

J. Guiochet et al. (Eds.): SAFECOMP 2023, LNCS 14181, pp. 41–54, 2023.
https://doi.org/10.1007/978-3-031-40923-3_4

Multiple strategies have been proposed and studied in N-version programming, such as "natural" independent development, where different programmers develop versions according to the same requirements without communicating to each other [6, 7], using different languages [8], different algorithms [9], and requirements written in different ways [10].

An interesting phenomenon has been consistently observed: programmers make the same mistakes, even though the "external factors" (languages, tools, processes, etc.) are different. For example, Eckhardt and Lee [6] demonstrated that truly independently developed versions can fail dependently. In Knight and Leveson's experiment [7], it was found that about half of the total faults were coincident faults when the programmers worked in isolation and developed programs independently. Avižienis, Lyu, and Schütz [8] found that identical faults were introduced despite the use of different programming languages. Similarly, Meulen and Revilla's experiment [11] on a large population of programs indicated that the benefit of language diversity was low. In Feldt's experiment [9], genetic programming was used to force algorithms to differ in various versions, and the test cases that caused versions to fail with high probabilities were located in specific areas. These studies suggest that in certain circumstances, programmers tend to make identical errors, although the underlying cognitive mechanisms are not well understood.

Programming involves significant cognitive activities on the part of programmers, and understanding the cognitive mechanisms behind how different programmers with similar backgrounds and using the same language and tool tend to make the same error is crucial to preventing coincident faults. To the best of our knowledge, there is no cognitive framework for modeling the likelihood of coincident faults that focuses on the contribution of the diversity of individual programmers in N-version programming.

This paper proposes a cognitive framework, the **Cogn**itive **F**ramework for **C**oincident **F**aults (**CognFCF**), for modeling the likelihood of different programmers introducing the same faults. We evaluated the framework through an experiment involving 200 student programmers with similar backgrounds. They independently developed program versions using using the same tool and language. Based on our findings, we derive an integrated framework for avoiding coincident faults from a human factors perspective.

2 The Cognitive Framework for Modeling Coincident Faults

2.1 Terminologies

Fault: An incorrect or missing step, process, or data definition in a computer program.

Error: An erroneous human behavior that leads to a software fault [12].

Occurrences (OC) of a fault: The number of programmers in an N-version programming study who introduced that fault.

The **Prevalence of Occurrence (POC)** of a fault: The percentage of programmers who introduced the fault *i*, defined as:

$$\mathrm{POC_i} = \mathrm{OC_i}/\mathrm{P} \quad (1)$$

where *P* is the total number of programmers who submitted code for the task. POC describes how common a fault is, that is, how likely it is to be introduced by different programmers. POC in an experiment is an estimator for the probability of the fault being inserted by a randomly chosen programmer from the population sampled for the experiment. For an N-version programming experiment in which P is the same for each fault, comparing the Occurrences of faults is equivalent to comparing the POCs of faults.

Coincident Fault: A fault whose Occurrences is two or more in N-version programming, i.e., that was introduced by at least two programmers.

2.2 The Cognitive Framework for Coincident Faults

The proposed cognitive framework for coincident faults (CognFCF) includes four cognitive error levels, based on Rasmussen's performance framework [13] and Byrne and Bovair's Post-Completion Error [14].

There have been many cognition and human error models used in diverse contexts, for instances, Palanque et al. [15] proposed a taxonomy that covers fault categories inducing operator (e.g. aircraft pilots) errors, Mohanani et al. [16] summarized a set of "cognitive biases" in software engineering, Huang et al. proposed human error taxonomies for defect root cause analysis [17] and defect early forecasting [18]. These different taxonomies are proposed for specific application contexts that differ from that of this paper, or many of them (e.g. in [16] and [17]) can be traced back to human error modes in classic psychological works, i.e. Reason's human error taxonomy [12]. As the first cognitive framework for modelling coincident defect, we prefer a classic fundamental psychological model that has been widely accepted by psychologists and applied in diverse contexts, while has an appropriate level of granularity that is possible for application by practitioners who seek software fault diversity.

Rasmussen's Performance Framework

Rasmussen's performance framework has been widely accepted as a fundamental theory in psychology [12]. For instance, J. Reason, one of the most well-acknowledged psychologists in the field of human errors, uses Rasmussen's framework to integrate various human error modes discovered by many psychologists [12]. Rasmussen's performance framework [13] classifies cognitive activities into three levels: skill-based, rule-based and knowledge-based level. Different performance levels have different cognitive characteristics, and are associated with different error modes [12].

Skill-based (SB) performance follows from the statement of an intention, "rolls along" automatically without conscious control. Skill-based activities in programming include typing a text string, compiling a program by pressing a button in the programming environment.

Skill-based errors are the human errors occurring in skill-based performances. Examples of skill-based errors in the psychology domain include such as perceptual confusions, e.g. taking an object that looks like another. In software development, typos and entering a wrong letter which looks similar to the correct one (e.g. taking 0 for o) are typical examples of skill-based errors [18].

Rule-based (RB) performance is applicable for tackling familiar problems. It is typically controlled by stored rules that have been derived from a person's experiences. The mind matches the situation at hand to the preconditions for such stored rules, allowing quick selection of actions. In programming, there are many rule-based performances, such as printing of a string line, and defining a variable in one's familiar programming language.

Rule-based errors are the errors occurring in rule-based performances. A typical rule-based error is "strong-but-now-wrong" error: one tends to use a rule that has been frequently used, but not necessarily matches well with the current situation [12].

Knowledge-based (KB) performance comes into play when individuals face novel situations, and no rules are available from previous experiences. At this level, actions must be planned using an analytical process. Errors at this level can arise from resource limitations and incomplete or incorrect knowledge. In programming, cognitive performances such as constructing the mental model of the system to understand a specific programming task and attempting to figure out a solution for a novel problem are considered knowledge-based performances.

Knowledge-based errors are human errors that occur in knowledge-based performances. Typical knowledge-based errors include "problems with complexity", and "confirmation bias"–the tendency of people to seek evidence that verifies their hypotheses rather than refuting them [12].

Cross-Level Errors

Additionally, we introduce Cross-level Errors to include the scenarios that may involve a combination of multiple levels of performances, though the three levels of performances are considered orthogonal.

A representative Cross-level errors is the **Post-Completion Error (PCE)** proposed by Byrne and Bovair [14]. Post-completion error refers to the tendency of people to forget a sub-goal in a specific situation. This occurs when there is an ultimate goal composed of several sub-goals, a sub-goal is to be carried out at the end of the task, and it is not a necessary condition for achieving the main sub-goal. Forgetting to attach a file when sending an email is a typical PCE in daily life.

Post-Completion Error was initially thought to be an error at the skill-based level [14]. However, recent studies found that it is an error that can occur at all of the three performance levels. For instances, Li et al. [19] conducted an experiment in the field of Human-Computer Interface, in which the participants were asked to solve river-crossing math problems and send the transport vessel back to the other side of the river as the last step of the solutions. The experiment showed that PCE can also occur in problem-solving tasks, i.e. math problems. Huang [20] conducted a controlled experiment on PCE in software development. The experiment showed that software developers are statistically prone to forgetting a requirement and/or design when the PCE scenario presents.

In summary, CognFCF includes four levels of cognitive errors: skill-based errors, rule-based errors, knowledge-based errors, and post-completion errors. These four levels of cognitive errors originate from four levels of cognitive performances shown in Fig. 1: skill-based, rule-based, knowledge-based, and cross-level performances, respectively. It is worth noting that post-completion errors are conceptually a subset of cross-level errors, and theoretically, there may be more cross-level error modes yet to be discovered by us and psychologists. Since our study only found post-completion errors at cross-level performances, we use post-completion errors specifically in the following sections.

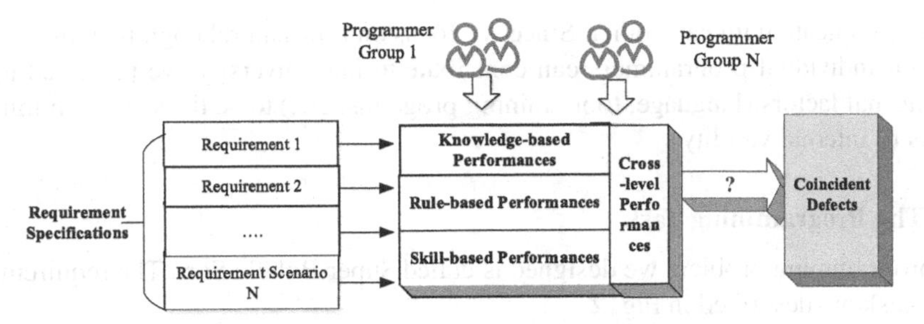

Fig. 1. The Cognitive Framework for Coincident Faults

3 Research Questions and Hypotheses

The aim of this study is to investigate whether the likelihoods of different programmers introducing the same faults differ across the four cognitive levels. The following research questions and hypotheses are proposed.

Research Question (RQ) 1: Do the likelihoods of a fault being a coincident fault differ across various cognitive levels? RQ1 addresses the overall trend of whether CognFCF has successfully captured the cognitive features that underlie coincident faults. RQ1 is answered by testing H1:

H1: The likelihoods of a fault being coincident are equal across the Skill-based, Rule-based, Knowledge-based, and Post-completion levels.

Research Question (RQ) 2: Do the Occurrences of coincident faults at skill-based, rule-based, knowledge-based, and post-completion errors differ?

The likelihood of different programmers introducing the same fault is measured by Occurrences (OC)–the number of programmers who introduced that fault in an N-version programming study. RQ2 is further divided into six sub-questions, corresponding to six hypotheses:

H2.1: The OC of skill-based faults is equivalent to that of rule-based faults.

H2.2: The OC of skill-based faults is equivalent to that of knowledge-based faults.

H2.3: The OC of skill-based faults is equivalent to that of post-completion faults.

H2.4: The OC of rule-based faults is equivalent to that of knowledge-based faults.

H2.5: The OC of rule-based faults is equivalent to that of post-completion faults.

H2.6: The OC of knowledge-based faults is equivalent to that of post-completion faults.

4 The Experimental Study

To test our hypotheses, we conducted an experiment in which we recruited 200 participants with similar backgrounds. They independently solve the same problem using the same language (C language) and tool (an Online Judge platform). It is worth noting that our experimental setting is not typical of N-version programming in industry, where programmers may use different languages and/or tools to develop versions and may or may not communicate with each other. Since we focus on how natural cognitive differences between individual programmers can contribute to fault diversity, we preferred to set the external factors (language, tool, training programs, etc.) to be the same to minimize threats to internal validity.

4.1 The Programming Task

The programming problem we designed is called Super Bubble Sort. The requirements of the task are described in Fig. 2.

The requirements of Bubble Sort Problem

For the set of integers to be sorted, we give a key number m, please sort by the distance to the key number m; that is, for two numbers a, b in the set, if $|a - m| < |b - m|$, then a should be in front of b; if the distances from a and b to m are equal, the smaller of a and b should be in front.

Input

The input consists of two lines. The first line contains two space-separated integers n, m, where n is the number of elements in the integer set to be sorted, and m is the sorting key. The second line contains n space-separated integers, where n_i is the ith element in the set. The input is guaranteed to be in the range $1 \leq n \leq 1000$, n_i and m in the range of int, i.e. $-2^{31} \leq n_i, m \leq 2^{31} - 1$.

Output

The output contains a total of n lines, each line outputs an integer representing the sorted result of the super bubble.

Sample Input

6 3
1 2 3 4 5 6

Sample Output

3
2
4
1
5
6

Hint

The value of $|n_i - m|$ may be outside the int range.

Fig. 2. The requirements specification of the task

4.2 The Participants

Our study involved 200 undergraduate Computer Science students from Beihang University who had completed the C Programming course. All participants signed an agreement to allow us to use their anonymized source code for scientific research, in accordance with the ethical guidelines of the university.

4.3 Data Collection

We included the Bubble Sort problem as one of the eight programming problems in a 90-min programming exam. The students were encouraged to choose the problems they felt most confident in solving and to correctly solve as many as possible. They were ranked based on the number of problems they correctly solved, and their rank was then transformed into credits, e.g., with the top 10% of participants receiving an A. A total of 1033 students participated the exam, while we randomly selected the programs submitted by 200 students who solved the Bubble Sort Problem. There was no specific time limit for each problem, but we analyzed the submission logs and found that the 200 participants spent an average of 28 min (Min = 9, Max = 62, Standard Deviation (SD) = 23.6) on the Bubble Sort Problem. The programming tasks were completed in C language only.

In addition to the Dev C++ installed in all the computers used by the participants, an Online Judge (OJ) system was used to submit their program versions. The OJ system is a reliable online system that supports multiple programmers working on the same problem concurrently and provides feedback to the programmers when they submit a version. This system has been used for over 10 years in programming education and contests at Beihang University. The feedback from the OJ system included the following six types:

Accepted (AC): The output of the program is the same as the standard answer, which means the program is correct.

Wrong Answer (WA): The output of the program is incorrect.

Runtime Failure (RF): The program performs an illegal operation, resulting in failure. Division by zero and out-of-bounds are examples of run-time failures.

Compilation Problem (CP): Source code fails during compiling.

Presentation Problem (PP): The data output by the program is correct, but the format does not conform to the requirements.

Time Limit Exceeded (TLE): The program runs for longer than the maximum time allowed by the online judge system.

The feedback types were obtained by running a total of ten groups of test cases. These test cases included tests that included normal and exceptional situations. The OJ system judged whether a program version passed each of these ten groups of test cases. Only programs that passed all test cases were accepted by the system. The complete set of test cases can be accessed at https://github.com/CognFCF/data/tree/main/testcases.

4.4 Data Analysis

Out of the 200 students, 154 students successfully solved the Super Bubble Sort problem, and their final code was judged to be "accepted" by the Online Judge system. Twenty-nine students didn't introduce any faults, while 171 students introduced one or more faults. The 200 participants submitted a total of 706 versions, since each participant could submit more than one version. The first versions submitted by all participants had an average of 39 lines of code (LOC) (Min = 18, Max = 71, SD = 9), and their Cyclomatic Complexity was an average of 9 (Min = 4, Max = 14, SD = 1.7).

Since a participant could revise his/her program based on feedback from the OJ system, for each participant, we have a series of versions showing the history of how he/she introduced and fixed faults.

We then analyzed the collected programs in two steps. The first step was code inspection to identify all the faults introduced by the participants. The identified faults were then integrated, and for each fault, we counted the number of participants who introduced that fault -- occurrences. The second step was to perform cognitive error cause analysis by classifying each fault with a cognitive error level.

Code Inspection. The first and third authors of the paper performed code inspections on the 706 versions of programs (submitted by the 200 participants, accessible through: https://github.com/CognFCF/data/tree/main/sourcecode), each inspected 100 participants' programs. Some participants submitted more than one version as they debugged and improved their programs if a submission was not "accepted" by the OJ.

The code inspections were performed in the following procedures: First, the code inspectors checked all versions of the code and recorded the faults in each version of their code. Then, for each fault, the number of participants who submitted the fault was counted. In cases where a fault appeared in multiple versions submitted by a programmer, the fault counted only once towards the metric "Occurrences".

The fault list was then integrated, discussed and consensus reached between the two inspectors, and finally summarized. Due to page limits, we only include a sample of the faults in this paper, shown in the first three columns of. The full set of faults is openly available here: https://github.com/CognFCF/data/tree/main/faultlist. In total, the code inspectors identified 70 faults, of which 31 were coincident faults.

Classifying Cognitive Error Levels. In this study, classifying the cognitive levels of faults was a crucial step. To minimize researcher bias, two professors who were knowledgeable in C programming and independent of the study performed the classification. It is worth noting that employing two raters for classification or coding is a common practice in social and psychological studies [21]. The classification process proceeded as follows:

First, the first author presented skill-based, rule-based and knowledge-based performances, as well as post-completion errors described in Sect. 2.2, to the professors. Next, the first author explained the programming task and each fault in the fault list (with a sample shown in) and ensured that both professors comprehended the materials thoroughly. Afterwards, the professors independently assigned a cognitive level category to each of the faults (Table 1).

We compared their classifications and discovered that 10 defects out of 70 were assigned different cognitive levels. The initial interrater reliability was calculated using by Cohen's kappa coefficient [21], which was 0.80, indicating substantial agreement [21]. Finally, the professors discussed the inconsistent classifications until a perfect agreement was reached, resulting in a final interrater reliability of 1.00. The final classification results are presented in the 4th column of.

Table 1. A sample of the experiment data

Fault ID	Fault Description	OC	CL	OJ
2	Data overflow	122	RB	WA
4	The variable is not defined before it is used	3	RB	WA
7	Mistook the "%d\n" for "\n%d"	1	SB	PP
19	Ignored the rule that the smaller number comes first when the distances are equal	66	PCE	WA
34	In the Bubble Sort algorithm, only the adjacent elements of the array holding the value (e.g. value[i] and value [i-1]) are exchanged, whereas the adjacent elements of the distance array are not exchanged (should be exchanged)	15	KB	WA
61	" = =" is written as " =" by mistake	11	RB	CP
62	A statement is followed by an extra '2'	1	SB	CP
64	A "}" is missing	5	PCE	CP

OC: Occurrences; CL: Cognitive Level; OJ: Online Judging feedback; SB: Skill-based error; RB: Rule-based error; KB: Knowledge-based error; PCE: Post-Completion Error.

5 The Results Analysis

We conducted Pearson's chi-squared test on H1. The chi-square test is an appropriate statistical method for determining whether there is a significant difference between expected frequencies and observed frequencies in one or more categories of the contingency table, as shown in Table 2.

Table 2. The contingency table for Chi-square test

Cognitive levels	Whether coincident		Total
	Non-coincident	Coincident	
Skill-based error	16	0	16
Rule-based error	10	18	28
Knowledge-based error	13	9	22
Post-completion Errors	0	4	4
Total	39	31	70

The largest category of faults is rule-based, constituting 40% (28/70) of all faults, of which 64% (18/28) are coincident faults. Knowledge-based faults account for 31.4% of all faults, of which 41% are coincident faults. Post-completion errors cause fewest faults (4) but all of them are coincident faults. Skill-based faults make up 23% of all faults, but there are no coincident fault (0%) at this level.

H1: The likelihoods of a fault being coincident are equal across the Skill-based, Rule-based, Knowledge-based, and Post-completion levels. H1 is rejected based on the results of the chi-square test on the data of the Bubble Sort Problem, χ^2 (df = 3, N = 70) = 22.39, p = 0.000. The test shows that the likelihoods of a fault being coincident fault are significantly different at various cognitive levels.

Finding A: The proposed CognFCF has overall captured the cognitive factors underlying fault diversity, as the likelihoods of a fault being coincident at various cognitive levels in CognFCF are statistically significant different.

Next, we analyzed the differences in fault occurrences across cognitive levels by testing hypotheses H2.1–2.6. We employed Mann-Whitney test, a nonparametric test used to examine whether the means of two populations differ significantly. Since each participant's fault only counts once, our fault samples are independent. The Mann-Whitney test is appropriate when the samples are independent and the populations are not normally distributed. We conducted the tests using SPSS Statistics 26. The statistics for fault occurrences at each cognitive level are presented in Table 3.

Table 3. The Statistics for the occurrences of faults at each cognitive error level

Cognitive Level	N	Mean	Std. Deviation	Min	Max
Skill-based	16	1.00	0.00	1	1
Rule-based	28	7.75	22.96	1	122
Knowledge-based	22	2.82	3.80	1	15
Post-completion error	4	19.00	31.36	2	66
Total	70	5.30	16.52	1	122

H2.1: The Occurrences (OCs) of skill-based faults are equivalent to that of rule-based faults. H2.1 is rejected based on the data of the Bubble Sort Problem (U = 368.00, N = 44, p = 0.000). The Occurrences of faults at the rule-based level are significantly higher than that at skill-based level ($p \leq 0.01$).

H2.2: The OCs of skill-based faults are equivalent to that of knowledge-based faults. H2.2 is rejected based on the data of the Bubble Sort Problem (U = 248.00, N = 38, p = 0.004). The occurrences of a fault at the knowledge-based level are significantly higher than that at the skill-based level ($p \leq 0.01$).

H2.3: The OCs of skill-based faults are equivalent to that of post-completion faults. H2.3 is rejected based on the data of the Bubble Sort Problem (U = 64.00, N = 20, p = 0.000). The occurrences of a post-completion fault are significantly higher than that at the skill-based level ($p \leq 0.01$).

Finding B: The occurrences of skill-based faults are significantly lower than that of faults at any other cognitive levels. We identified a total of 16 faults at skill-based performances in our experiment; all of them were unique faults (Occurrence=1).

The psychological explanation for Finding B is that skill-based errors are caused by attention and random situational factors, such as interruption. Typical skill-based performances in programming include typing and running a program by clicking a button in a programming environment that one is very familiar with. In N-version programming, where different programmers independently produce various versions of programs, they are unlikely to be interrupted at the same time or make the same typing errors.

H2.4: The OCs of rule-based faults are equivalent to that of knowledge-based faults. H2.4 is retained based on the data of the Bubble Sort Problem ($U = 237.50$, $N = 50$, $p = 0.145$). The occurrences of rule-based faults are not significantly different from that of knowledge-based faults.

H2.5: The OCs of rule-based faults are equivalent to that of post-completion faults. H2.5 is retained based on the data of the Bubble Sort Problem ($U = 81.50$, $N = 32$, $p = 0.137$). The occurrences of rule-based level faults are not significantly different from that of post-completion faults.

Rule-based coincident faults constitute 26% (18/70) of all the faults, ranking 1st among various categories in the contingency table, shown in Table 2. Combining the Chi-square test results (H1), we can see that, the risk of different programmers committing the same error on a function point corresponding to rule-based performances is highest. This suggests that software practitioners should pay special attention to those function points corresponding to rule-based performances.

Finding C: The occurrences of rule-based faults are not significant different from that of knowledge-based and post-completion faults, however, programmers are most likely to introduce coincident faults in rule-based performances.

H2.6: The OCs of knowledge-based faults are equivalent to that of post-completion faults. H2.6 is rejected based on the data of the Bubble Sort Problem ($U = 73.00$, $N = 22$, $p = 0.027$). The occurrences of a fault due to post-completion error are significantly higher than that at the knowledge-based level ($p < 0.05$).

Post-completion faults are an interesting category with a low probability of occurring (6%), but once a post-completion fault occurs, it is most likely to be repeatedly introduced by another programmer: 100% of the post-completion faults are coincident faults. Statistical tests also show that the occurrences of post-completion faults are significantly higher than that of faults at skill-based level (H2.3) and knowledge-based level (H2.6).

Finding D: Once a post-completion fault is introduced by a programmer, it is most likely to be repeated by another programmer (the conditional probability is the highest).

6 Discussions

6.1 Implications for Avoiding Coincident Faults in N-version Programming

The results and insights gained from this experimental study provide a guideline for practitioners on how to select various diversity-seeking strategies while reducing costs. Previous experimental or empirical studies have focused on single strategies such as independent development [7], use of different programming languages [8], and different algorithms [9], all of which have been shown to improve fault diversity but still resulted in coincident faults. An intuitive strategy is to combine multiple approaches, which was supported by empirical evidence in [22] that combining language diversity and program structural diversity can improve reliability. However, implementing multiple "forced diversity" strategies on every function of the requirement can be prohibitively expensive.

The results of this study suggest which parts of the requirements need forced diversity and which do not. Practitioners can first classify the functions in the requirements into three performance levels, and then, assess which parts need forced diversity, and which parts do not need "forced diversity" at all.

For skill-based performances, coincident errors (such as typos and using programming tools) are almost not a concern, so the cost of forcing programmers to use different programming tools can be saved. Independent development should be highly effective in preventing coincident faults in skill-based performances.

For knowledge-based performances, natural independent development is recommended, and no extra strategies are implied for seeking fault diversity.

For rule-based performances, forced diversity is recommended, and using different programming languages could be especially effective since it constitutes a large proportion of rule-based performances. A new strategy derived from this study is to assign two programmers at different expertise levels, e.g., one at the rule-based level and the other at the knowledge-based level, if such a pair of programmers available.

Another new strategy gained from this study is for post-completion faults. Once a post-completion scenario is presented in the requirements or design, it is highly likely to introduce coincident faults, furthermore, which are difficult to prevent using independent development alone. Extra cognitive strategies such as highlighting the last step of the task or changing the steps of the process [18] can be used to prevent coincident faults due to post-completion errors.

6.2 Limitations and Future Studies

Two limitations exist concerning the extent to which the findings can be generalized. The experimental study is limited to one single task, i.e. the super bubble sort problem.

While the use of a single task has covered all cognitive levels for the participants, it may limit the generalizability of our findings to other tasks that software developers engage in, such as refactoring, augmentative and corrective maintenance. The second limitation is that the participants in this study were undergraduate students.

In the future, we plan to empirically evaluate the derived strategies for avoiding coincident faults on more programming tasks with experienced programmers. Another interesting study would be to formally model the prevalence of errors, the relationships between different types of errors and faults, and faults and failures. This will enable us to incorporate the proposed cognitive framework into fault diversity modeling for pursuing fault tolerance of software systems.

7 Conclusion

In this paper, we have presented a cognitive framework for modeling coincident faults from a human factors perspective. Our experimental study involved 200 student programmers with similar training backgrounds who were asked to solve a Super Bubble Sort problem using the same language and tool. Our study provides insights into the contribution of individual diversity of programmers to fault diversity, i.e., the likelihood of different types of cognitive activities in programming introducing the same faults, while the external factors (i.e. language, tool and training backgrounds) are the same. Our results show that skill-based performances are unlikely to introduce coincident faults, and independent development is recommended for knowledge-based performances. Forced diversity and a new programmer task assigning strategy are recommended for preventing rule-based coincident faults. We also proposed a new cognitive strategy to avoid post-completion errors, which are prevalent and was proposed to avoid such errors. These findings enable us to derive an integrated framework for avoiding coincident faults that suggests which proportions of requirements should employ different diversity seeking strategies. This framework has the potential to reduce costs compared to full-scoped forced diversity while still achieving the same degree of fault tolerance in critical software systems. Future research can further evaluation this framework on more diverse tasks in industrial settings.

References

1. Lyu, M.R.: Handbook of Software Reliability Engineering. IEEE Computer Society Press, California (1996)
2. Littlewood, B., Popov, P., Strigini, L.: Modeling software design diversity: a review. ACM Comput. Surv. **33**, 177–208 (2001)
3. Lyu, M.R., Chen, J.-H., Avizienis, A.: Experience in metrics and measurements for N-version programming. Int. J. Reliab. Qual. Saf. Eng. **1**, 41–62 (1994)
4. Xu, H., Chen, Z., Wu, W., Jin, Z., Kuo, S.-y., Lyu, M.: NV-DNN: towards fault-tolerant DNN systems with N-version programming. In: 49th Annual IEEE/IFIP International Conference on Dependable Systems and Networks Workshops (DSN-W), pp. 44–47. IEEE (2019)
5. Levitin, G., Xing, L., Xiang, Y.: Optimal early warning defense of N-version programming service against co-resident attacks in cloud system. Reliab. Eng. Syst. Saf. **201**, 106969 (2020)

6. Eckhardt, D.E., et al.: An experimental evaluation of software redundancy as a strategy for improving reliability. IEEE Trans. Software Eng. **17**(7), 692–702 (1991). https://doi.org/10.1109/32.83905
7. John, C.K., Leveson, N.G.: An experimental evaluation of the assumption of independence in multi-version programming. IEEE Trans. Software Eng. **12**, 96–109 (1986)
8. Avzenis, A., Lyu, M.R., Schutz, W.: In search of effective diversity: a six-language study of fault-tolerant flight control software. In: Proceedings of the 18th International Symposium on Fault-Tolerant Computing, pp. 15–22. (1988)
9. Feldt, R.: Generating diverse software versions with genetic programming: an experimental study. IEE Proc., Softw. **145**, 228–236 (1998)
10. Yoo, C.S., Seong, P.H.: Experimental analysis of specification language diversity impact on NPP software diversity. J. Syst. Softw. **62**, 111–122 (2002)
11. Meine, J.P., van der Meulen, M.A.R.: Correlations between internal software metrics and software dependability in a large population of small C/C++ programs. In: 18th IEEE International Symposium on Software Reliability Engineering, pp. 203–208. IEEE Computer Society (2007)
12. Reason, J.: Human Error. Cambridge University Press, Cambridge, UK (1990)
13. Rasmussen, J.: Skills, rules, and knowledge; signals, signs, and symbols, and other distinctions in human performance models. IEEE Trans. Syst. Man Cybern. **13**, 257–266 (1983)
14. Byrne, M.D., Bovair, S.: A working memory model of a common procedural error. Cogn. Sci. **21**, 31–61 (1997)
15. Palanque, P., Cockburn, A., Gutwin, C.: A classification of faults covering the human-computer interaction loop. In: 39th International Conference Computer Safety, Reliability, and Security, pp. 434–448. Springer (2020)
16. Mohanani, R., Salman, I., Turhan, B., Rodríguez, P., Ralph, P.: Cognitive biases in software engineering: a systematic mapping study. IEEE Trans. Software Eng. **46**, 1318–1339 (2018)
17. Huang, F., Liu, B., Huang, B.: A taxonomy system to identify human error causes for software defects. In: The 18th international conference on reliability and quality in design, pp. 44–49. International Society of Science and Applied Technologies (2012)
18. Huang, F., Strigini, L.: HEDF: a method for early forecasting software defects based on human error mechanisms. IEEE Access **11**, 3626–3652 (2023)
19. Li, S.Y., Blandford, A., Cairns, P., Young, R.M.: Post-completion errors in problem solving. In: Proceedings of the Twenty-Seventh Annual Conference of the Cognitive Science Society. Citeseer (2005)
20. Huang, F.: Post-completion error in software development. In: The 9th International Workshop on Cooperative and Human Aspects of Software Engineering, ICSE 2016, pp. 108–113. ACM (2016)
21. Gisev, N., Bell, J.S., Chen, T.F.: Interrater agreement and interrater reliability: key concepts, approaches, and applications. Res. Social Adm. Pharm. **9**, 330–338 (2013)
22. Popov, P., Stankovic, V., Strigini, L.: An empirical study of the effectiveness of "forcing" diversity based on a large population of diverse programs. In: 23rd International Symposium on Software Reliability Engineering, pp. 41–50. IEEE (2012)

A Taxonomy of Software Defect Forms for Certification Tests in Aviation Industry

Fuqun Huang[1], Bing Huang[2], Yikun Wang[3], and Yichen Wang[4](✉)

[1] Department of Computer Science, Western Washington University, Bellingham, WA 98225, USA
huangf2@wwu.edu

[2] School of Computer Science and Engineering, Nanyang Technological University, Singapore, Singapore
bing.huang@ntu.edu.sg

[3] Antares Testing LLC, Beijing, China
echo@antares-testing.com

[4] School of Reliability and Systems Engineering, Beihang University, Beijing 100191, China
wangyichen@buaa.edu.cn

Abstract. The analysis of real software defects data not only helps to identify problems in a project, company, or industry but also enables continuous improvement in technologies and processes. In the Chinese aviation industry, a rigorous software quality assurance procedure is implemented, consisting of following a national process standard, conducting internal tests, expert milestone reviews, and formal documentation of all activities and results. After these quality assurance activities are completed, software systems undergo certification tests conducted by independent third-party centers. Discovering patterns of defects found in certification tests has significant practical implications for the industry on continuous improvement. This paper presents a taxonomy of defect forms based on the real defects found in 10 software systems, using Grounded Theory. The defect forms contain detailed information on how the defects are formed, compared to existing defect types. We then applied and validated this taxonomy on 9 additional software projects, using five software certification engineers independent of this paper. The results demonstrate that the developed taxonomy can describe the forms of 98% of defects found in certification tests. We recommend this taxonomy for process improvement and defect prevention in the aviation industry.

Keywords: Software Defect · Defect Form · Software Reliability · Aviation Industry

1 Introduction

How to appropriately classify software defects is important for safety-critical industries. A well-defined defect classification is not only necessary for effective communication, developers, and quality assurance engineers), but also can promote continuous process improvement. For instance, defect classification is a necessity for defect prevention [1].

J. Guiochet et al. (Eds.): SAFECOMP 2023, LNCS 14181, pp. 55–63, 2023.
https://doi.org/10.1007/978-3-031-40923-3_5

There are several well-known defect classifications methods. Orthogonal Defect Classification (ODC) [1] is the most popular defect classification method found in the literature, in addition to the IEEE Standard 1044–2009 on software Anomalies [2]. These defect classifications have been widely applied in academic research, e.g. defect prediction [3]. Leszak et al. [4] proposed 18 defect types in three categories (Implementation, Interface and External) for retrospective analysis of software defect projects, such as Data design/usage, functionality, and development environment. There have been several reports on applications of ODC on real safety-critical systems in aerospace domains. For instance, Lutz and Mikulski [5] used ODC to analyze residual defects on seven spacecraft systems, and identify the most frequent types appearing in those systems were "procedures" problems. Silva and Vieira [6] applied ODC on 4 aerospace systems and found that about one third of defects were difficult to be assigned to one of the defect types in ODC.

The Chinese aviation industry implements a rigorous process for assuring software quality [7]. The widely reported classification methods are not used in such certification tests as they were not suitable for the needs of the domain. Software certification tests are conducted by qualified and independent third-party software testing centers in the last phase of software quality assurance for safety-critical software systems. The industry requires the consistency between documentations and programs, and has a demanding standard for clearly describing the defects, and requires the classification can be easily understood and made consensus between multiple stakeholders, e.g. user representatives (usually air force administers), developers, project managers, domain experts, and software certification testing engineers. Thus, a defect classification method that works for the defects found in software certification tests is vital.

We approach this problem in an empirical manner by deriving the taxonomy of defect patterns from real historical data using Grounded Theory, rather than developing a taxonomy in a predetermined way. Then, we validate the taxonomy by professional testing engineers independent of this study on another set of real data.

2 Background and Concepts

In the Chinese aviation industry, software certification tests are usually performed by independent third-party centers. These centers must have proven their capability and obtained the relevant national qualifications. To certify a software system, a variety of certification tests must be performed, including system testing, code review, document review, etc. The defects found in these tests are formally documented, reviewed, and acknowledged by multiple stakeholders, including the software system's designers and developers, the project manager, the representative of the development institute, and the representatives of potential users.

Before entering such certification tests, a software system is supposed to have gone through a complete process of quality assurance, including requirement review carried out by the development institute. Every software project should follow the same national process standard, perform internal tests including requirements review, code review, unit test, integration test, and formally document these milestone reviews. Such an "internal" quality assurance process is guided by national standards and is mandatory [7] for a software system to be used in a Chinese aircraft.

The concepts used in this paper are defined as follows:

Defect: An incorrect or missing step, process, or data definition in software, including computer programs and documentation (adapted from the definition of "fault" in the IEEE Standard Glossary of Software Engineering Terminology [8]).

Defect Form (DF): The way how a snippet of software (including computer programs and documentations) is being a defect. Note that the Defect Form here is not the same as the "Defect Type" concept commonly used in software engineering, e.g. in "Orthogonal defect classification" [1]. Defect Form describes "what" a defect is, in more detail than "Defect Type".

3 Methodologies

We used the Coding method in Grounded Theory to derive the defect forms from historical defects identified in the certification tests. Subsequently, software certification testing engineers, who were independent of this study applied the taxonomy of the defect forms to another 9 projects. An overview of the methodologies and process employed in this study is depicted in Fig. 1.

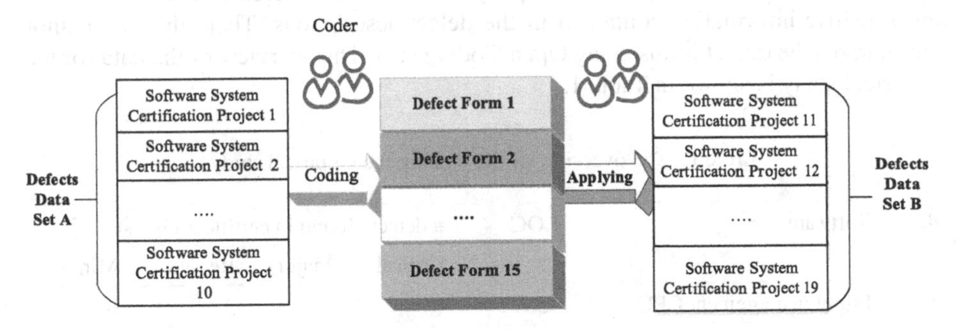

Fig. 1. The overview of the methodology and process of the study

3.1 Grounded Theory and Coding

Grounded Theory, initially developed by Glaser and Strauss [9], is a methodology used to build new theories based empirical data. The methodology involves a set of strategies, such as coding and memoing, to analyze data. Grounded Theory is widely used in various domains, including software engineering [10, 11]. Grounded theory employs an inductive and iterative process to generate a new theory from empirical data. Inductive research aims to identify categories and relations from empirical data and build a theory to explain a phenomenon. The theory-building process is an interactive one, and researchers continuously assimilate and integrate new information into the existing mental model, leading to the evolution of the theory during the research process.

To construct a new theory using Grounded Theory, major categories of concepts that can later be used to build the theory must be identified. This process is called Open Coding [12]. Codes may be derived from the text or created by researchers to provide meaning to the text. In this study, we used the Open Coding [12] method to derive defect forms from detailed descriptions of defects found in certification tests.

3.2 The Coder and Data Set

At the time of coding, the first author had over seven years of industrial experience in software systems verification, validation, and certification in the Chinese aviation industry, over ten years of experience in scientific research on software defects and human errors, and sufficient experience in coding [13].

In the first step, the last author reviewed the original data stored in our partner company, which is an independent third-party center for software verification, validation and certification. They recorded the following information: all defects for each software system, defect descriptions (not publishable due to a non-disclosure agreement), how each defect was fixed, the severity of each defect, and the defect type. The defect type was originally classified by software testing engineers and reviewed in formal project reviews by various independent stakeholders not involved in this study. The defect type includes four categories: Documentation, Program, Design, and Others.

In the second step, the partner company performed data desensitization to remove any sensitive information contained in the defect descriptions. Then, the first author summarized the defect forms using Open Coding [12]. The overview of the data for the empirical study is shown in Table 1.

Table 1. The overview of the data for the empirical study

#	Software	LOC	# defects found in certification tests			
			Critical	Major	Regular	Minor
1	Flight management CPU	75572	0	1	5	0
2	Flight Management MIO	9692	1	0	6	0
3	Inertial Navigation Software	26406	0	1	13	0
4	Inertial Attitude Software	18361	0	3	1	2
5	Anti-jamming all-in-one	21683	0	0	6	9
6	Radio Altimeter Software	7572	1	2	0	0
7	Task management software	20353	0	2	35	3
8	Power management	21947	0	4	3	0
9	Integrated monitoring	49450	0	0	18	3
10	Portable Maintenance Aid	27314	0	3	2	0
Total		278350	2	16	89	17
			124			

3.3 The Coding Results

Using the coding process, the coder extracted 13 defect forms that fully described the 124 defects in these 10 software systems. The defect forms, along with the number of defects for each form and examples are described as follows:

- **DF1-Useless requirements specification** (13 defects, 10%): A piece of requirements specified in the requirements specification document, but it is not implemented in the program. The defect is fixed by deleting this piece of requirements, and the program remains unchanged.
- **DF2-Missing requirements specification** (8 defects, 6%): A piece of function in a program that has no descriptions in the requirements document. The defect is fixed by adding description in the requirements document, and the program remains unchanged.
- **DF3-Inappropriately organizing requirement specifications** (2 defects, 2%): The requirements are inappropriately organized. The defect is fixed by adjusting the structure of the relevant specifications or putting them in another section, but no change is made to the text.
- **DF4-Incorrect requirements specification** (1 defect, 1%): The requirements specification is incorrect, causing the program implementation to be incorrect. The defect is fixed by changing both the requirement specification and the program.
- **DF5-Inconsistency between requirements specification and program function (I)** (19 defects, 15%): Inconsistency is found between the requirements specification and program function, and the defect is fixed by changing the requirements specification document.
- **DF6-Inconsistency between program function and requirements specification (II)** (35 defects, 28%): Inconsistency is found between the requirements and program function, and the defect is fixed by changing the program, while not changing the requirements. For instance, the interface protocols specify that the valid atmosphere altitude is -500 ~ 5000, while in the program it is set as "-500 ~ 6000". The defect is fixed by simply changing 6000 to 5000 in the program.
- **DF7-Ambiguous requirements specification** (2 defects, 2%): The description of a piece of requirements is not clear. The defect is fixed by clarifying the texts in the requirements specification document.
- **DF8-Calculation error** (9 defects, 7%): Some details of a function for calculating data is incorrect, causing the output of the program to be incorrect. The defect is fixed by changing the program. For instance, the accuracy of calculating the aircraft heading angle does not meet the requirements.
- **DF9-Algorithm error** (11 defects, 9%): A significant step or the logic of an algorithm is missing or incorrect, causing the output of the program to be incorrect. The defect is fixed by changing the program.
- **DF10-Assignment error** (2 defects, 2%): Inappropriately creating variables or assigning values to variables.
- **DF11-Missing function** (10 defects, 8%): A function or step is missing in the program, and the cause is unclear (not including post-completion error).
- **DF12-Exception handling error** (10 defects, 8%): Lack of mechanisms for handling exceptional conditions or data out of normal range. For instance, missing watchdog in a searching algorithm in the Radio Altimeter Software.

- **DF13-Annotation error** (2 defects, 2%): This defect form pertains to instances where the annotations or comments in the program are inaccurate or don't match with the code. The defect is fixed by modifying annotations without any changes to the code.
- **DF14**-Others. DF14 was added for validation purposes only. This defect form encompasses any defects that cannot be categorized under the above 13 forms.

As can be seen from the above list, a defect form contains more information than a defect type. Existing defect types (i.e. documentation, program and others) are very simple and provide little information on what's going on or what needs to be done for further improvement.

4 Application and Validation

In this section, we describe the application and validation of the developed taxonomy, which was offered to software engineers currently involved in software certification tests in the Chinese aviation industry.

4.1 Participants and Procedures

We recruited five experienced software certification testing engineers, all of whom held team leader positions at the time of participating the study. The defects data used in the study were obtained from the projects these engineers were actively involved in, and they were the most familiar with the defects under classification. The severity level and defect types were determined by the teams led by the participants. All of the participants were independent of our study.

We generated a translated version of the defect pattern list and descriptions for each defect form. We also added another category, DF14-Others, to the list to identify any defects that could not be classified using the 13 forms.

The participants used the defect form list to reclassify the defects they had found in the projects they were actively involved in. They were encouraged to raise any issues if there were any unclear definitions or any defects that could not be classified by the taxonomy. In particular, we asked them whether they had encountered any defect that could be classified by more than one defect form to evaluate the degree of mutual exclusivity between the defect forms.

4.2 The Validation Data

The 5 participants independently applied the taxonomy of defect forms to 9 different software systems that had completed certification tests. A total of 169 defects were found in these systems, of which 36 (21%) were Critical defects, 4 (2%) were Major, 111 (66%) were Regular, and 18 (11%) were Minor. Table 2 shows a summary of the defect data used for the validation of the taxonomy.

A participant assigned one defect form to each defect found in his/her certification test project. Table 3 summarizes the distribution of defects of different severity levels across various forms.

In total, 166 out of the 169 defects (98%) were accurately described by one of the 13 defect forms in the taxonomy. Only three defects (2%) were classified as DF14-other. The participants found that 10 out of the 13 defect forms provided a complete and clear description for the 166 defects.

Of the 169 defects, 55% (93) were originally classified as "Program" type, while 41% related to documentation issues. The remaining 4% were categorized as "other" defects. The distribution of defects across types and forms is presented in Table 3.

Table 2. The Defects for validating the taxonomy

Softw	Cr#	Mj#	Re#	Mi#	Total
CJGZKZ	4	0	6	0	10
JZCL	5	0	0	0	5
ZKGL	9	0	16	6	31
TXCL	2	0	5	0	7
HJGL	1	0	13	1	15
BFCC	3	0	2	1	6
IISS	0	4	1	2	7
DXX	5	0	5	5	15
JT	7	0	63	3	73
Total	36	4	111	18	169

Table 3. Defects across Types and Forms

Defect Forms	Doc	Prog	Other	Total #	#%
DF2	3	-	-	3	1.8
DF5	64	-	-	64	37.9
DF6	2[a]	38	-	40	23.7
DF8	-	2	-	2	1.2
DF9	-	4	-	4	2.4
DF10	-	30	-	30	17.8
DF11	-	8	-	8	4.7
DF12	-	9	-	9	5.3
DF13	-	-	6	6	3.6
DF14	1	2	-	3	1.8
Total	70	93	6	169	100

[a]. Comments in the program are inconsistent with requirements

5 Results Analysis and Discussions

The evaluation of software defect classification methods typically involves several criteria, such as clear definition and examples for each category, completeness, and non-overlapping categories.

To assess the clarity and non-overlapping nature of our taxonomy, we included an open-ended question in the questionnaire regarding the participants' experiences with assigning defect forms, e.g. misunderstanding the defect forms, not sure of which defect form to be assigned to a defect, or a defect can be assigned with more than one defect form. Participants reported no issues with the taxonomy.

We estimated the completeness of the taxonomy in describing defects found in the Chinese aviation industry's certification tests using Eq. (1):

$$Completeness = \frac{\sum_{1}^{13} DF_i}{\sum_{1}^{14} DF_i} \times 100\% \qquad (1)$$

where DF_{14} represents a defect was assigned to "other". That is, Completeness is estimated based on the number of defects assigned to the 13 defined forms out of all defects.

Results demonstrate that the taxonomy effectively covered 98% (166/169) of defects identified in the nine software certification projects. Moreover, participants utilized only 9 out of the 13 available defect forms, with their distribution presented in the sixth column of Table 3.

As such, we recommend the following 9 defect forms for use in the Chinese aviation and other safety-critical industries, ranked from highest to lowest frequency:

1) Inconsistency between requirements specification and program function, fixed by changing requirement documentation, 2) Inconsistency between program function and requirements specification, fixed by changing program, 3) Assignment error, 4) Exception handling error, 5) Missing function, 6) Annotation error, 7) Algorithm error, 8) Missing requirements specification, 9) Calculation error.

6 Conclusion

This paper proposed a new concept "defect form" to describe the patterns of defects found in certification tests in the Chinese aviation industry. We developed a taxonomy consisting of 13 defect forms derived from 10 software systems using Grounded Theory. The taxonomy were applied and validated by 5 independent professional certification testing engineers on another 9 software systems certification projects. Results show that nine defect forms were able to describe 98% defects. We recommend these defect forms for various stakeholders to consider in their future projects and strategy making for the whole Chinese aviation industry. These defect forms could also have implications for certification tests for safety-critical domains in other countries.

References

1. Chillarege, R., et al.: Orthogonal defect classification-a concept for in-process measurements. IEEE Trans. Software Eng. **18**, 943–956 (1992)
2. IEEE Standard Association: IEEE 1044–2009—IEEE Standard Classification for Software Anomalies (2009)
3. Patil, S., Ravindran, B.: Predicting software defect type using concept-based classification. Empir. Softw. Eng. **25**(2), 1341–1378 (2020). https://doi.org/10.1007/s10664-019-09779-6
4. Leszak, M., Perry, D.E., Stoll, D.: Classification and evaluation of defects in a project retrospective. J. Syst. Softw. **61**, 173–187 (2002)
5. Lutz, R.R., Mikulski, I.C.: Empirical analysis of safety-critical anomalies during operations. IEEE Trans. Software Eng. **30**, 172–180 (2004)
6. Silva, N., Vieira, M.: Experience report: orthogonal classification of safety critical issues. In: IEEE 25th International Symposium on Software Reliability Engineering, pp. 156–166. IEEE (2014)
7. Huang, F., Liu, B., Wang, S., Li, Q.: The impact of software process consistency on residual defects. J. Softw. Evol. Process **27**, 625–646 (2015)
8. IEEE: IEEE Standard Glossary of Software Engineering Terminology, vol. 1 EEE Std 610.121990. The Institute of Electrical and Electronics Engineers, New York (1990)
9. Glaser, B.G., Strauss, A.L.: The discovery of grounded theory: Strategies for qualitative research. Aldine, Chicago (1967)
10. Sjøberg, D.I.K., Dybå, T., Anda, B.C.D., Hannay, J.E.: Building theories in software engineering. In: Shull, F., Singer, J., Sjøberg, D.I.K. (eds.) Guide to advanced empirical software engineering, pp. 312–336. Springer London, London (2008). https://doi.org/10.1007/978-1-84800-044-5_12
11. Coleman, G., O'Connor, R.: Using grounded theory to understand software process improvement: a study of Irish software product companies. Inf. Softw. Technol. **49**, 654–667 (2007)
12. Saldaña, J.: The coding manual for qualitative researchers. Sage (2012)
13. Huang, F., Smidts, C.: Causal mechanism graph – a new notation for capturing cause-effect knowledge in software dependability. Reliab. Eng. Syst. Saf. **158**, 196–212 (2017)

Constraint-Guided Test Execution Scheduling: An Experience Report at ABB Robotics

Arnaud Gotlieb[1(✉)], Morten Mossige[2], and Helge Spieker[1]

[1] Simula Research Laboratory, Kristian Augusts Gate 23, 0164, Oslo, Norway
{arnaud,helge}@simula.no
[2] ABB Robotics, Bryne, Norway
morten.mossige@uis.no

Abstract. Automated test execution scheduling is crucial in modern software development environments, where components are frequently updated with changes that impact their integration with hardware systems. Building test schedules, which focus on the right tests and make optimal use of the available resources, both time and hardware, under consideration of vast requirements on the selection of test cases and their assignment to certain test execution machines, is a complex optimization task. Manual solutions are time-consuming and often error-prone. Furthermore, when software and hardware components and test scripts are frequently added, removed or updated, static test execution scheduling is no longer feasible and the motivation for automation taking care of dynamic changes grows. Since 2012, our work has focused on transferring technology based on constraint programming for automating the testing of industrial robotic systems at ABB Robotics. After having successfully transferred constraint satisfaction models dedicated to test case generation, we present the results of a project called DynTest whose goal is to automate the scheduling of test execution from a large test repository, on distinct industrial robots. This paper reports on our experience and lessons learned for successfully transferring constraint-based optimization models for test execution scheduling at ABB Robotics. Our experience underlines the benefits of a close collaboration between industry and academia for both parties.

1 Introduction

Continuous integration (CI) has been adopted by many companies all around the world in order to ensure better end-user product quality [3]. As part of CI, automated testing is crucial to get quicker feedback on the detected defects or regressions of a system under test. When a complete industrial system is tested under CI, a challenge arises if it relies on hardware and software components, because they can hardly be tested in isolation. Besides, additional challenges include the

List of authors is given in alphabetical order.

J. Guiochet et al. (Eds.): SAFECOMP 2023, LNCS 14181, pp. 64–72, 2023.
https://doi.org/10.1007/978-3-031-40923-3_6

requirement to generate tests with environmental hazards, the combinatorial explosion of the number of potential test cases due to parameter interactions, the automation of test execution scheduling which ensures proper coverage and diversity of test cases and agents.

This paper reports on our experience in deploying a constraint-guided test execution scheduling method as part of a CI process at ABB Robotics. By co-developing an automated testing process named DynTest through an industrial-academic partnership, the authors have explored the transfer of advanced constraint programming[1] models composed of global constraints and rotational diversity [9] in a highly automated industrial testing process [2]. Since 2012, multiple models for test case generation [4,5] and selection, test prioritization [8] and eventually test execution scheduling [6] have been explored, evaluated and transferred. Our experience underlines the benefits of a close collaboration between industry and academia for both parties in the area of automated testing.

2 Test Execution Scheduling at ABB Robotics

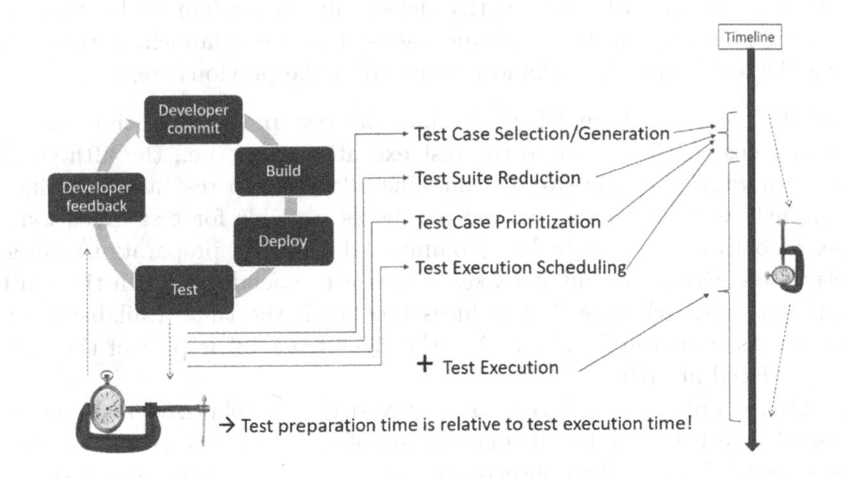

Fig. 1. Overview of the CI cycle and the challenge related to time management

ABB Robotics is an industrial robot supplier and manufacturer company operating in more than 50 countries around the world. A key objective of the company is to deliver high-quality products (thus involving an increased focus on testing robots for reliability and performance) for the benefice of its customers. Initially, robot testing was done mostly manually and using human-eyes visual control for checking the results of hand-crafted tests. This restricted the possible

[1] Constraint Programming is a declarative programming framework which uses relations among logical variables and search procedures to find solutions of combinatorial problems [7].

testing time to human-worked hours of test engineers (besides long-running tests, which could use nighttime and weekends) and did not use available robot to its full test capability. To reduce the time-to-market of new products and also improve the quality of these products, the testing process had to be much more automated. To start with, the test automation process had to be placed within a Continuous Integration (CI) process.

As shown in Fig. 1, a typical CI cycle includes software developer commit actions which automatically trigger build, deploy and test activities. The test results are then passed back to the developers to provide them with feedback. Typically, the test activity includes the following five steps:

1. **Test Case Selection and Generation:** Tests are either extracted from an existing repository or automatically generated from specific requirements;
2. **Test Suite Reduction:** Test suites that achieve a given objective (e.g., full requirement coverage) are pruned to eliminate spurious test cases;
3. **Test Case Prioritization:** Tests are ordered to provide a quick feedback by using either pre-determined or dynamically-computed priority values;
4. **Test Execution Scheduling:** Test plans are distributed on different robots, in a specific order according to a pre-computed test schedule;
5. **Test Execution:** Tests are then eventually executed according to the specified schedule, in order to identify defects in the system under test. This activity is clearly the most demanding as it requires launching the system with the test cases selected and prioritized in the previous steps.

It is worth noting that, in CI, controlling the test preparation time (i.e., the four first steps) with respect to the test execution time (i.e., the fifth step) is crucial. Knowing that the overall time-line allocated to test activities has to be bounded, we have to keep as much time as possible for test execution. Of course, an optimized test schedule (computed during test preparation) can lead to better test execution, but it makes no sense to spend too much time in the computation of a schedule if it reduces too much the time available for test execution. As shown in Fig. 1, finding the right trade-off is part of the testing challenges faced at ABB.

In ABB's context, a *test case* aims at verifying a robotized task, which is performed by a robot under the observation of some sensors. A test can either fail or succeed; it fails when observations reveal a misfunction, and it succeeds when no misfunction is observed. A test case is associated with some metadata, consisting of its average duration, previous execution times, results, and targeted robots. Each test case execution is non-preemptive, that is, it cannot be interrupted by another test or transferred to another robot during execution. All test cases are independent, without any dependency on the order in which they are executed. Still, they can be ordered by using their static priority, which is decided by the test engineers, and their dynamic priority, which is based on a combination of their effectiveness to reveal defects in earlier CI cycles and the time since their last execution. Test cases have furthermore hardware requirements, meaning that they can only be executed on certain robots.

Test cases are executed by *test agents*, which are software components that capture the various schedules computed for each CI cycle. Each test agent has a limited amount of time available per cycle and a set of compatible test cases, which it can execute. Computing a test schedule requires to vary the assignment between test cases and test agents between cycles to achieve a full coverage of all possible combinations between tests and hardware over time. Fulfilling this objective balances the confidence in the stability of certain features on different hardware, while giving room for executing many test cases and not executing the same tests multiple times during a cycle.

3 Automated Testing Process

Here, we present the approach to automate the testing process within the CI environment. The testing process is sequential with distributed components, orchestrated by a central test controller. Starting with *data initialization and acquisition* (Sec. 3.1), the process computes the priorities over test cases (Sec. 3.2) and the test schedules (Sec. 3.3). Test execution is performed by distributing test plans to each robot (Sec. 3.4) and eventually test execution reporting takes place (Sec. 3.5). The central test controller, referred to as DynTest, manages the process, acquires and distributes the necessary data from other sources, and provides the interface towards the automated testing process. Other components include a module for test case prioritization, selection, and scheduling, and a module for controlling test agents executing the test cases.

3.1 Data Initialization and Acquisition

The process is set up with available test cases and agents. Some test cases and agents are filtered out to exclude scripts and robots under maintenance or having incompatible hardware requirements. For the remaining test cases, historical meta-data is extracted from the central data repository. This data includes the most recent test execution results, their runtime, and previous test agents they were executed on, etc.

3.2 Test Case Prioritization

The prioritization step is initially designed with a simple approach, to ease the setup of automation and definition of assigned priorities during integration by the test engineers. The process iterates through all executable test cases and assigns each a priority, which is a weighted sum of the time since the last run, the test case duration, and the most recent last results The weights and number of considered historical test results are manually chosen during integration, but that process could be replaced by a self-adaptive method in the future.

3.3 Selection and Scheduling

Selection and scheduling focus on taking the test cases with the highest priorities and distributing them to the test agents until all time available for testing is used. Although test case selection and scheduling are often regarded as two separate

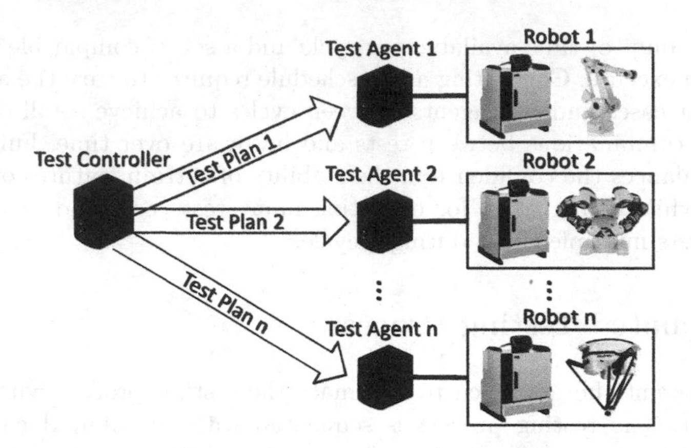

Fig. 2. The test controller distributes individual test plans to each test agent, which controls, in turn, one robot and records log information and test outcomes.

tasks in the literature, in practice, we closely integrate both steps. Selection means to take those test cases from the set of prioritized test cases, which are most desirable to execute. Because the execution of test cases is constrained, this selection has to consider which subset of test cases can actually be executed and at the same time maximizes the available resources (as we want to avoid idle times). The selection and scheduling step receives a set of prioritized test cases and a set of available test agents as inputs. During this step, DynTest creates an execution schedule, where each test case is assigned to one test agent for execution while preferring to assign high-priority test cases over low-priority test cases. During selection, test cases which are marked as obligatory to be run, are always included in the final schedule, regardless of their calculated priority.

We now approach this scheduling task by using Constraint Programming (CP), even if, in its initial version, only a simple greedy first-fill algorithm was used. This heuristic algorithm's first ordered the test cases by descending priority. Then, successively for each test agent, the test case with the highest priority was assigned to the test agent until the maximum time limit was reached. However, we quickly discovered that this too-simplistic approach was not suitable to ensure sufficient diversity in the selection of test cases and agents. We then developed a refined model based on CP. CP is a paradigm in which a problem is not modelled as a sequence of steps to achieve a desired solution, i.e., an algorithm, but relations between variables are described to formulate properties of a desired solution (see [7]). CP and its associated optimization methods are efficient and well-performing techniques for modelling strictly constrained problems, such as planning and scheduling problems [1]. Using CP for scheduling enables precise control over the execution time and the trade-offs made between time looking for a solution and the solution's quality. We replaced the initial scheduling method with a dedicated constraint optimization model which further optimizes the schedules by ensuring that the assignment between test cases

Fig. 3. Visualization of a test schedule with interactive access to test results.

and test agents changes between test cycles. We called this process rotational diversity and used global constraints to develop it. Full details on this constraint model are available in [9].

3.4 Distribution and Execution

Once the test schedule is created, DYNTEST transforms it into separate, individual test plans and sends them to the corresponding test agent. An example test plan is shown in Fig. 3. Each test agent executes all assigned test cases independently, as there is no interdependency between test cases and robots. The test agent records all test results and log files from the test cases and returns them to the test controller.

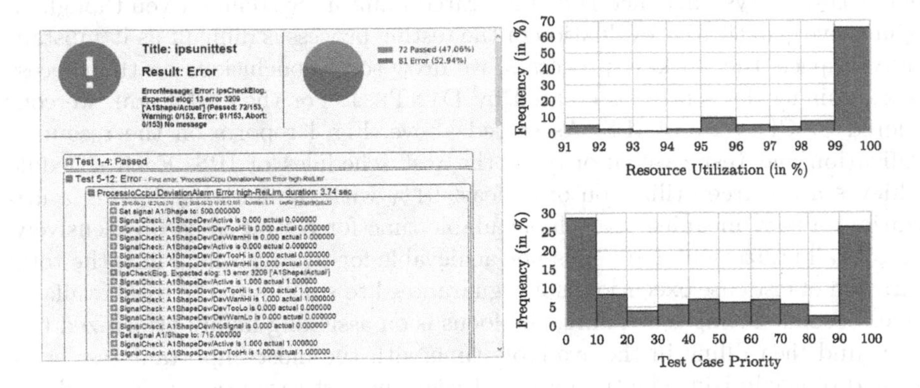

Fig. 4. Visualization of a test report for the IPS project and distribution of resource usage and test case priority among CI cycles for a one-month period.

3.5 Reporting

Reporting aims to communicate the test execution results back to the developer for failure analysis. An example of a test report is shown in Fig. 4. The report summarizes the results of a test cycle, allows us to navigate into lower levels of the test hierarchy and access specific details of single test executions. This hierarchical structure makes the report accessible to different user groups and is the first step of debugging and failure analysis. Another goal of the reporting step is to gather and visualize information about the testing process itself. A visual report of the scheduling outcome is created as a means for individual run analysis and communication (see Fig. 4). It is built on web technologies and enables interactive exploration, including access to test case information and results of recent executions. Exhaustive reporting and data collection enables better long-term evaluation of the system's behavior as well as impact evaluation onto software development, which is an important aspect for tuning the process in the future.

Besides the reporting of individual test results, monitoring the overall behaviour of DynTest is performed. Figure 4 shows examples of two such monitoring metrics. The resource utilization monitors how efficiently the available resources are filled by the test case scheduling algorithm, here most plans should show a high utilization of close to 100% to make the best use of the available resources. The distribution of test case priorities shows the variation in relevance of test cases. Here, there is a large block of highly important test cases with high priority but also chunks with low priority as well as average priority, indicating a good overall balance of priorities.

4 Empirical Evaluation

After a development phase where the integration of all steps of the automated testing process was realized, we performed a one-month empirical evaluation of an existing subsystem called IPS (Integrated Painting Systems). Even though an exhaustive quantitative evaluation of the testing process is difficult as it substantially impacts the working processes, we drew some conclusions on the process by examining the schedules created by DynTest. For the evaluation, we considered 87 CI cycles of IPS. As stated above, Fig. 4 reports on the resources utilization and test case priority of the test schedules of IPS. Each schedule achieves a resource utilization of at least 91 % with the majority having a utilization of 99 % meaning that the available time for testing is used extensively. An overall utilization of 100 % is not achievable for two reasons. First, the total duration of test case execution is not guaranteed to sum up to the total available time. Second, during scheduling, the focus is on assigning highly prioritized test cases and then filling in the available time with the most important test cases instead of maximizing the time usage. Regarding test case priority, Fig. 4 shows that the test cases are spread among the spectrum of possible priorities, with two noticeable clusters at the lower and upper bound of the spectrum. Having a similar number of high- and low-priority test cases stems from the fact, that high-priority test cases, once they have passed their last execution, tend to receive a low priority during the next cycle. This behavior distinguishes from test cases which have not failed during the observed period. After having not been executed for a while, the priority grows again and these test cases become likelier to be executed again.

5 Lessons Learned

We report on three lessons learned while developing test automation process.

Automated test scheduling through CI is crucial to improve robot software/hardware quality. Automated testing through CI allows us to detect at an early stage hardware/software defects on robots and avoid the propagation of failure at customer sites. It also reveals regression issues when the specification of a new product is not yet finalized. This approach significantly improves the overall product quality;

Incremental co-development is relevant when complex constraint optimization models have to be developed. We co-developed a test execution scheduling component as part of DynTest. Starting from a simple version (based on an inefficient greedy-based scheduling approach), we developed a complex constraint optimization model based on global constraints and rotational diversity incrementally. This approach was key to fostering the adoption and maintenance of this complex model by people who do not necessarily have the expertise to maintain advanced constraint models;

Industry-academic co-development. The outcomes of this co-development were beneficial for both sides. On one hand, ABB Robotics benefited from the academic expertise in constraint-based scheduling, which was required to develop test execution scheduling models. On the other hand, scientists took advantage of the industrial experience of the test engineers in the test automation processes, to publish advanced research results with empirical results. Finally, thanks to this co-development, the transferability of the method was easier.

6 Conclusion

This paper reports on an experience to transfer constraint-based models for automated test execution scheduling at ABB Robotics. In this work, advanced constraint-based scheduling models using global constraints and rotational diversity were developed and empirically evaluated, and industrialized as part of a complete CI process. Further work includes refinement in the description of test cases to handle specific globally-shared external equipment.

References

1. Bartak, R., Salido, M.A., Rossi, F.: Constraint satisfaction techniques in planning and scheduling. J. Intell. Manuf. **21**(1), 5–15 (2010)
2. Gotlieb, A., Marijan, D., Spieker, H.: Testing Industrial Robotic Systems: A New Battlefield! In: Software Engineering for Robotics, pp. 109–137. Springer, Cham (2021). https://doi.org/10.1007/978-3-030-66494-7_4

3. Klotins, E., Gorschek, T., Sundelin, K., Falk, E.: Towards cost-benefit evaluation for continuous software engineering activities. Empir. Soft. Eng. **27**, 157(2022) https://doi.org/10.1007/s10664-022-10191-w
4. Mossige, M., Gotlieb, A., Meling, H.: Using CP in Automatic Test Generation for ABB Robotics' Paint Control System. In: O'Sullivan, B. (ed.) CP 2014. LNCS, vol. 8656, pp. 25–41. Springer, Cham (2014). https://doi.org/10.1007/978-3-319-10428-7_6
5. Mossige, M., Gotlieb, A., Meling, H.: Testing robot controllers using constraint programming and continuous integration. Inf. Softw. Technol. **57**, 169–185 (2015)
6. Mossige, M., Gotlieb, A., Spieker, H., Meling, H., Carlsson, M.: Time-aware test case execution scheduling for cyber-physical systems. In: Principles and Practice of Constraint Programming (CP). Springer LNCS, vol. 10416 (2017)
7. Rossi, F., Beek, P.V., Walsh, T.: Handbook of Constraint Programming (Foundations of Artificial Intelligence). Elsevier Science Inc. (2006)
8. Spieker, H., Gotlieb, A., Marijan, D., Mossige, M.: Reinforcement learning for automatic test case prioritization and selection in continuous integration. In: Proceedings of the 26th ACM SIGSOFT International Symposium on Software Testing and Analysis(ISSTA). pp. 12–22 (2017)
9. Spieker, H., Gotlieb, A., Mossige, M.: Rotational diversity in multi-cycle assignment problems. In: Proceedings of the AAAI Conference on Artificial Intelligence. pp. 7724–7731 (2019)

Neural Networks Robustness and Monitoring

A Low-Cost Strategic Monitoring Approach for Scalable and Interpretable Error Detection in Deep Neural Networks

Florian Geissler[1(✉)], Syed Qutub[1], Michael Paulitsch[1], and Karthik Pattabiraman[2]

[1] Intel Labs, Munich, Germany
{florian.geissler,qutub.syed,michael.paulitsch}@intel.com
[2] University of British Columbia, Vancouver, Canada
karthikp@ece.ubc.ca

Abstract. We present a highly compact run-time monitoring approach for deep computer vision networks that extracts selected knowledge from only a few (down to merely two) hidden layers, yet can efficiently detect silent data corruption originating from both hardware memory and input faults. Building on the insight that critical faults typically manifest as peak or bulk shifts in the activation distribution of the affected network layers, we use strategically placed quantile markers to make accurate estimates about the anomaly of the current inference as a whole. Importantly, the detector component itself is kept algorithmically transparent to render the categorization of regular and abnormal behavior interpretable to a human. Our technique achieves up to ∼96% precision and ∼98% recall of detection. Compared to state-of-the-art anomaly detection techniques, this approach requires minimal compute overhead (as little as 0.3% with respect to non-supervised inference time) and contributes to the explainability of the model.

1 Introduction

Deep neural networks (DNNs) have reached impressive performance in computer vision problems such as object detection, making them a natural choice for problems like automated driving [1]. However, DNNs are known to be highly vulnerable to faults. For example, even small changes to the input such as adding a customized noise pattern that remains invisible to the human eye, can stimulate silent prediction errors [8]. Similarly, modifying a single out of millions of network parameters, in the form of a bit flip, is sufficient to cause severe accuracy drops [14].

Because DNNs are being deployed in safety-critical applications such as autonomous vehicles (AVs), we need efficient mechanisms to detect errors that cause such silent data corruptions (SDC). Beyond the functional part, trust in the safety of the application requires that the error detectors are interpretable by the user, so that he/she can develop an intuitive understanding of the regular

J. Guiochet et al. (Eds.): SAFECOMP 2023, LNCS 14181, pp. 75–88, 2023.
https://doi.org/10.1007/978-3-031-40923-3_7

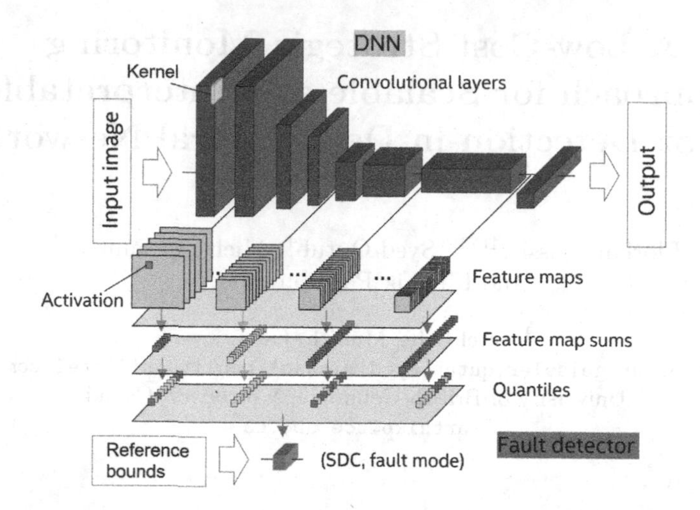

Fig. 1. Monitoring architecture for quantile shift detection.

and irregular behavior of the network [2]. In an AV, for example, a user who does not trust an automated perception component due to its opaque decision-making, will not trust a black-box fault monitor either. Therefore, it is important to build interpretable error detectors for DNNs.

The goal of error detection is to supervise a small, yet representative subset of activations - during a given network inference - for comparison with a previously extracted fault-free baseline. This leads to three key challenges: **(1)** How can one compress the relevant information into efficient abstractions? **(2)** How can one efficiently perform the anomaly detection process, for complex patterns? **(3)** Can the anomaly detection decision be understandable to a human, so that insights are gained about the inner workings of the network?

Unfortunately, no existing approach satisfactorily addresses all three of the above challenges (Sect. 2). This paper presents an solution using a monitoring architecture that taps into the intermediate activations only at selected strategic points and interprets those signatures in a transparent way, see Fig. 1. Our approach is designed to detect SDC-causing errors due to input corruptions *or* hardware faults in the underlying platform memory. Our main observation that underpins the method is that an SDC occurs when a fault injection (FI) *either* increases the values of a few activations by a large margin (referred to here as an *activation peak shift*), or the values of many activations each by a small margin (*activation bulk shift*). As Fig. 2 shows, the former is observed typically for platform faults, while the latter is observed for input faults. We then use discrete quantile markers to distill the knowledge about the variation of the activation distribution in a given layer. Conceptually, within a faulty layer, we can expect a large change of only the top quantiles for activation peak shifts, and small changes of the lower and medium quantiles for bulk shifts (Fig. 2). This idea allows us to produce discriminative features for anomaly detection from a small number of monitored elements, with a single detector.

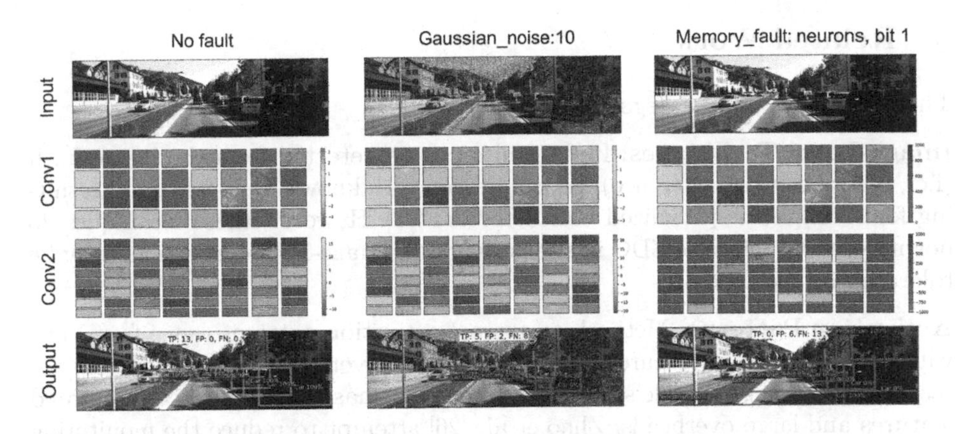

(a) Feature map visualization, the color code represents activation magnitudes. The final prediction result is given as inset in the output row (TP: True positives, FP: false positives, FN: false negatives).

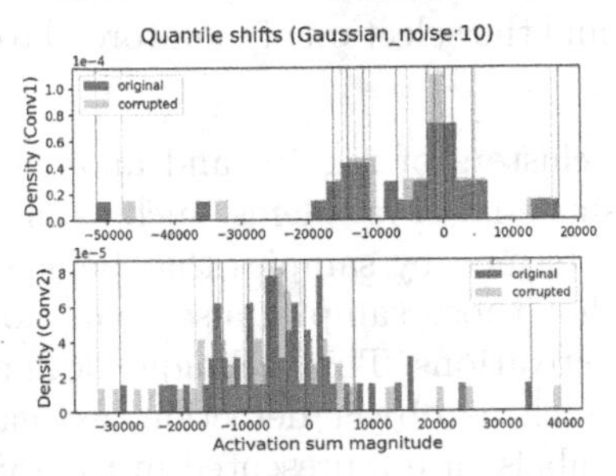

(b) Original image vs. noise corruption of (a). We show both activation sums (bars) and quantile values (overlayed vertical lines).

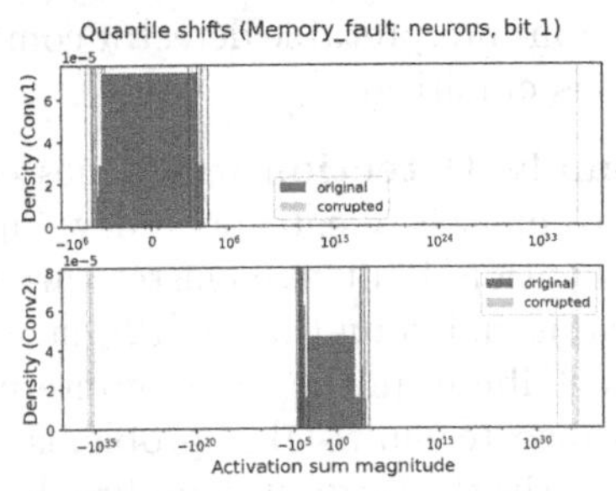

(c) Original image vs. memory fault of (a) (symmetric log scale). Large corruption magnitude ranges lead to histogram densities $<10^{-30}$.

Fig. 2. **(a)** The feature map appearance is slightly changed with noise and massively affected by the memory FI. **(b)** Noise causes a small shift of multiple quantiles from the affected layer onwards (activation bulk shift). **(c)** The layer with the memory FI shows a large shift of the maximum quantile (activation peak shift), which then propagates to other quantiles.

In summary, we make the following contributions in this paper:

- We demonstrate that even for complex object detection networks, we can identify anomalous behavior from quantile shifts in only a few layers.
- We identify minimal sets of relevant features and discuss their universality across models.
- We efficiently differentiate input and hardware fault classes with a single detector.
- We show that the anomaly detection process can be achieved with algorithmically transparent components, such as decision trees.

The article is structured as follows: Sect. 2 discusses related work, while Sect. 3 describes our experimental setup. We present our method in Sect. 4, and the results of our evaluation in Sect. 5.

2 Related Work

There are three main categories of related work.

Image-Level Techniques: Input faults can be detected from the image itself (i.e., before network inference), in comparison with known fault-free data, resulting for example in specialized blur detectors [15]. However, these techniques do not necessarily relate to SDC in the network, as image-level corruptions may be tolerated by the model.

Activation Patterns: Methods to extract activation patterns range from activation vectors [5] to feature traces [24,25]. However, these techniques do not scale well to deeper models as they result in a massive number of monitored features and large overheads. Zhao et al. [26] attempt to reduce the monitoring effort by leveraging only activations from selected layers and compressing them with customized convolution and pooling operations. This leads to a rather complex, non-interpretable detector component, and the selection of monitored layers remains empirical.

Anomaly Detection techniques establish clusters of regular and anomalous data to efficiently categorize new input. In single-label problems, such as image classification, fault-free clusters are typically formed by samples that belong to the same individual label [12], suggesting that those samples also share common attributes in the space of intermediate activations. This technique does not generalize to multi-label problems though, such as object detection, as many objects (in the form of bounding boxes and labels) are represented in the same image. More abstracted clustering rules such as the maximum activation range per layer have been proposed [4,18]. However, these detectors omit more subtle errors within the activation spectrum, for example resulting from input faults. In other work [24–26], a secondary neural network is trained to perform the detection process. This comes at the cost that the detector then does not feature algorithmic transparency [2] and hence the anomaly decision is not understandable to a human. The same limitations are found in the context of detector subnetworks that are trained to identify adversarial perturbations [19].

Summary: We see that none of the prior techniques satisfactorily address the challenges outlined earlier. We present a new technique to overcome this problem in this paper.

3 Experimental Setup and Preliminary Study

Models and Datasets: We use the three classic object detection networks Yolo (v3), Single Shot Detector (SSD), and RetinaNet from the *open-mmlab* [3] framework, as well as the two standard image classification networks ResNet50 and AlexNet from *torchvision* [21]. Object detection networks are pretrained on Coco [20] and were retrained on Kitti [6], with the following AP50 baseline performances: Yolo+Coco: 55.5%, Yolo+Kitti: 72.8%, SSD+Coco: 52.5%, SSD+Kitti: 66.5%, RetinaNet+Coco: 59.0%, RetinaNet+Kitti: 81.8%. Image classification models were pretrained on ImageNet [17], providing accuracies of 78.0% (ResNet) and 60.0% (AlexNet) for the test setup. The data was split in a ratio of 2:1 for detector training and testing. All models are run in *Pytorch* with the IEEE-standard FP32 data precision [16].

Fault Modes: Input faults are modeled using *torchvision* [21] transform functions and are applied in three different magnitudes to the raw RGB images. We select three perturbation patterns that are popular in computer vision benchmarks such as ImageNet-C [11] for our analysis: i) *Gaussian noise* due to low lighting conditions or noise in the electronic signal of the sensor device. Low (0.1), medium (1), and high (10) noise is tested. ii) *Gaussian blur*, reflecting for example a camera lens being out of focus. We choose a kernel size of $(5, 9)$ and a symmetric, variable standard deviation $(0.3, 1, 3)$. iii) *Contrast reductions* simulate poor lighting conditions or foggy weather. We adjust the contrast by a factor between zero (no contrast, gray image) and one (original image). The selected models have different vulnerabilities to input faults, for example, the two image classification models ResNet and AlexNet are highly sensitive to contrast adjustments, but are rather robust to noise and blur faults. For the remaining models, the trend is reversed.

Hardware faults are modeled as single bit flips in the underlying memory and injected using *PytorchAlfi* [9]. Such flips can occur randomly either in the buffers holding temporary activation values (*neuron* fault), or in dedicated memory which holds the parameters of the network (*weight* faults). We group both neuron and weight faults into a single class *memory* fault. This approach is in line with previous work [4,7,13,18,23–25]. We target all convolutional layers.

Fault Metrics: First, detectable uncorrectable errors (DUE) can occur when invalid symbols such as *NaN* or *Inf* are found among the activations at inference time. During fault injection, we observe DUE events only for memory faults, with rates <1% across all models. DUEs can be generated also at the detector stage, in the process of adding up feature map sums that contain platform errors. The rates for such events vary between 0.2% and 5.1% with our method. While DUE errors may affect the system's availability, they are considered less critical as they are readily detectable and there is no need for further activation monitoring [7].

In this article, we are concerned therefore only with silent data corruption (SDC), events that lead to a silent alteration of the predicted outcome. For image classification networks, this is represented by a change in the top-1 class prediction. For object detection systems, we use an asymmetric version of the IVMOD metric [23] as SDC criterion, i.e., an image-wise increment in the FP or FN object numbers is counted as SDC. Each experiment was done with a subset of 100 images of the test data set, running 100 random FIs on each image individually. For hardware faults, SDC rates are typically low (~1–3%) since drastic changes will result only from bit flips in the high exponential bits of the FP32 data type [7,18]. Therefore, an additional 500 epochs with accelerated FI only into the three highest exponential bits are performed for both flavors of memory faults. Overall, the faulty and fault-free data is found to be balanced at a ratio of about 2:1.

4 Model

Notational Remarks: We use the range index convention, i.e., a vector is given as $\mathbf{x} = (x_i) = (x^i)$, a matrix reads $\mathbf{A} = (A_{ij})$, and similarly for higher-dimensional tensors.

Monitoring Approach: Let us denote a four-dimensional activation tensor that represents an intermediate state of a convolutional neural network as $\mathbf{T} = (T_{n,c,h,w}) \in \mathbb{R}^{N \times C \times H \times W}$, where N is the sample number, C the number of channels, H the height, and W the width. We list n as running global sample index, where samples may be further grouped in batches. An output tensor of a specific layer $l \in [1, \ldots L]$ shall be given as $\mathbf{T}^l$, with L being the total number of monitored layers of the model. Subsets of a tensor with fixed n, c are called feature maps. Our monitoring approach first performs the summation of individual feature maps and subsequently calculates quantile values over the remaining kernels, see Fig. 1,

$$(F_{n,c})^l = \sum_{h,w} (T_{n,c,h,w})^l, \tag{1}$$

$$(q_n)_p^l = \left(Q_p((F_{n,c})^l)_n\right). \tag{2}$$

Here Q_p is the quantile function for the percentile p which acts on the n-th row of $(F_{n,c})^l$. In other words, Q_p reduces the kernel dimensions c to a set of discrete values where we use the 10-percentiles, i.e., $p \in [0, 10, 20, 30, \ldots, 90, 100]$. The result is a quantile value set, q_p, for a given image index n and layer l. Note that both the summation and the quantile operations (and hence the detector) are invariant under input perturbations such as image rotations.

Supervised Layers: We intercept the output activations of all convolutional layers, as those layers provide the vast majority of operations in the selected computer vision DNNs. Yet, the same technique can be applied to any neural network layer.

Reference Bound Extraction: Applied to a separate data set D_{bnds}, the above technique is used pre-runtime to extract reference bounds which represent the minimum and maximum feature sums during fault-free operation:

$$\begin{aligned} q_{p,\min}^l &= \min_{n \in D_{\text{bnds}}} \left((q_n)_p^l\right), \\ q_{p,\max}^l &= \max_{n \in D_{\text{bnds}}} \left((q_n)_p^l\right), \end{aligned} \tag{3}$$

For D_{bnds}, we randomly select 20% of the training data [4].

Anomaly Feature Extraction: For a given input during runtime, Eqs. (1) to (2) are used to obtain the quantile markers of the current activation distribution. Those are further processed to a so-called *anomaly feature vector* which quantifies the similarity of the observed patterns with respect to the baseline references of Eq. 3,

$$q_p^l \rightarrow \frac{1}{2}\left(f_{\text{norm}}(q_p^l, q_{p,\min}^l, q_{p,\max}^l) + 1\right). \tag{4}$$

Here, f_{norm} normalizes the monitored quantiles to a range of $(-1, 1)$ by applying element-wise ($\epsilon = 10^{-8}$ is a regularization offset)

$$f_{\text{norm}}(a, a_{\min}, a_{\max}) = \begin{cases} \tanh\left(\frac{a - a_{\max}}{|a_{\max}| + \epsilon}\right) & \text{if } a \geq a_{\min}, \\ \tanh\left(\frac{a_{\min} - a}{|a_{\min}| + \epsilon}\right) & \text{if } a < a_{\min}. \end{cases} \tag{5}$$

Intuitively, the result of Eq. 5 will be positive if a is outside the defined minimum ($a_{\min}$) and maximum ($a_{\max}$) bounds (approaching +1 for very large positive or negative values). The function is negative if a is within the bounds (lowest when barely above the minimum), and will become zero when a is of the order of the thresholds. In Eq. 4, a shift brings features to a range of $(0, 1)$ to facilitate the interpretation of feature importance. Finally, all extracted features are unrolled into a single anomaly feature vector $\mathbf{q} = ((q^l)_p) = [q_0^1, q_0^2, \dots q_0^L, q_{10}^1, \dots q_{100}^L]$, that will be the input to the anomaly detector component.

Anomaly Detector: We use a decision tree [10] approach to train an interpretable classifier, leveraging the *sklearn* package [22]. The class weights are inversely proportionally to the number of samples in the respective class to compensate for imbalances in the training data. As a measure of the split quality of a decision node we use the Gini index [10]. To avoid overfitting of the decision tree, we perform cost-complexity pruning [22] with a factor varying between 1×10^{-5} and 2×10^{-5}, that is optimized for the respective model.

To investigate fault class identification, we study three different detector modes with varying levels of fault class abstractions and quantify each mode $x \in \{cls, cat, sdc\}$ by precision, $\text{P}_\text{x} = \text{TP}_\text{x}/(\text{TP}_\text{x}+\text{FP}_\text{x})$ and recall $\text{R}_\text{x} = \text{TP}_\text{x}/(\text{TP}_\text{x}+\text{FN}_\text{x})$. Here we abbreviated true positives (TP), false positives (FP), and false negatives (FN). In the class mode (*cls*), we consider only those detections as true positives where the predicted and actual fault modes (see Sect. 3) coincide exactly. Cases where SDC is detected correctly but the fault class does not match will be counted as either FP or FN in this setting. In the category mode (*cat*), those SDC detections are considered true positives where the predicted and actual fault class fall into the same category of either *memory fault* or *input fault* $= \{noise, blur, contrast\}$. That means, fault class confusions within a category will not reduce the performance in this mode. The final precision and recall values for the class and category mode are given as the average over all classes or categories, respectively. Finally, in the mode *sdc*, we consider all cases as true positives where SDC was correctly identified regardless of the specific fault class. This reflects a situation where one is only interested in the presence of SDC overall, rather than the specific fault class.

5 Results

5.1 Detector Performance

Error Detection: Table 1 shows the precision, recall, and decision tree complexity for the studied detectors and models. When all extracted features are

Table 1. Precision (P), Recall (R), and decision tree (DT) complexity - given as the number of used features (N_{ft}) and monitored layers (N_l) - for different setups. Every detector was retrained 10 times with different random seeds and the averages across all runs are given. We list both the classifiers making use of all extracted quantiles (*full*) and the averaged reduced (*red*) detector models, where guided feature reduction was applied, see Fig. 3. Best-in-class detectors are highlighted in each column.

Model	P(%)			R(%)			DT
	P_{cls}	P_{cat}	P_{sdc}	R_{cls}	R_{cat}	R_{sdc}	N_{ft}/N_l
Yolo+Coco							
full	95.8	96.4	96.1	98.2	98.6	98.4	825/75
red (avg)	93.3	94.6	93.4	97.4	96.3	96.7	2/2
Yolo+Kitti							
full	97.3	97.5	97.4	**99.1**	99.3	99.2	825/75
red (avg)	92.6	92.1	92.0	97.3	96.4	96.8	3/2
SSD+Coco							
full	96.6	97.2	96.6	98.2	98.5	98.3	429/39
red (avg)	95.2	96.3	94.9	96.5	94.5	95.9	3/3
SSD+Kitti							
full	96.0	97.1	96.2	98.4	98.7	98.6	429/39
red (avg)	92.8	94.6	92.1	98.0	97.7	98.2	2/2
RetinaNet+Coco							
full	96.6	95.7	96.9	97.1	94.9	98.0	781/71
red (avg)	**96.6**	96.6	96.5	97.0	94.6	98.2	2/2
RetinaNet+Kitti							
full	**97.5**	97.3	97.5	98.6	98.2	98.7	781/71
red (avg)	96.2	96.6	95.9	**98.6**	97.8	98.9	2/2
ResNet+Imagenet							
full	93.9	**98.3**	**97.6**	98.1	**99.6**	**99.4**	583/53
red (avg)	92.1	**97.6**	**96.7**	98.3	**99.6**	**99.5**	3/3
AlexNet+Imagenet							
full	96.1	**98.3**	97.3	98.4	99.2	99.0	55/5
red (avg)	93.2	96.8	95.0	98.0	99.0	98.8	4/3

leveraged by the decision tree classifier (referred to as *full* model), the average class-wise detection precision varies between 93.9% (ResNet) and 97.5% (RetinaNet+Kitti), while the recall is between 97.1% (RetinaNet+Coco) and 99.1% (Yolo+Kitti). If only the fault category needs to be detected correctly, we find $P_{cat} > 95\%$ and $R_{cat} > 94\%$. Correct decisions about the presence of SDC only are done with $P_{sdc} > 96\%$ and $R_{sdc} \geq 98\%$. Across models, we observe (not shown in Table 1) that the most common confusion are false positive noise detections, leading to a reduced precision in the individual *noise* class (worst

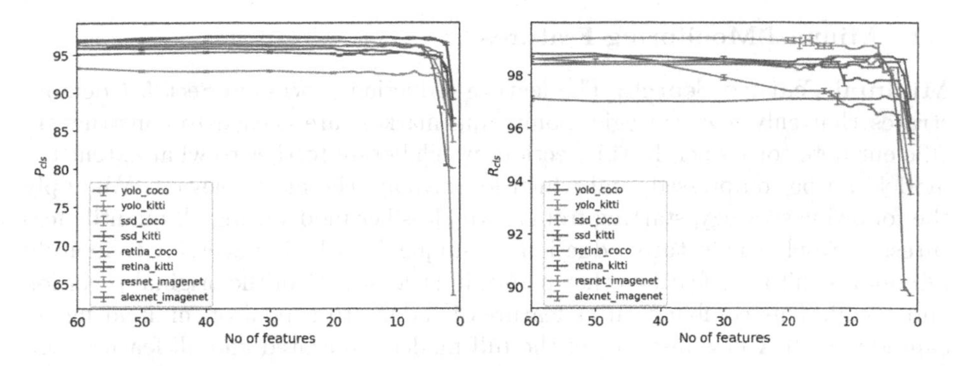

Fig. 3. Precision and recall of class-wise SDC detection when reducing the number of monitored features (average of 10 independent runs).

case 75.8% for ResNet). The recall is most affected by memory faults (lowest individual class recall 90.6% for RetinaNet+Coco).

The detection rates of the full model in Table 1 outperform the ones reported in the comparable approach of Schorn et al. [25] (using feature map tracing) and the blur detection in Huang et al. [15] in terms of precision and recall. When using alternative metrics (not shown in Table 1) for comparison with other detector designs, we find that our method achieves class-wise misclassification rates ranging between 0.7% and 2.0%, depending on the model, which is on par with the results for example in Cheng et al. [5]. Similarly, the calculated class-wise true negative rates vary between 99.6% and 99.8%, reaching or exceeding the classifier performance in Zhao et al. [26]. Note that all mentioned references are limited to image classification networks.

Feature Reduction: The number of monitored features can be drastically reduced without significantly affecting the detection performance. This means that many quantiles represent similar information and further distillation can be applied. For feature reduction, we follow two steps: First, all quantile features of the full model are ranked according to their Gini importance [22] in the decision tree. Then, we retrain the classifier with a successive number of features, starting from the most important one only, to the two most important ones, etc. A reduced model is accepted as efficient if it recovers at least 95% of both the precision and recall performance of the original model with all features.

Figure 3 shows the results of the feature reduction. Inspecting performance trends from larger to smaller feature numbers, we observe that the detection rate stagnates over most of the elimination process, before dropping abruptly when the number of used features reduces beyond a limit. On average, the number of monitored features and layers that are required to maintain close-to-original performance (as defined above) are as few as 2 to 4 and 2 to 3, respectively. For a model like Yolo, this means that only 2 out of the 75 convolution layers have to be supervised. The average characteristics of the resulting detector models is shown in Table. 1 as reduced (*red*) model.

5.2 Minimal Monitoring Features

Minimal Feature Search: The feature reduction process in Sect. 5.1 demonstrates that only few strategic monitoring markers are needed to construct an efficient detector model. In this section, we elaborate further to what extent the model can be compressed, and which features are the most relevant. We apply the following strategy, starting from a full classifier model using all quantile features: 1) Apply the feature reduction technique described in Sect. 5.1 to identify minimal monitoring features that maintain at least 95% of the original precision and recall. This combination of features is added to a pool of minimal model candidates. 2) A new instance of the full model is initiated and all feature candidates from the pool are eliminated. Return to the first step to find alternative candidates until a certain search depth (we choose 24) is exhausted.

Universal Trends: The identified minimal feature combinations are shown in Fig. 4. We find that just 2 features from 2 different layers are sufficient to constitute an efficient error detector for all studied models except for AlexNet (4 features from 3 layers).

Almost universally, one of the monitored layers needs to be among the very last layers of the deep neural network. Since memory faults are injected randomly across the network, errors in the last layers would go unnoticed otherwise. Only for SSD models, it turns out that most of the SDC faults occur in earlier layers, so that a supervision of the last layers is less crucial to achieve a similar statistical detection performance. We observe that it is favorable to supervise a higher percentile (e.g., q_{100}) in the later layers, especially in more shallow networks (AlexNet and SSD). This is because in shallow networks, peak shifts have a shorter propagation path and hence it is more important to intercept faults directly. This can only be achieved by the highest percentiles. In models with ReLU activation functions (all except Yolo here), the minimum quantile does not serve as a meaningful peak shift monitor as negative activations are clipped automatically.

A second monitoring marker should to be set in the first half of the network layer stack. This helps to identify input faults (which are interceptable from the very first layer) and discriminate them from memory faults. Either a low or high percentile can be chosen for supervision.

Explainability: Given the above generalizable trends and the fully transparent nature of the classifier, we can make statements about the inner workings of the DNN that correlate a given input with an anomalous or regular outcome. Those statements can be interpreted intuitively by a human as a proxy of a decision, and hence qualify as an explanation [2].

5.3 Overhead

We measure the average inference time per image when running the supervised model on random input, using the *Torch profiler* [21]. The profiled overall self compute time in Fig. 5 is shared between CPU and GPU. Compared to the

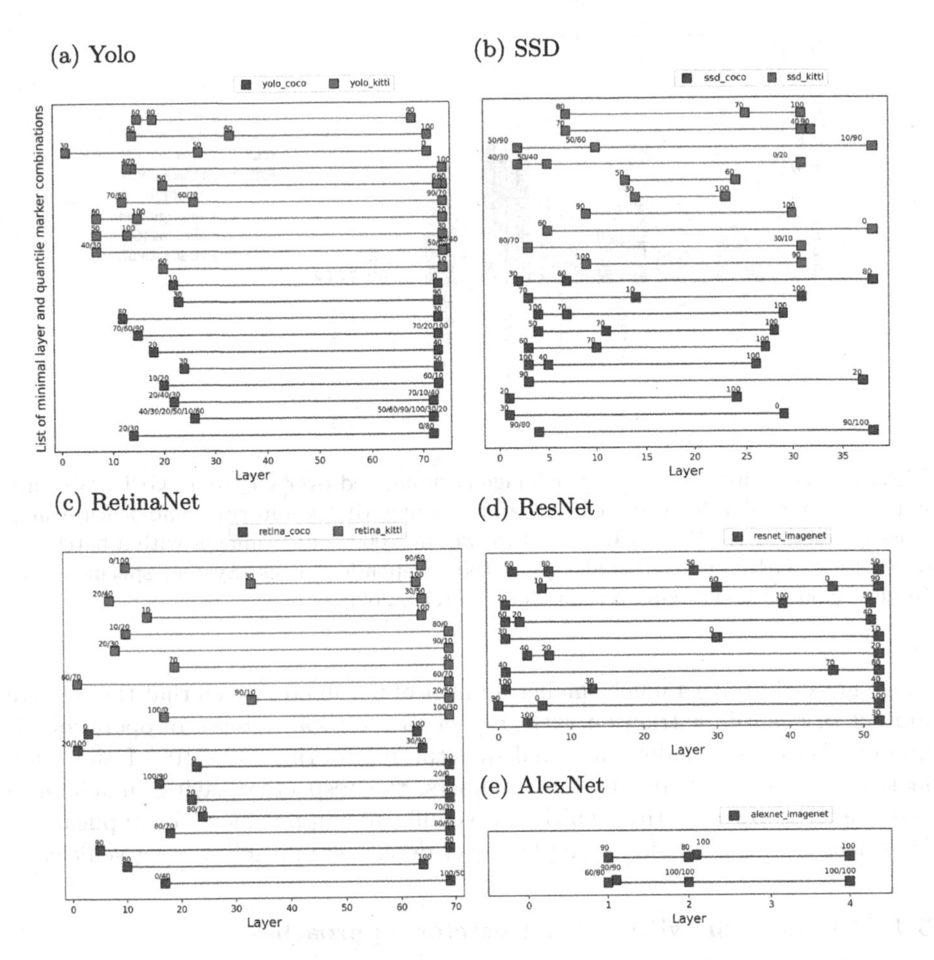

Fig. 4. Minimal combinations of features as identified by the search process in Sect. 5.2. All combinations in (a)–(e) constitute a reduced classifier model with at least 95% of the performance of the respective full model. Inset numbers designate the percentile numbers (or combinations thereof if multiple combinations are equally valid).

feature map tracing method of Schorn et al. [24,25], the quantile operation introduces additional compute, but at the same time saves the time of storing large tensors, due to the compression of many feature sums into only a few quantiles.

Between these two opposing trends, full quantile monitoring turns out to be *faster* than feature map tracing for all the studied models except for the shallow AlexNet, as shown in Fig. 5. If only selected layers are monitored to create a reduced classifier, the overhead can be decreased significantly. We find that the impact of minimal quantile monitoring on the overall inference time is between +0.3% and +1.6% for all studied object detection DNNs. For the image classification networks, on the other hand, quantile monitoring imposes a more significant overhead of +10.7% (ResNet) and +53.8% (AlexNet). This is because

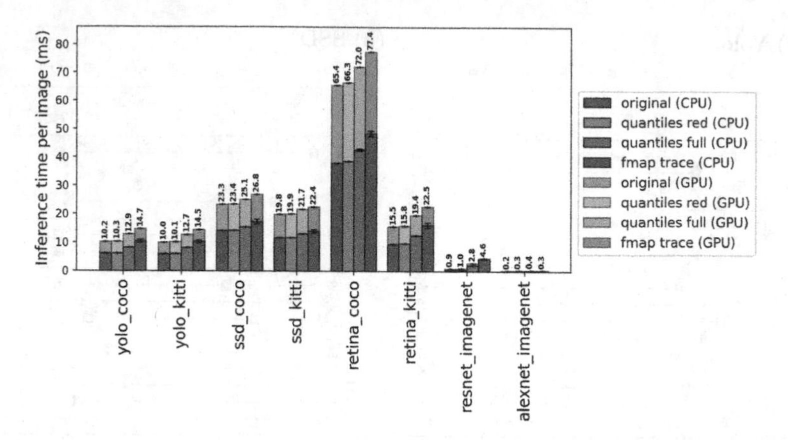

Fig. 5. Average inference time per image accumulated over CPU and GPU. We compare the original inference, reduced and full quantile monitoring, and feature map tracing (method of [24,25]). In the setup, we run 100 random images with a batch size of 10 (with GPU enabled) and repeat 100 independent runs. System specifications: Intel® Core^{TM} i9-12900K, Nvidia GeForce RTX 3090.

those networks have a much smaller number of parameters such that the relative impact of quantile extraction with respect to the total number of operations is higher. Across all models, minimal quantile monitoring is $>$ 10% faster than feature map tracing. In absolute numbers, the respective saving in inference time can be up to $\sim$10 ms, which is a significant improvement for applications operating at real-time, for example object detection in a self-driving vehicle.

5.4 Comparison with Other Detector Approaches

Alternative to a decision tree, we can deploy a linear machine learning model for error detection (similar to [24]). We study the feasibility of doing so in this section. For this setup, we select Yolo+Kitti to train a classifier for 1000 epochs using the Adam optimizer and cross entropy loss. A batch size of 100 and learning rates optimized between 1×10^{-4} and 5×10^{-3} were chosen. In the simplest form, with a multi-layer-perceptron, the algorithmic transparency is preserved and we find P_{cls} = 86.0% and R_{cls} = 95.7%. If more hidden linear layers are added, higher detection rates can be achieved at the cost of explainability. For example, including one extra hidden layer with 64 neurons [24], we find a performance of P_{cls} = 88.9% and R_{cls} = 96.3%, with three such extra layers we obtain P_{cls} = 91.7% and R_{cls} = 95.1%. Compared to decision trees, however, this strategy suffers from more complex hyperparameter tuning and large training times. Therefore, decision trees are a better fit for our use case.

6 Summary and Future Work

In this paper, we show that critical silent data corruptions in computer vision DNNs (originating either from hardware memory faults or input corruptions) can be efficiently detected by monitoring the quantile shifts of the activation distributions in specific layers. In most studied cases, it is sufficient to supervise two layers with one quantile marker each to achieve high error detection rates up to ~96% precision and ~98% recall. We also show that the strategic monitoring location can be associated with the concept of intercepting bulk and peak activation shifts, which gives a *novel, unifying perspective on the dependability of DNNs.* Due to the large degree of information compression in this approach, the compute overhead of the approach is in most cases only between 0.3% and 1.6% compared to the original inference time, and outperforms the comparable state of the art. In addition, we show that the method contributes to the model's explainability as the error detection decision is interpretable and transparent. For future work, we can further guide the search for optimized minimal feature combinations, for example, by taking into account specifics of the model architecture.

Acknowledgement. We thank Neslihan Kose Cihangir and Yang Peng for helpful discussions. This project has received funding from the European Union's Horizon 2020 research and innovation programme under grant agreement No 956123. This work was partially funded by the Federal Ministry for Economic Affairs and Climate Action of Germany, as part of the research project SafeWahr (Grant Number: 19*A*21026*C*), and the Natural Sciences and Engineering Research Council of Canada (NSERC).

References

1. Balasubramaniam, A., Pasricha, S.: Object Detection in Autonomous Vehicles: Status and Open Challenges (2022)
2. Barredo Arrieta, A., et al.: Explainable artificial intelligence (XAI): concepts, taxonomies, opportunities and challenges toward responsible AI. Inf. Fusion **58**, 82–115 (2019)
3. Chen, K., et al.: MMDetection: open MMLab detection toolbox and benchmark. arXiv:1906.07155 (2019)
4. Chen, Z., Li, G., Pattabiraman, K.: A low-cost fault corrector for deep neural networks through range restriction. In: Proceedings - 51st Annual IEEE/IFIP International Conference on Dependable Systems and Networks, DSN (2021)
5. Cheng, C.H., Nührenberg, G., Yasuoka, H.: Runtime monitoring neuron activation patterns. In: Proceedings of the 2019 Design, Automation and Test in Europe Conference and Exhibition, DATE (2019)
6. Geiger, A., Lenz, P., Stiller, C., Urtasun, R.: Vision meets robotics: the KITTI dataset. Int. J. Robot. Res. **32**(11) (2013)
7. Geissler, F., et al.: Towards a safety case for hardware fault tolerance in convolutional neural networks using activation range supervision. In: CEUR Workshop Proceedings, vol. 2916 (2021)

8. Goodfellow, I.J., Shlens, J., Szegedy, C.: Explaining and harnessing adversarial examples. In: 3rd International Conference on Learning Representations, ICLR 2015 - Conference Track Proceedings (2015)
9. Graefe, R., Geissler, F., Syed, Q.: Pytorch application-level fault injector (pytorch-Alfi) (2022). https://github.com/IntelLabs/pytorchalfi
10. Hastie, T., Tibshirani, R., Friedman, J.: Springer Series in Statistics, vol. 27 (2009)
11. Hendrycks, D., Dietterich, T.: Benchmarking neural network robustness to common corruptions and perturbations. In: 7th International Conference on Learning Representations, ICLR (2019)
12. Henzinger, T.A., Lukina, A., Schilling, C.: Outside the box: abstraction-based monitoring of neural networks. Front. Artif. Intell. Appl. **325** (2020)
13. Hoang, L.H., Hanif, M.A., Shafique, M.: FT-ClipAct: resilience analysis of deep neural networks and improving their fault tolerance using clipped activation. In: Proceedings of the 2020 Design, Automation and Test in Europe Conference and Exhibition, DATE (2020)
14. Hong, S., Frigo, P., Kaya, Y., Giuffrida, C., Dumitras, T.: Terminal brain damage: exposing the graceless degradation in deep neural networks under hardware fault attacks. In: Proceedings of the 28th USENIX Security Symposium (2019)
15. Huang, R., Feng, W., Fan, M., Wan, L., Sun, J.: Multiscale blur detection by learning discriminative deep features. Neurocomputing **285** (2018)
16. IEEE: 754–2019 - IEEE Standard for Floating-Point Arithmetic. Technical report (2019). https://doi.org/10.1109/IEEESTD.2019.8766229
17. Deng, J., Dong, W., Socher, R., Li, L.-J., Li, K., Fei-Fei, L.: ImageNet: a large-scale hierarchical image database. In: IEEE CVPR (2009). https://doi.org/10.1109/cvprw.2009.5206848
18. Li, G., et al.: Understanding error propagation in Deep Learning Neural Network (DNN) accelerators and applications. In: Proceedings of the International Conference for High Performance Computing, Networking, Storage and Analysis, SC (2017)
19. Metzen, J.H., Genewein, T., Fischer, V., Bischoff, B.: On detecting adversarial perturbations. In: Proceedings of International Conference on Learning and Representation (2017)
20. Microsoft: Coco 2017 dataset (2017). https://cocodataset.org/github.com/cocodataset/cocoapi
21. Paszke, A., et al.: PyTorch: an imperative style, high-performance deep learning library. In: Advances in Neural Information Processing Systems, vol. 32 (2019)
22. Pedregosa, F., et al.: Scikit-learn: machine learning in Python. J. Mach. Learn. Res. (2011)
23. Qutub, S., et al.: Hardware faults that matter: understanding and estimating the safety impact of hardware faults on object detection DNNs. In: Safecomp (2022)
24. Schorn, C., Gauerhof, L.: FACER: a universal framework for detecting anomalous operation of deep neural networks. IEEE ITSC (2020)
25. Schorn, C., Guntoro, A., Ascheid, G.: Efficient on-line error detection and mitigation for deep neural network accelerators. In: Safecomp (2018)
26. Zhao, F., Zhang, C., Dong, N., You, Z., Wu, Z.: A uniform framework for anomaly detection in deep neural networks. Neural Process. Lett. **54**(4) (2022)

Are Transformers More Robust? Towards Exact Robustness Verification for Transformers

Brian Hsuan-Cheng Liao[1,2(✉)], Chih-Hong Cheng[3], Hasan Esen[1], and Alois Knoll[2]

[1] DENSO AUTOMOTIVE Deutschland GmbH, 85386 Eching, Germany
{h.liao,h.esen}@eu.denso.com
[2] Technical University of Munich, 85748 Garching, Germany
knoll@in.tum.de
[3] Fraunhofer IKS, 80686 Munich, Germany
chih-hong.cheng@iks.fraunhofer.de

Abstract. As an emerging type of Neural Networks (NNs), Transformers are used in many domains ranging from Natural Language Processing to Autonomous Driving. In this paper, we study the robustness problem of Transformers, a key characteristic as low robustness may cause safety concerns. Specifically, we focus on Sparsemax-based Transformers and reduce the finding of their maximum robustness to a Mixed Integer Quadratically Constrained Programming (MIQCP) problem. We also design two pre-processing heuristics that can be embedded in the MIQCP encoding and substantially accelerate its solving. We then conduct experiments using the application of Land Departure Warning to compare the robustness of Sparsemax-based Transformers against that of the more conventional Multi-Layer-Perceptron (MLP) NNs. To our surprise, Transformers are not necessarily more robust, leading to profound considerations in selecting appropriate NN architectures for safety-critical domain applications.

Keywords: NN verification · Robustness · Transformers · Lane Departure Warning · Autonomous Driving

1 Introduction

Over the past decade, Neural Networks (NNs) have been widely adopted for many applications, including automated vehicles (AVs) [2]. Lately, as an emerging type of NNs, Transformers [29] are often found to be the most effective models, compared to the more conventional Multi-Layer Perceptrons (MLPs) or their convolutional and recurrent variants [6], thereby gradually replacing them in these applications. For instance, Tesla and Cruise use Transformers in their perception units [5,27]. However, most of the studies and discussions focus on evaluating NNs' accuracy. Parallel research has shown that NNs often lack robustness against input changes such as adversarial attacks or domain shifts, hence hindering the overall dependability of the NN-based applications [11,14,26].

The above background naturally prompts a question of whether Transformers are more robust than MLPs, given the often better accuracy and wide applications.

This project has received funding from the European Union's Horizon 2020 research and innovation programme under grant agreement No 956123 - FOCETA.

J. Guiochet et al. (Eds.): SAFECOMP 2023, LNCS 14181, pp. 89–103, 2023.
https://doi.org/10.1007/978-3-031-40923-3_8

To answer this question, we study the maximum robustness of the NNs against local input perturbations commonly modeled by l_p-distances, i.e., exact robustness verification (where p can be $1, 2, ..., \infty$). Ultimately, the goal is to hold a direct comparison of the robustness of the two kinds of NNs, Transformers and MLPs, and gain insights from the results.

In the literature, research efforts have been made to enable exact robustness verification for MLPs, particularly ones with feed-forward layers and ReLU activation function [4,7,19,28]. The main approach is to employ an optimization framework, encode the NNs' architecture into the constraints, and calculate the exact robustness within some admissible input perturbation region [4,28]. However, there is still a gap for Transformers' exact robustness verification due to the more complex operations in this kind of NNs, namely the dot product between variables and the activation function in the Multi-Head Self-Attention (MSA) block [29]. The existing (small volume of) works handle these operations with approximations during verification yet at the expense of verification precision [3,25] (more details can be found in Sect. 2).

Our work attempts to close the gap towards exact robustness verification for Transformers but provides merely an interim solution. To elaborate, having formulated a Mixed Integer Programming (MIP)-based optimization problem, we focus on the Transformers that use Sparsemax (instead of Softmax) for MSA activation. This allows for precisely encoding the NN into the MIP, or more particularly, Mixed Integer Quadratically Constrained Programming (MIQCP) due to the remaining quadratic terms. We provide a comparison study to show that the Sparsemax-based Transformers perform similarly to their Softmax-based counterparts. Then, to faster solve the MIQCP, we devise two pre-processing heuristics that lead to a total speedup of an order of magnitude. Notably, these heuristics are not restricted to our work but can be applied to related studies.

We perform the experiments using a Lane Departure Warning (LDW) application, which is widely adopted in AVs. Such an LDW application can be used for human driving assistance or run-time monitoring on separate automated driving functions. Essentially, an LDW application is a time-series classification and regression task. The embedded model, usually an NN, has to predict the direction of and the time to a potential lane departure, given a sequence of past driving information such as the ego vehicle velocity and estimated time to collision against adjacent vehicles. Our methodology and experimental results, though limited (similar to the exact robustness verification works focusing on ReLU-based MLPs), demonstrate that Sparsemax-based Transformers tend to be less robust than similar-sized MLPs despite generally higher accuracy. Resonating the government publications [8] and industrial guidelines [23], our findings suggest that conducting thorough studies and providing rigorous guarantees on metrics beyond accuracy is crucial before deploying an NN-based application. In summary, our contributions include the following:

- To implement exact robustness verification for Sparsemax-based Transformers;
- To propose two accelerating heuristics for related robustness verification studies;
- To benchmark ATN and MLP accuracy and robustness with an industrial application (i.e., LDW).

The rest of the paper is organized in the following way. Section 2 browses the relevant literature emphasizing verification methods for general NNs and Transformers;

Sect. 3 introduces the branch of Transformers concerned in this paper. Section 4 details the problem formulation and our heuristics for robustness verification, whose effectiveness and efficiency are demonstrated and discussed in Sect. 5. Lastly, Sect. 6 concludes with a few final remarks.

2 Related Works

This section overviews related works, focusing on robustness verification for ReLU-based MLPs (i.e., piece-wise linear feed-forward NNs) and Transformers.

2.1 Robustness Verification for Neural Networks

Following the common categorization [28], we introduce two main branches of verification methods: complete and incomplete. To illustrate the difference, we assume an *adversarial polytope* to be the exact set of NN outputs resulting from the norm-bounded perturbation region. To assert the robustness of the NN within the perturbation region, complete methods handle the adversarial polytope directly, attaining an adversarial example or a robustness certificate for each query when given sufficient processing time. These methods usually apply Mixed Integer Programming (MIP) [4,19,28] or Satisfiability Modulo Theory (SMT) [7,16,17], which in turn utilizes Linear Programming (LP) or Satisfiability (SAT) solvers with accelerating techniques such as interval analysis [4,7,28] or region partitioning [9] in a Branch-and-Bound (BnB) fashion [30]. By contrast, incomplete methods reason upon an outer approximation of the adversarial polytope. Such reasoning typically results in faster verification time, yet possibly some robust queries being evaluated non-robust due to the over-approximation. Common methods in this branch include duality [31], abstract interpretation [10] and Semi-Definite Programming (SDP) [30]. For more details, interested readers are referred to the survey paper [15].

2.2 Robustness Verification for Transformers

Transformers are typically more challenging to verify because they contain more complex operations than other NNs, such as MLPs. We are only aware of two lines of existing works [3,25], both conducted with sentiment classification in Natural Language Processing. In [25], the authors calculate linear intervals for all operations in a Transformer to find the lower bound of the difference between the Softmax values of the ground-truth class and the most-probable-other-than-ground-truth class. If the computed lower bound is larger than zero (within an admissible perturbation region), the model is guaranteed robust since the prediction remains unchanged. As mentioned, the effectiveness of such approaches depends heavily on how well the lower bound approximates the actual difference of the Softmax values. Holding a similar strategy, the most recent work [3] applies abstract interpretation and suggests techniques, such as noise symbol reduction and Softmax summation constraint, to achieve better speed and precision during verification.

Our work differs from these verification frameworks in two ways. First, we find Transfomers' maximum robustness (i.e., minimum adversarial perturbation) through

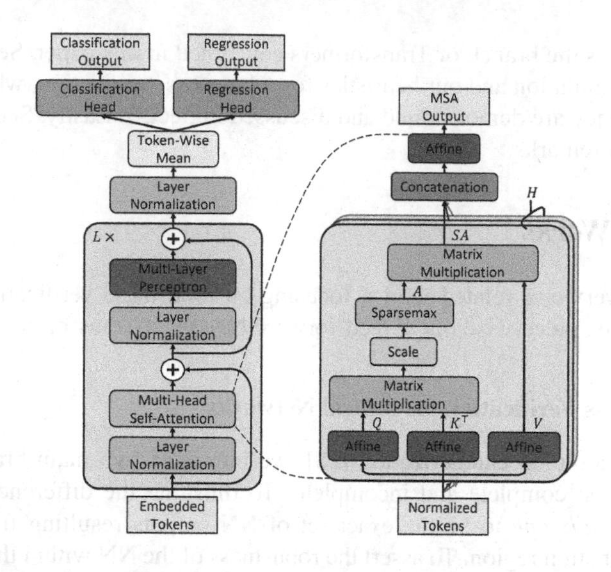

Fig. 1. The Sparsemax-based Transformer under exact robustness verification [6,29].

precise MIP encoding and fasten the procedure with novel tactics, allowing us to compare the robustness with common MLPs. Second, we conduct experiments with LDW, an industrial application not used before in robustness verification studies.

3 Preliminaries

This section provides a brief description of the Transformer under verification. For a more elaborate illustration of general Transformers, readers are referred to [6,29].

As shown in Fig. 1, a Transformer typically processes an array of embedded tokens with alternating blocks of MSA and MLP, which are both preceded by Layer Normalization (LN) and followed by residual connections[1]. Then, after another LN and token-wise mean extraction, the network is appended with suitable affine heads for downstream predictions such as classification (CLS) and regression (REG). Mathematically, given an input $\mathbf{x} \in \mathbb{R}^{N \times D}$, in which N is the number of tokens and D the dimension of features, we write:

$$\mathbf{z}^0 = \mathbf{x} \tag{1}$$

$$\widehat{\mathbf{z}^\ell} = \mathsf{MSA}(\mathsf{LN}(\mathbf{z}^{\ell-1})) + \mathbf{z}^{\ell-1}, \tag{2}$$

$$\mathbf{z}^\ell = \mathsf{MLP}(\mathsf{LN}(\widehat{\mathbf{z}^\ell})) + \widehat{\mathbf{z}^\ell}, \tag{3}$$

$$f^{\mathsf{CLS}}(\mathbf{x}) = \mathsf{CLS}(\widetilde{\mathbf{z}^L}) \in \mathbb{R}^C, \tag{4}$$

$$f^{\mathsf{REG}}(\mathbf{x}) = \mathsf{REG}(\widetilde{\mathbf{z}^L}) \in \mathbb{R}, \tag{5}$$

[1] Placing LN before MSA and MLP is found to give better network performance than placing it after residual addition [32].

in which $\widetilde{\mathbf{z}^L} \in \mathbb{R}^D$ is the token-wise mean of $\mathsf{LN}(\mathbf{z}^L) \in \mathbb{R}^{N\times D}$, C is the number of predefined classes and $\ell = 1,\ldots,L$ is the layer index. We define the functions in the following.

As introduced, we consider Sparsemax-based MSA with the definition:

$$[\mathbf{Q}_h, \mathbf{K}_h, \mathbf{V}_h] = \mathbf{z}\mathbf{W}_h^{\mathsf{QKV}}, \tag{6}$$

$$\mathbf{A}_h = \mathsf{Sparsemax}\left(\mathbf{Q}_h\mathbf{K}_h^\top / \sqrt{D_H}\right), \tag{7}$$

$$\mathsf{SA}_h(\mathbf{z}) = \mathbf{A}_h\mathbf{V}_h, \tag{8}$$

$$\mathsf{MSA}(\mathbf{z}) = [\mathsf{SA}_1(\mathbf{z}),\ldots,\mathsf{SA}_H(\mathbf{z})]\,\mathbf{W}^{\mathsf{MSA}}, \tag{9}$$

where $\mathbf{z} \in \mathbb{R}^{N\times D}$ is a general input matrix, $\mathbf{Q}_h, \mathbf{K}_h, \mathbf{V}_h \in \mathbb{R}^{D_H}$ are the query, key and value matrices for the h-th self-attention head, H is the number of self-attention heads, D_H is the dimension of each head, $\mathbf{W}_h^{\mathsf{QKV}} \in \mathbb{R}^{D\times(D_H\cdot 3)}$ and $\mathbf{W}^{\mathsf{MSA}} \in \mathbb{R}^{(D_H\cdot H)\times D}$ are the trainable weights, and $h = 1,\ldots,H$ is the head index. Essentially, Sparsemax projects an input vector $\mathbf{u} \in \mathbb{R}^D$ to an output vector $\mathbf{p} \in \mathbb{R}^D$, where $\mathbf{p}_1 + \cdots + \mathbf{p}_D = 1$ and $\mathbf{p}_i \geq 0$ for $i = 1,\ldots,D$ (i.e., a probability simplex)[2]. Technically, it can serve as a piece-wise linear approximation to Softmax [21]. Algorithm 1 and Fig. 2 provide the calculation steps for a closed-form solution and the visualization of a 2D input-output relation.

To proceed with the Transformer architecture, MLP is a two-layer NN with ReLU activation, written as $\mathsf{MLP}(\mathbf{v}) = \mathsf{ReLU}(\mathbf{v}\mathbf{W}_1^{\mathsf{MLP}} + \mathbf{b}_1^{\mathsf{MLP}})\mathbf{W}_2^{\mathsf{MLP}} + \mathbf{b}_2^{\mathsf{MLP}} \in \mathbb{R}^D$, where $\mathbf{v} \in \mathbb{R}^D$ is the input vector and $\mathbf{W}_1^{\mathsf{MLP}} \in \mathbb{R}^{D\times D_{\mathsf{MLP}}}$, $\mathbf{b}_1^{\mathsf{MLP}} \in \mathbb{R}^{D_{\mathsf{MLP}}}$, $\mathbf{W}_2^{\mathsf{MLP}} \in \mathbb{R}^{D_{\mathsf{MLP}}\times D}$, $\mathbf{b}_2^{\mathsf{MLP}} \in \mathbb{R}^D$ the trainable parameters, and $+$ the addition with broadcasting rules in

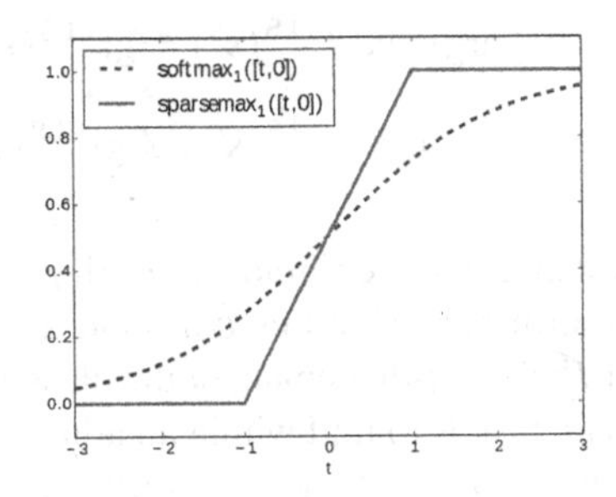

Fig. 2. Softmax vs. Sparsemax, given an input vector $\mathbf{u} = [t, 0] \in \mathbb{R}^2$ (adapted from [21]).

Algorithm 1. Calculate Sparsemax activation [21]

1: **Input:** $\mathbf{u} \in \mathbb{R}^D$
2: Sort $\mathbf{u}$ into $\hat{\mathbf{u}}$, where $\hat{\mathbf{u}}_1 \geq \cdots \geq \hat{\mathbf{u}}_D$
3: Find $s(\hat{\mathbf{u}}) := \max\left\{k \in [1,D] \mid 1 + k\hat{\mathbf{u}}_k > \sum_{j=1}^{k} \hat{\mathbf{u}}_j\right\}$
4: Define $\tau(\hat{\mathbf{u}}) = \frac{\left(\sum_{j=1}^{s(\hat{\mathbf{u}})} \hat{\mathbf{u}}_j\right) - 1}{s(\hat{\mathbf{u}})}$
5: **Output:** $\mathbf{p} \in \mathbb{R}^D$, where $\mathbf{p}_i = \max(\mathbf{u}_i - \tau(\hat{\mathbf{u}}), 0)$

[2] We write here in vector form (c.f. (1)-(3)) as the operations are applied vector-wise essentially.

common NN implementation libraries (see footnote 5). Similarly, we have the two affine heads, $\mathsf{CLS}(\mathbf{v}) = \mathbf{v}\mathbf{W}^{\mathsf{CLS}} + \mathbf{b}^{\mathsf{CLS}} \in \mathbb{R}^C$ and $\mathsf{REG}(\mathbf{v}) = \mathbf{v} \cdot \mathbf{W}^{\mathsf{REG}} + b^{\mathsf{REG}} \in \mathbb{R}$, where $\mathbf{v} = \widetilde{\mathbf{z}}^L \in \mathbb{R}^D$ and $\mathbf{W}^{\mathsf{CLS}} \in \mathbb{R}^{D\times C}$, $\mathbf{b}^{\mathsf{CLS}} \in \mathbb{R}^C$, $\mathbf{W}^{\mathsf{REG}} \in \mathbb{R}^D$ and $b^{\mathsf{REG}} \in \mathbb{R}$. Finally, we note that a linearized variant of LN is used, i.e., $\mathsf{LN}_i(\mathbf{v}) = \mathbf{w}_i \times (\mathbf{v}_i - \mu_{\mathbf{v}}) + \mathbf{b}_i$, where $\mathbf{v} \in \mathbb{R}^D$ is the input vector, $\mu_{\mathbf{v}} \in \mathbb{R}$ its mean, $\mathsf{LN}_i(\mathbf{v})$ the i-th element for $i = 1,...,D$, and $\mathbf{w} \in \mathbb{R}^D$ and $\mathbf{b} \in \mathbb{R}^D$ the trainable parameters[5]. Such modification avoids the division over relatively small input variance $\sigma_{\mathbf{v}}$ in the quadratic LN definition, thereby allowing the variables to be bounded more tightly [25].

4 Methodology

Having defined the Sparsemax-based Transformer, we now give the robustness property for verification. Subsequently, we highlight our MIQCP encoding steps and two heuristics that shall accelerate the solving of the encoded MIQCP.

4.1 Problem Formulation

We formalize the problem of robustness verification as follows: Let $f(\cdot) : \mathbb{R}^M \to \mathbb{R}^N$ denote the NN under verification, $\mathbf{x} \in \mathbb{R}^M$ the original data point on which the NN is being verified, and $\mathbf{x}' \in \mathbb{R}^M$ a perturbed input which tries to deceive the NN, we write:

$$\min_{\mathbf{x}'} \mathscr{D}_p(\mathbf{x}', \mathbf{x}) \tag{10}$$

$$\text{subject to } \mathbf{x}' \in \mathscr{B}_p(\mathbf{x}), \tag{11}$$

$$\arg\max_i(f_i^{\mathsf{CLS}}(\mathbf{x})) = \mathrm{gt}^{\mathsf{CLS}}(\mathbf{x}), \tag{12}$$

$$\arg\max_i(f_i^{\mathsf{CLS}}(\mathbf{x}')) \neq \mathrm{gt}^{\mathsf{CLS}}(\mathbf{x}), \tag{13}$$

where $\mathscr{D}_p(\cdot,\cdot)$ is the l_p-distance with commonly used $p \in \{1,2,\infty\}$ and $\mathscr{B}_p(x) = \{\mathbf{x}' \mid \|\mathbf{x}' - \mathbf{x}\|_p \leq \varepsilon\}$ the l_p-norm ball of radius ε around $\mathbf{x}$, $\mathrm{gt}^{\mathsf{CLS}}(x) \in \{1,\ldots,C\}$ the ground-truth class label and f_i^{CLS} the i-th element of the classification head output. Conceptually, the optimizer's main task is to find within an admissible perturbation region a perturbed data point closest to the original one and fulfills the misprediction constraints. Due to the space limit, we write only the classification model here, but regression cases can be derived similarly, as suggested by [3].

4.2 MIQCP Encoding

In the following, we highlight how to encode Sparsemax (i.e., Algorithm 1, Line 2- 4) into the optimization problem (which will essentially be a MIQCP problem). Encoding methods for other terms in the Transformer (e.g., affine transformation and ReLU) can be seen in [4, 19, 28].

For sorting (Algorithm 1, Line 2), we introduce a binary integer permutation matrix $\mathbf{P} \in \{0,1\}^{D\times D}$ and encode the following constraints:

$$\sum_{i=1}^{D} \mathbf{P}_{ij} = 1, \quad \text{for } j = 1,\ldots,D; \tag{14}$$

$$\sum_{j=1}^{D} \mathbf{P}_{ij} = 1, \quad \text{for } i = 1,\ldots,D; \tag{15}$$

$$\hat{\mathbf{u}} = \mathbf{P}\mathbf{u}, \tag{16}$$

$$\hat{\mathbf{u}}_i \geq \hat{\mathbf{u}}_{i+1}, \quad \text{for } i = 1,\ldots,D-1, \tag{17}$$

where $\mathbf{u} \in \mathbb{R}^D$ is the (perturbed) input vector at Sparsemax and $\hat{\mathbf{u}}$ the sorted output.

For calculating the support (Algorithm 1, Line 3), we first define a vector $\rho \in \mathbb{R}^D$, where $\rho_k = 1 + k\hat{\mathbf{u}}_k - \sum_{j=1}^{k} \hat{\mathbf{u}}_j$ $(k = 1,\ldots,D)$, and then introduce another binary integer vector $\zeta \in \{0,1\}^D$ such that:

$$\zeta_k = \begin{cases} 1, & \text{if } \rho_k > 0; \\ 0, & \text{otherwise.} \end{cases} \tag{18}$$

It can be seen that finding the support is equivalent to summing up the vector as $s(\hat{\mathbf{u}}) = \sum_{k=1}^{D} \zeta_k$. However, we actually need to implement the step function in (18) in the MIQCP. For this, we introduce the Big-M method [12] with large positive constants $M_k^{+/-}$ and a small positive constant η (e.g., 10^{-6}) such that:

$$\rho_k \leq M_k^+ \times \zeta_k, \tag{19}$$

$$-\rho_k + \eta \leq M_k^- \times (1 - \zeta_k). \tag{20}$$

We now provide two lemmas to explain how $M_k^{+/-}$ are set.

Lemma 1. *For all $k = 1,\ldots,D$, the smallest value for the Big-M encoding in* (19) *is* ${}^{opt}M_k^+ = 1$.

Proof. We first rewrite $\rho_k = 1 + k\hat{\mathbf{u}}_k - \sum_{j=1}^{k} \hat{\mathbf{u}}_j = 1 + \sum_{j=1}^{k}(\hat{\mathbf{u}}_k - \hat{\mathbf{u}}_j)$. Now, with the sorting result (i.e., $\hat{\mathbf{u}}_1 \geq \hat{\mathbf{u}}_2 \geq \ldots \geq \hat{\mathbf{u}}_D$), it follows that $1 = \rho_1 \geq \rho_2 \geq \cdots \geq \rho_D$, hence the lemma. ∎

Lemma 2. *For all $k = 1,\ldots,D$, let the input of Sparsemax be bounded as $\mathbf{u}_k \in [\underline{\mathbf{u}_k}, \overline{\mathbf{u}_k}]$ and $\eta = 10^{-6}$. We first define $\lambda \in \mathbb{R}^D : \lambda_k = 1 + (k-1)(\underline{\mathbf{u}} - \overline{\mathbf{u}})$, where $\underline{\mathbf{u}} = \min(\underline{\mathbf{u}_1},\ldots,\underline{\mathbf{u}_D})$ and $\overline{\mathbf{u}} = \max(\overline{\mathbf{u}_1},\ldots,\overline{\mathbf{u}_D})$. Then, the smallest value for the Big-M encoding in* (20) *is ${}^{opt}M_k^- = |\lambda_k| + \eta$, if $\lambda_k \leq 0$; otherwise,* (20) *needs not be implemented.*

Proof. After sorting on $\mathbf{u}$, we can only rely on vector-wise bounds for estimating $\hat{\mathbf{u}}$, i.e., $\hat{\mathbf{u}}_k \in [\underline{\mathbf{u}}, \overline{\mathbf{u}}]$. Considering the result of sorting (i.e., $\hat{\mathbf{u}}_1 \geq \hat{\mathbf{u}}_2 \geq \ldots \geq \hat{\mathbf{u}}_D$), we can then derive $\rho_k = 1 + k\hat{\mathbf{u}}_k - \sum_{j=1}^{k} \hat{\mathbf{u}}_j = 1 + \sum_{j=1}^{k}(\hat{\mathbf{u}}_k - \hat{\mathbf{u}}_j) \geq 1 + (k-1)(\underline{\mathbf{u}} - \overline{\mathbf{u}}) = \lambda_k$. Now, there are two cases: If $\lambda_k \leq 0$, then the smallest value for (20) is ${}^{opt}M_k^- = |\lambda_k| + \eta$. Otherwise, $\rho_k \geq \lambda_k > 0$ and (20) is not needed. ∎

Lastly, Algorithm 1 Line 4 can be encoded with a linear constraint for the summation term and a quadratic constraint for the division. As such, we arrive at a plain MIQCP encoding for quantifying the Sparsemax-based Transformer's robustness.

4.3 Acceleration Heuristics

As related works indicate, one usually needs several acceleration heuristics to solve an encoded MIP problem efficiently [4, 19, 28]. We present our proposals in this section.

Interval Analysis Interval analysis has been widely studied and proven effective in aiding MIP solving [4, 28]. The central idea is that with tight interval bounds propagated across the network, non-linear functions such as ReLU or max can be constrained to certain behaviors, thus avoiding undetermined variables. We extend this line of thought by finding a novel bounding technique over Sparsemax[3]. Denoting the activation function as σ, (perturbed) input vector as $\mathbf{z} \in \mathbb{R}^D$, output vector as $\mathbf{a}$, upper and lower bounds as upper and lower bars, we have:

$$\begin{aligned} \mathbf{a}_i \in [\underline{\mathbf{a}_i}, \overline{\mathbf{a}_i}] = &[\sigma_i(\overline{\mathbf{z}_1}, \ldots, \underline{\mathbf{z}_i}, \ldots, \overline{\mathbf{z}_D}), \\ &\sigma_i(\underline{\mathbf{z}_1}, \ldots, \overline{\mathbf{z}_i}, \ldots, \underline{\mathbf{z}_D})], \end{aligned} \tag{21}$$

where $i = 1, \ldots, D$. Essentially, this formula stems from the observation that the lower (upper) bound of the output of an element can be calculated by applying the activation to a vector consisting of this element's input lower (upper) bound and other elements' input upper (lower) bounds. In our evaluation, we generally see much tighter intervals (usually one-tenth of the simple probability interval $[0, 1]$) with this technique.

Progressive Verification with Norm-Space Region Partitioning. Region partitioning shares a similar goal to interval analysis, attempting to tighten variable bounds and generate faster solutions. Related works have focused on how to do partitioning and prioritizing [9]. Our work differs slightly by exploiting the formulation of the optimization problem and the observation that adversarial examples usually appear close to the clean inputs. Hence, we propose a progressive verification procedure based on norm-space region partitioning. To illustrate, given ε, the radius of the l_p-norm ball, and a partition step $0 < \varepsilon_{step} \leq \varepsilon$, we first create a sub-region with a lower bound $\varepsilon_{min} = 0$ and an upper bound $\varepsilon_{max} = \varepsilon_{min} + \varepsilon_{step}$ and then run verification on this sub-region. If the verifier cannot find a solution in the current sub-region, we move on to the next one by setting $\varepsilon_{min} \mathrel{+}= \varepsilon_{step}$ and $\varepsilon_{max} \mathrel{+}= \varepsilon_{step}$ until the entire admissible region is covered. As such, we generally obtain a tighter interval for the perturbed input variable (taking $p = 1$, for instance):

$$0 \leq \varepsilon_{min} \leq \|\mathbf{x}' - \mathbf{x}\|_1 = \|\delta\|_1 \leq \varepsilon_{max} \leq \varepsilon \tag{22}$$

$$\Rightarrow \quad \varepsilon_{min} \leq \sum_{d=1}^{D} |\delta_d| \leq \varepsilon_{max} \tag{23}$$

$$\Rightarrow \quad 0 \leq |\delta_d| \leq \varepsilon_{max} \tag{24}$$

$$\Rightarrow \quad -\varepsilon_{max} \leq \delta_d \leq \varepsilon_{max} \tag{25}$$

$$\Rightarrow \quad \mathbf{x}'_d \in \left[\underline{\mathbf{x}'_d}, \overline{\mathbf{x}'_d}\right] = \left[\mathbf{x}_d - \varepsilon_{max}, \mathbf{x}_d + \varepsilon_{max}\right], \tag{26}$$

[3] This technique also applies to Softmax.

where $d = 1, \dots, D$. The tightness of the variable intervals depends on the value of ε_{max}. If ε_{max} grows towards ε, the variable intervals shall fall back to the original regions. Nonetheless, considering adversarial examples often appear closely around the clean inputs, we conjecture that early sub-regions would already contain one, offering a high possibility for a quick solution.

We summarize the progressive procedure in Algorithm 2. Additionally, to prevent the verifier from executing unboundedly, we place a pre-defined time limit t_{limit} on the algorithm. We employ Gurobi 9.5 [13] to solve the progressively encoded MIQCPs $\mathcal{M}$. As shown in Line 3, for each call on a sub-region, the solver returns an interim tuple including the optimization status s (being optimal, timeout or infeasible), solution count n, objective obj, counterexample $\mathbf{x}'$, solving gap α and execution time t_{exec}. Based on the interim tuple, we check if the verification is done optimally (OPT), timed out with a sub-optimal solution (SAT), timed out with no solution (UNDTM), unsatisfied (UNSAT), or to be continued with the next sub-region. Each of these conditions will be appended with corresponding data from the interim tuple to form the final verification result. Notably, apart from fastening the verification process, another advantage of such a progressive procedure is that one can still attain a good lower bound of the optimal objective if a timeout occurs (i.e., Algorithm 2, Line 9).

Algorithm 2. Verify an NN model with norm-space region partitioning

```
1: procedure VERIFYMODEL(f, x, gt(x), ε, ε_step, ε_min, ε_max, t_limit)
2:     Encode f, x, gt(x) into an MIQCP M with ε_min and ε_max
3:     Solve M with Gurobi under t_limit and obtain an interim tuple (s, n, obj, x', gap, t_exec)
4:     if s is optimal then
5:         return (OPT, obj, x')
6:     else if s is timeout and n > 0 then
7:         return (SAT, obj, x', α)
8:     else if s is timeout and n = 0 then
9:         return (UNDTM, ε_min)
10:    else if s is infeasible then
11:        t_limit -= t_exec
12:        ε_min += ε_step
13:        ε_max = min(ε_max+ε_step, ε)
14:        if ε_min > ε then
15:            return (UNSAT)
16:        else
17:            return VERIFYMODEL(f, x, gt(x), ε, ε_step, ε_min, ε_max, t_limit)
18:        end if
19:    end if
20: end procedure
21: Input: f, x, gt(x), ε, ε_step, t_limit
22: Initialize: ε_min ← 0, ε_max ← ε_step
23: verification_result = VERIFYMODEL(f, x, gt(x), ε, ε_step, ε_min, ε_max, t_limit)
24: Output: verification_result
```

5 Experimental Results and Discussions

This section presents the experimental results of the proposed method and techniques. We first introduce the Lane Departure Warning (LDW) application, then show an accuracy benchmark, and finally present an ablation study on the heuristics and a robustness comparison.

5.1 Lane Departure Warning

The LDW application performs a time-series joint classification and regression task [20]. For training and evaluating the NNs, we utilize the High-D dataset[4] [18], a drone-recorded bird's-eye-view highway driving dataset. For each vehicle in the recordings, we process raw data into trajectory information, including the past ten steps of time-wise features $\mathbf{x} \in \mathbb{R}^{10\times14}$ (spanning across one second), the direction $\mathsf{gt}^{\mathsf{CLS}} \in \{0,1,2\}$ of the lane departure (where 0 is no departure, 1 a left departure, and 2 a right departure), and the time $\mathsf{gt}^{\mathsf{REG}} \in [0,1]$ to the lane departure. More specifically, the 14 features include left and right lane existence (2), ego distance to lane center (1), longitudinal and latitudinal velocities and accelerations (4), and time-to-collision to surrounding vehicles excluding the following one (7). Based on these features, the NN's task is to predict a potential lane departure direction (i.e., classification) and timing (i.e., regression) up to one second in the future.

5.2 Accuracy Benchmark

As a side experiment, we evaluate the accuracy of different NNs for LDW. For classification, the model is accurate if the predicted direction matches the ground-truth direction; for regression, it is accurate if the predicted timing falls within 0.1 s from the ground-truth timing. For the Transformers, we set $H = 2, D_H = 4, D_{\mathsf{MLP}} = 8$ (as per Sect. 3). For the MLPs, we replace the MSA within the Transformer with another MLP of hidden-layer dimension $D_{\mathsf{MLP}} = 16$, resulting in similar numbers of network parameters. We implement the NNs with PyTorch [22], train them using the Adam optimizer and a fixed learning rate of 0.003 for 50 epochs, and report the best results in Table 1.

As observed, even for a relatively small application (considering the variable dimensions), the Transformers generally perform better than MLPs regarding accuracy. Additionally, NNs with piece-wise linear activation functions (i.e., Sparsemax and ReLU) are on a par with, if not stronger than, the ones with Softmax or Tanh. This result corresponds well with the original Sparsemax paper [21] and justifies the piece-wise linear activation functions in exchange for better verifiability (as also suggested in [25]).

5.3 Ablation Study

We now conduct an ablation study on the acceleration heuristics described in Sect. 4.3, using the Sparsemax-based Transformer with $L = 1$ and linearized LN. During our verification, we only allow the final token of the input variable $\mathbf{x}$ to be perturbed, resulting

[4] The utilization of the High-D dataset in this paper is for knowledge dissemination and scientific publication and is not for commercial use.

Table 1. Accuracy of various NNs in LDW: Transformers tend to be more accurate than MLPs. L denotes the number of layers in the NN as defined in Sect. 3. For LN, 2 denotes the quadratic variant, 1 the linear variant, and 0 no layer normalization. Within each column, we mark the best model overall in bold and the best model by its type (i.e., Transformer or MLP) in italic fonts.

Network	Activation	LN	$L=1$		$L=2$	
			CLS	REG	CLS	REG
Transformer	Sparsemax	2	98.01%	87.24%	98.50%	91.49%
		1	98.31%	88.46%	98.52%	91.39%
		0	***98.34%***	***90.38%***	98.54%	***92.31%***
	Softmax	2	98.01%	87.11%	98.55%	91.88%
		1	97.95%	87.56%	98.61%	91.52%
		0	98.21%	88.03%	***98.64%***	91.21%
MLP	ReLU	2	97.62%	83.81%	97.81%	84.70%
		1	97.46%	83.55%	97.89%	85.04%
		0	97.68%	*84.27%*	97.92%	84.81%
	Tanh	2	*97.80%*	83.20%	*97.99%*	84.63%
		1	97.32%	82.93%	97.69%	84.81%
		0	97.23%	83.02%	97.85%	*85.14%*

Table 2. Ablation study on the proposed acceleration heuristics, including interval analysis without and with Sparsemax bounding (IA and IA-σ) and norm-space region partitioning with different epsilon steps (RP-*0.001*, RP-*0.005* and RP-*0.01*): Our two heuristics can give a total speedup of up to one order of magnitude. We set $\varepsilon = 0.05$ in all experiments. We mark the best-performing numbers of each sub-group in italic fonts if they are better than the control and further mark the best-performing numbers across all heuristics in bold fonts. For the three listed samples, we specify optimally solved cases as OPT and augment undetermined cases (UNDTM) with a lower bound of the minimum adversarial distortion (better if higher) and satisfied cases (SAT) with the MIQCP solution gap (better if lower).

Techniques	Time elpased (s)			Nodes explored			Random samples		
	Mean	Best	Worst	Mean	Best	Worst	No. 1	No. 2	No. 3
Control	2367.14	89.63	3602.50	9326275	118329	16395255	(UNDTM, 0.0)	(SAT, 1.0)	OPT
IA	1703.65	330.26	3605.29	5370138	911639	13103193	(UNDTM, 0.0)	OPT	OPT
IA-σ	*950.37*	127.33	*2936.72*	*3061462*	247916	*10298216*	OPT	OPT	OPT
RP-*0.001*	755.66	*7.08*	3609.83	849968	*11517*	4058680	(UNDTM, 0.005)	OPT	OPT
RP-*0.005*	*278.20*	15.41	*1276.86*	***276310***	13734	***1255818***	OPT	OPT	OPT
RP-*0.01*	793.76	51.39	1556.72	1825979	70243	3429947	OPT	OPT	OPT
IA-σ+RP-*0.001*	852.67	***5.36***	3608.60	1689582	14557	6573066	(UNDTM, 0.004)	OPT	OPT
IA-σ+RP-*0.005*	296.90	7.74	1413.32	505555	***7156***	2413044	OPT	OPT	OPT
IA-σ+RP-*0.01*	***222.68***	59.56	***644.10***	*494973*	99642	*1479260*	OPT	OPT	OPT

in an encoded MIQCP with roughly 4000 linear constraints, 700 quadratic constraints, 400 general constraints (e.g., max or absolute operations) and 2500 binary variables. We test the heuristics with five random samples from the curated dataset and summarize the results in Table 2. All experiments are run with an Intel i9-10980XE CPU @ 3.0GHz using 18 threads and 20 GB of RAM.

It is observed that the novel activation bounding technique is effective. Likewise, progressive verification with norm-space region partitioning further reduces verification time. However, the best epsilon step might vary among test cases as its magnitude does not necessarily correlate to the best-case verification speed. A reason behind this might be the excessive number of MIP instances created by smaller epsilon steps. Notably, combining interval analysis and progressive verification delivers a speedup of approximately an order of magnitude.

5.4 NN Robustness Comparisons

For robustness comparisons, we first verify and compare the robustness of the aforementioned Sparsemax-based Transformer and the similar-sized ReLU-based MLP. They respectively achieve 98.31% and 97.46% in accuracy for classification in Sect. 5.2[5]. We set $\varepsilon = 0.03$ and $p = 1$ for the admissible ℓ_p-ball[6] and enable IA-σ+RP-*0.01* from the previous section during verification. We collect results from 60 random data points (on which the Transformer and MLP predict correctly before perturbation) and plot the robustness comparison diagram in Fig. 3. Overall, excluding 2 points where they draw, the Transformer is more robust than the MLP on 26 points yet less robust on 32. The result shows that the Transformer is not necessarily more robust despite the higher accuracy.

We further train and verify the robustness of five Transformers and five MLPs, each on 20 data points, giving 100 samples for each NN type. For faster verification, we set $\varepsilon = 0.01$ as the threshold for being robust and surprisingly find that the MLPs are verified robust on all data points. In contrast, the Transformers are verified robust on 69 data points and have a mean robustness value of 0.0041 (excluding the robust ones). Our finding is a counterexample to several testing results on vision tasks, which suggest that Transformers are generally more robust [1,24]. Accordingly, it is believed that NNs may perform differently in diverse domain tasks, and it is vital to rigorously evaluate both accuracy and robustness before deploying NN-based applications.

[5] Verifying the MLP follows similar steps in Sect. 4 except that the encoding is relatively simple and can be solved by Mixed Integer Linear Programming (MILP).

[6] The admissible perturbation region can be derived from input feature values analytically for better physical interpretability. For example, we can set ε as the normalized value of ego car lateral acceleration, considering it a decisive feature for LDW. Perturbations in this context can stem from sensor noises or hardware faults. The binary features are not perturbed.

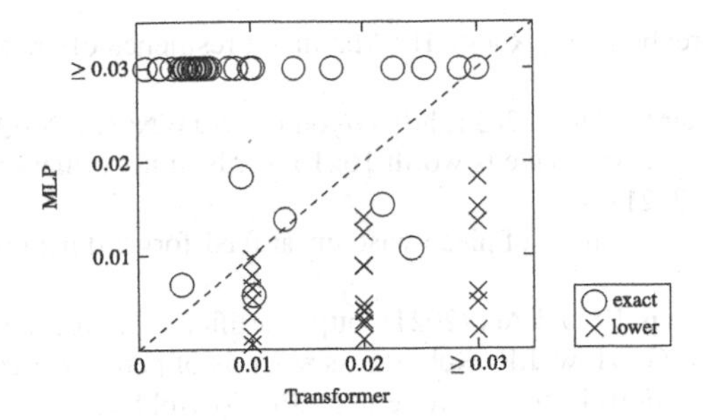

Fig. 3. Robustness of the Transformer and MLP on 60 random data points (better if larger). The dashed line highlights where they perform equally. There are 2 points on the dashed line, 26 in the lower triangle and 32 in the upper one. Since verifying the Transformer still takes much time, we report the lower bounds of the robustness values for the data points requiring more than one hour to verify. This means the points marked by "lower" can be further pushed to the right if the verifier is given more time. Nonetheless, this does not affect the overall observation as we ensure all such cases are points where the Transformer performs more robustly already.

6 Conclusion

This paper works towards exact robustness verification for ATNs. Specifically, we focus on Sparsemax-based ATNs, encode them into a MIQCP problem, and propose accelerating heuristics for solving the problem faster. When applied, our proposals fasten the verification process roughly one order of magnitude. We conduct experiments with a Lane Departure Warning application and find that ATNs are less robust than MLPs in our settings.

This initial study opens some interesting directions for further exploration. First, as we consider only Sparsemax-based ATNs, we are exploring further techniques to improve the bounds on Softmax that may facilitate its exact robustness verification. Second, we evaluate only small-scale networks (approximately 1600 neurons) on one dataset (with inputs of 14×10 dimensions). Whether our observations remain true for larger-scale networks and more datasets is yet to be explored. Third, our problem formulation only examines point-wise robustness verification for NNs. Such analyses can be combined with systematic sampling and testing methods to give formal and statistical guarantees on safety-critical applications. Lastly, our verification requires ground truths and works only in design time. How to utilize the studied techniques in a run-time setting remains an open question.

References

1. Bhojanapalli, S., Chakrabarti, A., Glasner, D., Li, D., Unterthiner, T., Veit, A.: Understanding robustness of transformers for image classification. In: ICCV (2021)
2. Bojarski, M., et al.: End to end learning for self-driving cars (2016)
3. Bonaert, G., Dimitrov, D.I., Baader, M., Vechev, M.: Fast and precise certification of transformers. In: PLDI (2021)

4. Cheng, C.H., Nührenberg, G., Ruess, H.: Maximum resilience of artificial neural networks. In: ATVA (2017)
5. Cruise: Cruise Under the Hood 2021, https://youtu.be/uJWN0K26NxQ?t=1342
6. Dosovitskiy, A., et al.: An image is worth 16x16 words: transformers for image recognition at scale. In: ICLR (2021)
7. Ehlers, R.: Formal verification of piece-wise linear feed-forward neural networks. In: ATVA (2017)
8. European Commission: EU AI Act (2021), https://artificialintelligenceact.eu/
9. Everett, M., Habibi, G., How, J.P.: Robustness analysis of neural networks via efficient partitioning with applications in control systems. IEEE Control Syst. Lett. **5**, 2114–2119 (2021)
10. Gehr, T., Mirman, M., Drachsler-Cohen, D., Tsankov, P., Chaudhuri, S., Vechev, M.: Ai2: safety and robustness certification of neural networks with abstract interpretation. In: SP (2018)
11. Goodfellow, I.J., Shlens, J., Szegedy, C.: Explaining and harnessing adversarial examples. In: ICLR (2015)
12. Grossmann, I.E.: Review of nonlinear mixed-integer and disjunctive programming techniques. Optim. Eng. **3**, 227–252 (2002)
13. Gurobi Optimization, LLC: Gurobi optimizer reference manual (2021)
14. Hu, B.C., Marsso, L., Czarnecki, K., Salay, R., Shen, H., Chechik, M.: If a human can see it, so should your system: Reliability requirements for machine vision components. In: ICSE (2022)
15. Huang, X., et al.: A survey of safety and trustworthiness of deep neural networks: Verification, testing, adversarial attack and defence, and interpretability. Comput. Sci. Rev. **37**, 100270 (2020)
16. Huang, X., Kwiatkowska, M., Wang, S., Wu, M.: Safety verification of deep neural networks. In: CAV (2017)
17. Katz, G., Barrett, C., Dill, D., Julian, K., Kochenderfer, M.: Reluplex: An efficient SMT solver for verifying deep neural networks. In: CAV (2017)
18. Krajewski, R., Bock, J., Kloeker, L., Eckstein, L.: The highD dataset: a drone dataset of naturalistic vehicle trajectories on German highways for validation of highly automated driving systems. In: ITSC (2018)
19. Lomuscio, A., Maganti, L.: An approach to reachability analysis for feed-forward relu neural networks (2017)
20. Mahajan, V., Katrakazas, C., Antoniou, C.: Prediction of lane-changing maneuvers with automatic labeling and deep learning. TRR J. **2674**, 336–347 (2020)
21. Martins, A.F.T., Astudillo, R.F.: From softmax to sparsemax: A sparse model of attention and multi-label classification. In: ICML (2016)
22. Paszke, A., et al.: PyTorch: an imperative style, high-performance deep learning library. In: NeurIPS (2019)
23. Poretschkin, M., et al.: AI assessment catalog (2023), https://www.iais.fraunhofer.de/en/research/artificial-intelligence/ai-assessment-catalog.html
24. Shao, R., Shi, Z., Yi, J., Chen, P.Y., Hsieh, C.J.: On the adversarial robustness of vision transformers. In: UCCV (2021)
25. Shi, Z., Zhang, H., Chang, K.W., Huang, M., Hsieh, C.J.: Robustness verification for transformers. In: ICLR (2020)
26. Su, J., Vargas, D.V., Sakurai, K.: One pixel attack for fooling deep neural networks. IEEE Trans. Evol. Comput. **23**, 828–841 (2019)
27. Tesla: Tesla AI Day 2022, https://www.youtube.com/live/ODSJsviD_SU?feature=share&t=4464
28. Tjeng, V., Xiao, K., Tedrake, R.: Evaluating robustness of neural networks with mixed integer programming. In: ICLR (2019)

29. Vaswani, A., et al.: Attention is all you need. In: NeurIPS (2017)
30. Wang, S., et al.: Beta-crown: efficient bound propagation with per-neuron split constraints for complete and incomplete neural network verification (2021)
31. Wong, E., Kolter, J.Z.: Provable defenses against adversarial examples via the convex outer adversarial polytope. In: ICML (2018)
32. Xiong, R., et al.: On layer normalization in the transformer architecture. In: ICLR (2020)

Model-Based Security and Threat Analysis

Model-Based Generation of Attack-Fault Trees

Raffaela Groner[1], Thomas Witte[1], Alexander Raschke[1](✉), Sophie Hirn[1], Irdin Pekaric[2,3], Markus Frick[2], Matthias Tichy[1], and Michael Felderer[2,4,5]

[1] Institute of Software Engineering and Programming Languages, Ulm University, Ulm, Germany
{raffaela.groner,thomas.witte,alexander.raschke,sophie.hirn, matthias.tichy}@uni-ulm.de

[2] Department of Computer Science, University of Innsbruck, Innsbruck, Austria
{irdin.pekaric,markus.frick,michael.felderer}@uibk.ac.at

[3] Department of Information Systems and Computer Science, University of Liechtenstein, Vaduz, Liechtenstein

[4] Institute for Software Technology, German Aerospace Center (DLR), Cologne, Germany

[5] Department of Mathematics and Computer Science, University of Cologne, Cologne, Germany

Abstract. Joint safety and security analysis of cyber-physical systems is a necessary step to correctly capture inter-dependencies between these properties.

Attack-Fault Trees represent a combination of dynamic Fault Trees and Attack Trees and can be used to model and model-check a holistic view on both safety and security. Manually creating a complete AFT for the whole system is, however, a daunting task. It needs to span multiple abstraction layers, e.g., abstract application architecture and data flow as well as system and library dependencies that are affected by various vulnerabilities.

We present an AFT generation tool-chain that facilitates this task using partial Fault and Attack Trees that are either manually created or mined from vulnerability databases. We semi-automatically create two system models that provide the necessary information to automatically combine these partial Fault and Attack Trees into complete AFTs using graph transformation rules.

Keywords: AFT · CPS · safety analysis · security analysis

1 Introduction

As cyber-physical systems (CPS) become more and more ubiquitous, safety and security analysis of such systems must take new and emerging problems into consideration. The proliferation of connected and smart devices, their interaction,

J. Guiochet et al. (Eds.): SAFECOMP 2023, LNCS 14181, pp. 107–120, 2023.
https://doi.org/10.1007/978-3-031-40923-3_9

and constantly changing software (e. g., due to over-the-air updates) leads to new safety and security problems. In particular, the increasing inter-connectivity of CPSs has a significant impact on their security. The amount of reported vulnerabilities increases year by year[1]. Each vulnerability can cause a failure of (parts of) the system, which in turn may affect its safety. In consequence, safety and security analysis in isolation, and without consideration of the specific environment the system is used in, usually cannot capture inter-dependencies between these concerns easily. The heterogeneity of CPSs in terms of hardware, but especially in terms of operating system versions with different installed package versions, requires that the deployed CPSs are analyzed and constantly monitored during operation.

Attack-Fault Trees (AFT) as a combination of Attack Trees and dynamic Fault Trees enables the joint analysis of safety and security properties in a single modeling formalism [1,7]. The AFT model can then be checked using existing model-checking techniques, e. g., critical path analysis or calculation of failure rates and probabilities. Generating large AFTs for realistic systems by hand, however, is error-prone and infeasible. It is much easier for safety experts to model Fault Trees on a system level and for security experts to create Attack Trees for used components [5].

These partial models are often on very different levels of abstraction. On the one side, a Fault Tree for a safety hazard ends on the level of logical system components and data channels. On the other side, vulnerabilities are reported on the level of packages and libraries and not on the level of components. The creation of an Attack Tree for a component thus requires intimate knowledge of the implementation of this component. Our approach attempts to bridge this gap of abstraction levels by deriving system dependency models from a running system and combining Fault Trees and Attack Trees into an AFT on this basis. Since we allow the extension of the generated models by manual models at each modeling level, we refer to our approach as semi-automatic.

In this paper, we present our toolchain (Fig. 1) and preliminary evaluation for the semi-automatic generation of AFTs, previously outlined in [22]: As an input, we use Attack and Fault Trees created by security and safety experts respectively as well as additional Attack Trees, that are automatically mined from vulnerability databases for components, tools and libraries found on the analyzed system. Compared to [22], which provided a vision of the approach, this paper provides a complete running pipeline.

In order to combine these partial models into a single AFT, we use information from two system models. Separate models for the logical system architecture and dataflow *(Dataflow Model)*, and the deployment and dependencies of these logical components *(Deployment Model)* are automatically derived from the running system and used to provide the necessary information to combine higher-level Fault Trees with system-level Attack Trees.

We then use graph transformation rules inferred from typical attack and fault propagation patterns to identify and generate missing AFT fragments to

[1] https://www.cve.org/About/Metrics.

bridge the abstraction gap between the partial attack and Fault Trees. To reduce the number of candidate Attack Trees to attach to the partial Fault Trees, we annotate basic events with impact requirements for potential attacks. These requirements are then propagated and matched with the impact scores of basic attack steps derived from vulnerability databases.

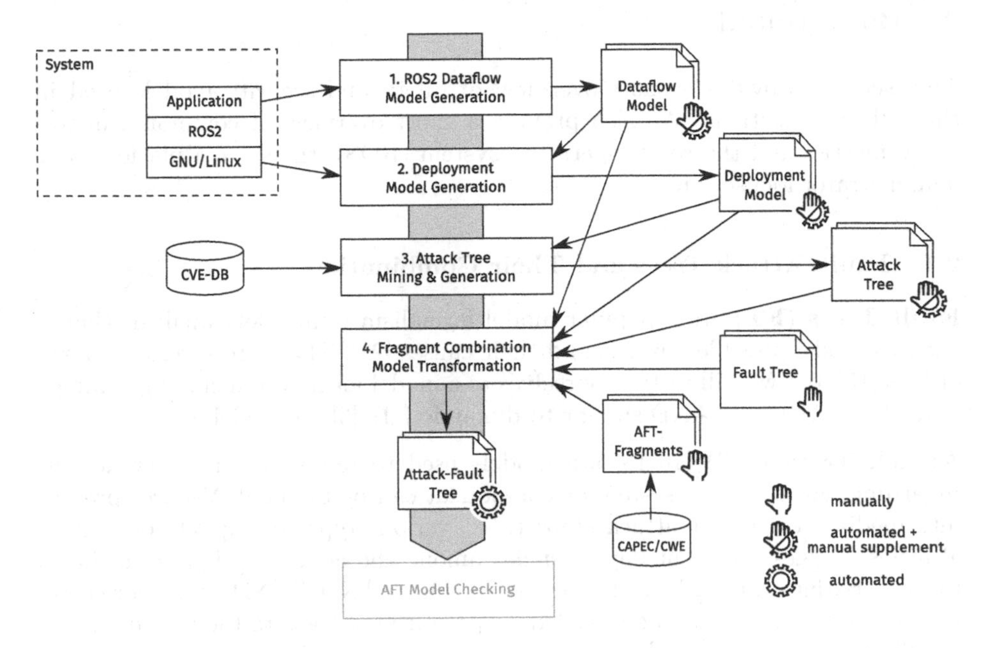

Fig. 1. Overview of the AFT generation toolchain and its models. Data sources for model generation are shown on the left and resulting models on the right.

We evaluate our implementation by analyzing a quadcopter system: the quadcopter pose is tracked using multiple cameras. Flight trajectories that avoid obstacles are generated on a PC and then sent to the quadcopter. Our toolchain connects to the system, automatically gathers dataflow and deployment information of the system, mines vulnerability databases for found library and system dependencies, and generates initial Attack Trees. Then it combines these models with manually created Fault Trees, using dataflow and deployment information to complete the Fault Trees to full AFTs that include potential system vulnerabilities. On the intentionally not updated operating system, our approach successfully found some possible attacks that might lead to the hazards modeled in the Fault Trees and created meaningful AFTs for this scenario.

However, we identified some possible improvements especially with respect to the mapping of software packages to vulnerabilities in CVE-databases and related the precise decision, which of the generated Attack Trees can be applied to an AFT to minimize the amount of false positives.

The outline of the paper is as follows: Sect. 2 shortly introduces besides necessary background on safety and security models, security metrics, and the robot operation system, a running example used in the remaining paper. In Sect. 3, we present our approach to combined safety and security modeling and analysis followed by a discussion of its chances and limitations in Sect. 4. Section 5 compares the presented approach to related ones before Sect. 6 concludes the paper.

2 Background

This section provides a short overview of safety and security models used in the following sections. We also provide a short overview of common vulnerability metrics and the robot operating system (ROS), the system platform our demonstrator focuses on.

2.1 Fault/Attack Trees and Their Combination

Fault Trees (FT) are a popular model formalism from safety analysis that is used to model possible hazards and their causes [21]. There are several variants of FTs [15], but we will restrict ourselves to a model formalism including complex gates like SAND or PAND similar to dynamic FTs [3] for modeling.

Attack Trees (AT) [10,18] are models used to represent adversary actions regarding how a certain system or component can be targeted. Vertices present intermediate targets or attack steps (i. e., various types of exploits or vulnerabilities). Edges represent dependencies among the actions and (intermediate) targets. By introducing logical gates (AND, OR, PAND, SAND, etc.) more complex attack scenarios can be modeled. A path from a leave to the root of an AT is called an attack path [8]. Moreover, weights can also be assigned to edges to include costs, probability, risks, or other metrics.

Automatic generation of (simple) ATs is a complex task that can be achieved by employing various existing databases that include vulnerability, library, attack, and severity data.

Attack-Fault Trees (AFT) [7,11] integrate Attack Trees within Fault Trees. This is achieved by redefining the basic events of a FT which now describe not only accidental events (e.g., the sudden failure of a component) but also the failure of a component due to the malicious actions of an attacker. While ATs usually describe a general attack pattern, the Attack Trees attached to FTs must be directed at a specific target. Therefore, it is necessary to make the definition of ATs more specific for this purpose, as described in [11].

In our approach, an attack event consists of a description of the attack, a reference to a model element from the deployment or the dataflow model (see Sect. 3.1) and a requirement in the form of minimum CIA values (see Sect. 2.2) that an attack on the linked component must possess in order to trigger the attack. Adapted from the external events from Vesely et al. [21], we use a house shape to represent our attack events graphically.

2.2 Vulnerability Metrics

In the following, we discuss vulnerability-related terms that are utilized in the proposed modeling approach. CVE[2] data is used to uniquely distinguish different vulnerabilities. CWE[3] entries represent specific higher-level groups to which CVEs are assigned in order to provide a hierarchy for vulnerability data. There exist two main hierarchies, which divide them into software and hardware weakness types. Most of the CVEs contain a CVSS[4] vector. It provides various types of qualitative scores related to the severity of the vulnerability. Besides some more detailed information assessing the severity, it contains information about the impact of a CVE, the so-called CIA triad [16].

CPEs[5] represent various systems, software, and platforms, which are represented using syntax for Uniform Resource Identifiers (URI) including a specific version of a software or library. This allows security engineers and researchers to exactly know which software is affected by a certain CVE. In cyber-security, attacks can also occur by exploiting multiple vulnerabilities, which form attack chains. Some of these chains are similar in a way that they address CVEs that belong to the same CWE or they have corresponding mechanics. In order to represent these similarities, CAPEC[6] entries were created, which allow an easy understanding of common attacker actions. The aforementioned databases present a solid foundation for the proposed approach since they include vulnerabilities, weaknesses, platforms, and attack patterns.

2.3 Robot Operating System (ROS)

While our safety and security analysis toolchain is technology and system agnostic, we implemented specific dataflow and deployment generators for the *Robot Operating System (ROS)*. ROS [9] is a middleware for component-based robotic applications. It consists of various helper libraries, e.g., for message transport, standardized interfaces, and tools. Components, called *Nodes*, communicate over named and typed channels (*Topics*) and via RPC (*Services*). Application components can use multiple implementation languages; 3rd party components provided by, e.g., hardware manufacturers might act as black boxes, showcasing heterogeneous systems in need of joint safety and security analysis of the system as a whole.

2.4 Running Example

As a running example, we present the following scenario: an autonomous drone might pose an injury hazard to a bystander in case of a collision. This can be

[2] Common Vulnerabilities and Exposures, https://cve.mitre.org/.
[3] Common Weakness Enumeration, https://cwe.mitre.org/.
[4] Common Vulnerability Scoring System, https://www.first.org/cvss/.
[5] Common Platform Enumerations, https://nvd.nist.gov/products/cpe.
[6] Common Attack Pattern Enumeration and Classification, https://capec.mitre.org/.

caused through a mechanical malfunction causing the accident, or as a result of an attack on the control system of the drone.

The drone control system in our quadcopter lab consists of a camera array for optical tracking, that calculates the exact position and orientation of the drone at a high frequency. This pose data is then sent to a ROS application, which consists of several components that implement trajectory planning, obstacle avoidance, and position control, among others. The quadcopter is connected via WiFi and control commands are sent to it using the AR.Drone SDK, closing the position control loop.

We create a Fault Tree to model the injury hazard: if one or more components or channels in the position control loop fail, a drone operator standing near the drone might be injured. Refining this FT and figuring out which software vulnerabilities, exploits or weaknesses might be applicable to trigger these fault events is done using our proposed AFT generation toolchain. An excerpt of the generated AFT is shown in Fig. 3. Two potential attacks that might lead to a failure of the position controller are identified and attached based on libraries used by this component.

3 SafeSec Attack-Fault Tree Generation Toolchain

We developed the *SafeSec Attack-Fault Tree Generation Toolchain (SAFT-GT)* (Fig. 1) to semi-automatically create and analyze AFT models for self-adaptive systems. Our toolchain uses *dataflow* (Sect. 3.1), and *deployment models* (Sect. 3.2) to capture the state of the system and uses this information to automatically combine generated *Attack Trees* (Sect. 3.3) and manually created *Fault Trees* that use different abstraction levels. A set of *combination rules* is used to find and connect these AFT fragments (Sect. 3.4). The complete toolchain including the models used for the running example can be downloaded here: https://www.uni-ulm.de/in/sp/research/projects/safesec/

3.1 Dataflow Model

Our dataflow model captures the logical components and dataflow of the system. We separate this logical view on the system from the actual implementation and deployment in the deployment model. The meta-model of the dataflow model is rather simple: it consists of *components* – entities that provide, transform or process data – and *channels* – ways for components to communicate, send messages or observe the state of other components.

This high level of abstraction is necessary to easily map basic fault events to system components and channels. A safety engineer manually creating Fault Trees for the system needs to specify the origin of basic fault events by annotating respective components and channels.

We designed the model to be simple to generate for systems using different middlewares or frameworks. Our model simplifies the ROS component meta-model similar to the abstract component meta-model in [4] in order to easily generate and integrate dataflow models from ROS systems as well as other systems. ROS nodes are mapped to components while ROS topics, services, and actions are mapped to channels. Additionally, the model is designed to be manually extensible: additional components and channels can be defined manually and interface with the rest of the model. For example, the camera system and infrastructure to optically track quadcopters might be manually added including an optical channel from the quadcopter to the camera components to model the optical tracking of the quadcopter position.

Our implementation includes a dataflow model generator for ROS2 systems. ROS nodes do not define their interface statically, but connect to topics and services dynamically. Therefore, our generator consists of a single ROS node, that can be triggered to collect architecture and dataflow data using ROS' introspection capabilities at runtime. Due to its design as a daemon that collects data at runtime, the generation of the dataflow model is fast and captures exactly the current state of the system. The generator can be triggered repeatedly in order to monitor the system for changes or architecture reconfigurations. Deriving the dataflow model from a static configuration/composition description instead would require complex static analysis and access to the source code of all nodes.

3.2 Deployment Model

In order to bridge the gap between the high-level dataflow model and the components that are deployed on a certain system, we introduce a so-called deployment model. This model contains the information which component is running on which system. Our toolchain automatically extends this initial information with the files and ultimately the libraries, a component depends on. Dependencies that cannot be derived automatically (e.g., because the platform a component runs on cannot be reached by our analysis tool) can be given manually.

Unlike other existing tools, such as snyk[7], we do not rely on component source code to obtain dependency information. Instead, our tool uses information about the open files of a component's running processes.

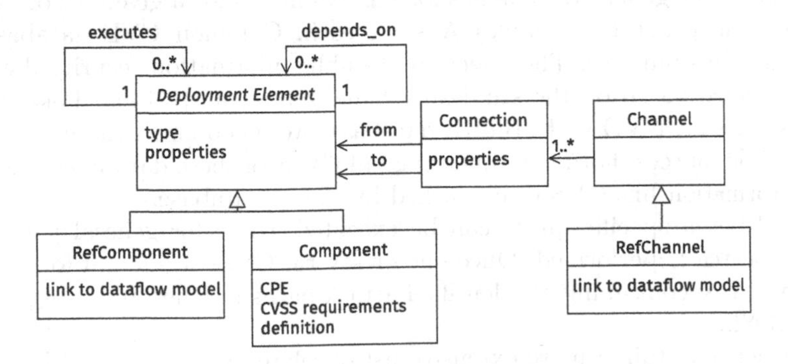

Fig. 2. Simplified meta-model of deployment model.

[7] https://snyk.io.

Figure 2 shows a slightly simplified meta-model of the deployment model. Due to limited space, the *Channels* are shown but not described further below. A *Deployment Element* is either a newly defined *Component* or a reference to a dataflow component (*RefComponent*). Each component has a type (e.g., File, Library, Package, Platform, OS, HW, Sensor, Actor, ...) and arbitrary properties (key/value-pairs whose interpretation depends on the type). A (low level) component might have a CPE entry or CVSS requirements. If these requirements are met by an appropriate attack, the component can be considered as corrupted (see Sect. 3.4).

Each deployment element can be executed on another element. In our running example most ROS nodes run on a PC called "rosbox". The properties of this "rosbox", including information how our analysis tool can reach this PC, are used to gather more detailed information about the components running on it. A deployment element can also depend on other elements. For instance, the component "default_FARFETCH_bebop_position_control" of our running example depends on library "fast_dds" in version 2.1.1 (compare Fig. 3). This dependency information is generated recursively by our analysis tool via the used files and libraries of a component returned by Unix tools like `lsof` and `ldd`. System specific package managers like `apt` and `dpkg` abstract this information into package names for which CVEs can be found. So far, the tool supports Ubuntu and Gentoo as platforms, but its architecture includes several abstraction layers to facilitate the integration of other platforms.

The next step is to find the corresponding CPE for each identified package. For this purpose, we use the tool CPEguesser[8] in combination with some heuristic preprocessing like shortening names, removing additional version information, etc. Both lists, CPEs and all packages for which no CPE entry can be found, are then passed to the Attack Tree Generator to find possible CVEs for these pieces of software.

3.3 Attack Tree Generation

The Attack Tree generator searches for vulnerabilities for a given set of software packages and generates (simple) ATs for each. Common CVE databases are utilized for this purpose. The selection of public information security databases was conducted based on the studies by Sauerwein et al. [17] and Pekaric et al. [13]. As a result, NVD and MITRE databases were chosen as the most current and credible sources. For faster querying, all CVEs of these databases (including meta-information like CVSS) are cached in a local database.

For CPEs, a specific query can be executed, while for general packages, a full-text search is performed. Once one or several CVEs are found for an entry, an Attack Tree containing all identified attack paths is generated using our self-defined DSL.

In order to obtain a more extensive list of related CVEs, the CWE data, its hierarchy, and especially their relationship information like *PeerOf*, *CanFollow*,

[8] https://github.com/cve-search/cpe-guesser.

and *CanPrecede* is considered in order to create attack chains in which multiple CVEs are linked using SAND, AND, and OR gates, telling the combination in which different CVEs must be exploited to conduct more complex attacks. Besides these automatically generated Attack Trees (ATs), more complex attack scenarios are only in a later step with the help of manually predefined AFT fragments.

3.4 Attack-Fault Tree Generation

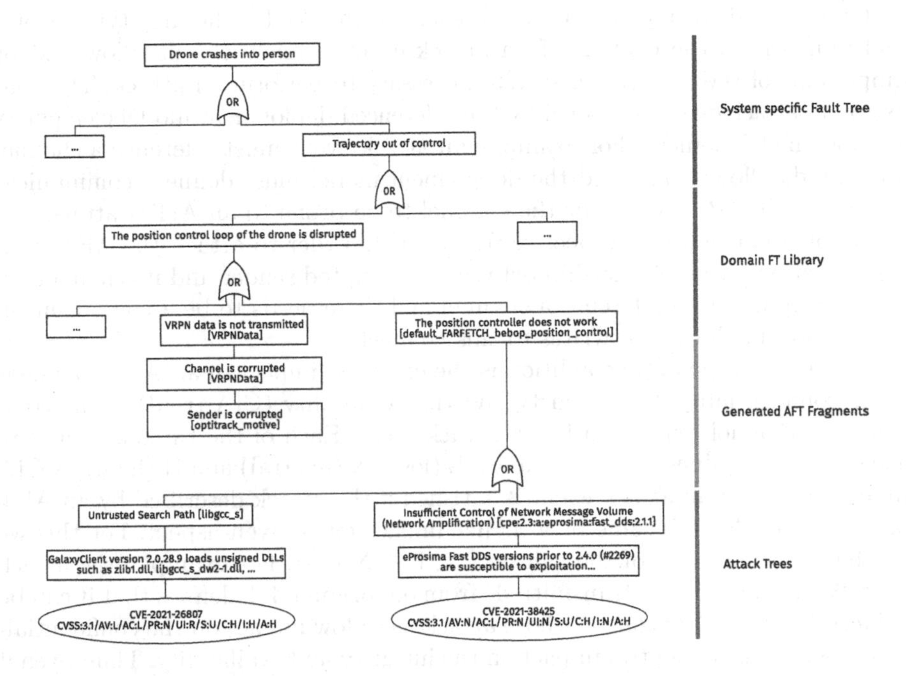

Fig. 3. Generated AFT for the running example (truncated for readability).

The AFT generation, as shown in Fig. 1, mainly consists of combining fragments using model transformations based on our different input models. Overall, the AFT generation can be divided into three phases. First, the FT that forms the upper part of an AFT is copied into a new AFT model. Second, the newly generated AFT model is extended using manually (pre-)defined AFT fragments, which represent generic, common attack patterns. Third, appropriate ATs are attached at the leaves of the created AFT model.

To decide whether an AFT fragment or an AT can replace an attack event and thus be attached to the AFT we use the information about the system provided by the dataflow and the deployment model. Since copying the FT into the AFT is trivial, we will present only the last two phases of AFT generation.

AFT Fragments-Phase: The goal of the second phase is to create a bridge between the more abstract FTs, which are based on events caused by corrupted logical data flow or components, and the very technical ATs, which model, e.g., attacks on individual hardware or protocols. This bridge is formed by our AFT fragments. These fragments represent different attack patterns that help to break down the more abstract attack events of the FT to the same level as the ATs. Such a fragment describes for example the individual steps, which are necessary to perform an adversary in the middle attack (AiTM) or another one describes the relationship that when a sender is corrupted its associated channel on which it transmits is also corrupted.

We have defined two types of preconditions for each AFT fragment, which must be fulfilled, to replace an attack event in the AFT. The first type of precondition defines the context of an attack event, namely the dataflow and/or components of the underlying system necessary to perform an attack. The context of an attack event is defined by the referenced deployment model element or dataflow model element. For example, an attack event must reference a channel from the dataflow model and the deployment model must define a communication over TCP/IP or UDP for the channel to be prone to an AiTM attack and thus allow to attach the respective fragment. In order to attach the AFT fragment describing the relationship between a corrupted sender and its channel, an attack event needs to reference a channel and there needs to be a component in the dataflow model which writes to this channel.

The second type of precondition is the expected impact of an AFT fragment to the confidentiality, the integrity and the availability (CIA triad) of the component or channel referenced by an attack event. Each of the three aspects can take one of four values, namely * (any), L (low), N (neutral) and H (high). A CIA value required by an attack event is satisfied if the attack described by an AFT fragment provides the same or a higher impact for a given aspect. For this we use the following order of the values: $* < L < N < H$. For example, the attack event "VRPN data is not transmitted" from our original FT defines that it can be replaced by an AFT fragment whose attack has a low impact on the confidentiality and a more than neutral impact on the integrity and availability. Thus, even if the context is correct, an AiTM attack which aims to obtain data cannot replace this attack event since the confidentiality value of this AiTM attack is high and its integrity and availability values are low. Our AFT fragment describing the relationship between a corrupted sender and its channel, defines that such an attack towards a sender has a high impact on the integrity and a neutral impact on the confidentiality and availability. Thus, this AFT fragment satisfies both types of preconditions specified by the attack event "VRPN data is not transmitted" from our initial FT and is attached to the AFT, as shown in Fig. 3.

AFT fragments can also introduce new attack events, which then can be further replaced with other AFT fragments or ATs. For example, "Sender is corrupted" in Fig. 3 is an attack event introduced by our AFT fragment describing the relationship between a corrupted sender and its channel.

In our proof of concept implementation, we manually defined five AFT fragments and their preconditions, which we have developed based on our expertise in modeling AFTs and ROS. We also analyzed the available list of CAPEC entries to identify entries that are suitable as a basis for AFT fragments and, e.g., used the CAPEC-94 [2] as basis for our AiTM AFT fragment.

Append AT-Phase: In the last phase of AFT generation, we attempt to replace attack events that were not replaced in the previous phase or are newly added by the AFT fragments with generated ATs. In order to decide whether an AT is suitable to replace an attack event or not we use the same two types of preconditions, namely the context and the CIA values defined the respective attack event.

Since we generate the ATs, their preconditions regarding their context are much simpler than for the AFT fragments we define manually. For an AT, the precondition for its context is already fulfilled if the context referenced by the attack event is mentioned in the description, the CPE or the name of the AT. If an attack event references a deployment model element, it is checked whether this referenced element or one of its subcomponents is affected by the attack described in the AT. If the attack event references a dataflow element, the deployment model is first searched for a more precise system description for the referenced dataflow model element. If a corresponding element is found in the deployment model, it is checked as before whether this element or one of its subcomponents is affected by the attack described in the AT.

To decide whether the values of the CIA triad are fulfilled by the attack we use the corresponding values from the CVSS vector of the generated AT and the same rules as already used to attach our AFT fragments. In our example, the attack event "The position controller does not work" defines that an attack must have at least one low impact on the confidentiality and at least a neutral impact on the integrity and availability. This requirement is fulfilled by two generated ATs since according to its CVSS vector the AT "Insufficient Control of Network Message Volume" contains the value high for confidentiality and availability and a neutral one for the integrity. The CVSS vector for the AT "Untrusted Search Path" even possesses a value high for all three CIA values. Since we rely on CIA values, our approach is limited to ATs that possess a CVSS vector.

4 Discussion

We have modeled two different FTs to demonstrate the feasibility of our approach. The first one describes the possible injury of a person by a drone, we have also used this FT in our running example. The second FT describes how a privacy violation by a drone can occur due to errors or malicious behavior. We performed our AFT generation for both FTs in the context of our quadcopter lab. In total, the three phases of the AFT generation took approx. 15 s for the first and 10 s for the second FT on a Intel Core i7-3770 CPU with 16GB RAM.

Based on these results and a manual review of the generated AFTs, we can conclude that the whole pipeline, starting with the automated dataflow extraction of a (intentionally not updated) running ROS system generates correct and useful AFTs in a reasonable time. The most time consuming part is the search for vulnerabilities, especially in the case of full-text search. To improve the performance, the intermediate results could be cached and only recalculated, if new CVEs are reported and/or new package/file (versions) are detected.

Unfortunately, some of the generated and attached attacks are false positives: The AT "Untrusted Search Path" fulfills all our preconditions, but the described attack does not apply to our system. The reason for this is that identifying software and software versions and mapping them to vulnerability data is hard. CPEs mitigate this problem by uniquely identifying affected software and software versions. However, to our knowledge, no mapping between arbitrary OS packages and CPEs exists and not all CVE entries contain affected CPE information. Our fallback to a full-text search may result in wrong or incomplete mappings due to different package names (e.g., software split into multiple packages, OS specific package naming), different versioning (e.g., due to additional applied patches), or renamed software projects. We only applied some simple heuristics, and thus the results can definitely be improved by using more sophisticated mining techniques such as NLP and/or by merging more detailed package databases. We think, this is an interesting research field that many other CVE-related research approaches could benefit from.

Moreover, our precondition consisting of the consideration of the three CIA values may be too vague to decide reliably whether an AFT fragment or an AT should be attached or not. Here, a solution could be to include the other metrics provided by a CVSS vector in the decision. In addition, we have defined only a few AFT fragments and we did not yet conduct a structured review of all CAPEC mechanisms and other related taxonomies of common attack patterns to derive and evaluate a complete set of general AFT fragments. We envision future iterations of our toolchain to provide a core set of general AFT fragments that can then be tailored toward specific domains and application areas by adding additional, more specific, AFT fragments.

5 Related Work

The research field of combined security and safety analysis in software and systems engineering is huge [14]. For this reason, we will concentrate on the aspect of Attack/Fault Tree generation in our discussion of related work and only briefly address other aspects.

The idea of generating fault- or attack graphs is not new. Swiler et al. [20] present a tool that generates attack graphs based on an assessment of security attributes and vulnerabilities in computer networks. Similar to our presented approach the authors use vulnerability scanning tools and attack templates. They mention the integration of an attacker profile might be interesting but also do not take this into account. Besides their focus on network attacks, the main difference to our approach is our attempt to combine the generated Attack Trees with Fault Trees. The configuration files introduced by Swiler et al. are similar to our dataflow and deployment models, but we generate these partially automatically.

Kotenko et al. [6] utilize in their approach for Attack Tree generation similar techniques as proposed by us: CPEs are used to identify CVEs for components and CAPECs are employed for the generation of more complex attack scenarios. However, the main difference is their focus on network scenarios. Therefore, they use network security detection tools to identify possible vulnerabilities. Instead, we build upon software packages of a running (ROS) system. Similarly, Ou et al. [12] generate attack graphs for network topologies using logic programming. One disadvantage of their approach is that all information of the system must be given manually in advance by "facts" in the logical programming language.

Our combination of FTs and ATs into AFTs follows the general approach stated by Steiner et al. [19] which is in accordance with the work of Fovino et al. [11]. They describe the introduction of so-called "security events" in FTs in contrast to the existing "safety events". In contrast, Stoelinga et al. [7] introduce new model elements in order to combine FTs with ATs.

6 Conclusion and Future Work

In this paper, we presented our automated tool pipeline for generating AFTs based on generated and manually supplemented models. To bridge the gap between high level Fault Trees and low-level Attack Trees, we introduce different intermediate models that describe the data flow between components and system dependencies. Using these models, we extend manually created FTs with generic and specific AFT fragments, and then attach generated ATs to the created AFT. The advantage of this combined approach is the possibility to add manually created (partial) models at any stage. This allows this approach to be used even if the level of automation in a particular environment is not yet very high.

At the moment, however, we see potential for improvement with respect to the mapping of used software packages and their vulnerabilities and the decision of whether an AT can be attached to the AFT or not. Here, we see several possibilities for improvement we plan to investigate in the future, from which also other research approaches can benefit from. Also the extension of our approach towards other operating systems and software platforms besides ROS2 is an interesting future research direction.

Acknowledgements. This work was partially supported by the Austrian Science Fund (FWF): I 4701-N and the German Research Foundation (DFG): 435878599.

References

1. André, É., Lime, D., Ramparison, M., Stoelinga, M.: Parametric analyses of attack-fault trees. Fund. Inform. **182**, 69–94 (2021). https://doi.org/10.3233/FI-2021-2066
2. CAPEC-94: Adversary in the Middle (AiTM). https://capec.mitre.org/data/definitions/94.html. Accessed 13 Feb 2023
3. Dugan, J., Bavuso, S., Boyd, M.: Dynamic fault-tree models for fault-tolerant computer systems. IEEE Trans. Reliab. **41**(3), 363–377 (1992). https://doi.org/10.1109/24.159800

4. Gherardi, L.: Variability modeling and resolution in component-based robotics systems. Ph.D. thesis (2013)
5. Giese, H., Tichy, M.: Component-based hazard analysis: optimal designs, product lines, and online-reconfiguration. In: Górski, J. (ed.) SAFECOMP 2006. LNCS, vol. 4166, pp. 156–169. Springer, Heidelberg (2006). https://doi.org/10.1007/11875567_12
6. Kotenko, I., Chechulin, A.: A cyber attack modeling and impact assessment framework. In: CYCON 2013, pp. 1–24 (2013)
7. Kumar, R., Stoelinga, M.: Quantitative security and safety analysis with attack-fault trees. In: HASE 2017, pp. 25–32 (2017). https://doi.org/10.1109/HASE.2017.12
8. Lallie, H.S., Debattista, K., Bal, J.: A review of attack graph and attack tree visual syntax in cyber security. Comput. Sci. Rev. **35**, 100219 (2020)
9. Macenski, S., Foote, T., Gerkey, B., Lalancette, C., Woodall, W.: Robot operating system 2: design, architecture, and uses in the wild. Sci. Robot. **7**(66), eabm6074 (2022)
10. Mauw, S., Oostdijk, M.: Foundations of attack trees. In: Won, D.H., Kim, S. (eds.) ICISC 2005. LNCS, vol. 3935, pp. 186–198. Springer, Heidelberg (2006). https://doi.org/10.1007/11734727_17
11. Nai Fovino, I., Masera, M., De Cian, A.: Integrating cyber attacks within fault trees. Reliab. Eng. Syst. Saf. **94**(9), 1394–1402 (2009). https://doi.org/10.1016/j.ress.2009.02.020
12. Ou, X., Boyer, W.F., McQueen, M.A.: A scalable approach to attack graph generation. In: 13th ACM Conference on Computer and Communications Security. CCS '06, pp. 336–345. Association for Computing Machinery, New York, NY, USA (2006). https://doi.org/10.1145/1180405.1180446
13. Pekaric, I., Felderer, M., Steinmüller, P.: VULNERLIZER: cross-analysis between vulnerabilities and software libraries. In: HICSS, pp. 1–10 (2021)
14. Pekaric, I., et al.: A Systematic Review on Security and Safety of Self-Adaptive Systems (2022). https://dx.doi.org/11.2139/ssrn.4029617, preprint at SSRN
15. Ruijters, E., Stoelinga, M.: Fault tree analysis: a survey of the state-of-the-art in modeling, analysis and tools. Comput. Sci. Rev. **15–16**, 29–62 (2015). https://doi.org/10.1016/j.cosrev.2015.03.001
16. Samonas, S., Coss, D.: The CIA strikes back: Redefining confidentiality, integrity and availability in security. J. Inf. Syst. Secur. **10**(3) (2014)
17. Sauerwein, C., Pekaric, I., Felderer, M., Breu, R.: An analysis and classification of public information security data sources used in research and practice. Comput. Secur. **82**, 140–155 (2019)
18. Schneier, B.: Modeling security threats. Dr. Dobb's J. **24**(12) (1999)
19. Steiner, M., Liggesmeyer, P.: Combination of safety and security analysis - finding security problems that threaten the safety of a system. In: SAFECOMP 2013. Workshops and Tutorials: CARS, SASSUR, DECS, ASCOMS (2013)
20. Swiler, L., Phillips, C., Ellis, D., Chakerian, S.: Computer-attack graph generation tool. In: DISCEX'01, vol. 2, pp. 307–321 (2001). https://doi.org/10.1109/DISCEX.2001.932182
21. Vesely, W.E., Goldberg, F.F., Roberts, N.H., Haasl, D.F.: Fault tree handbook. Technical report, Nuclear Regulatory Commission Washington DC (1981)
22. Witte, T., Groner, R., Raschke, A., Tichy, M., Pekaric, I., Felderer, M.: Towards model co-evolution across self-adaptation steps for combined safety and security analysis. In: SEAMS 2022, pp. 106–112 (2022). https://doi.org/10.1145/3524844.3528062

MBTA: A Model-Based Threat Analysis Approach for Software Architectures

Anas Motii(✉)

Mohammed VI Polytechnic University, Ben Guerir, Morocco
anas.motii@um6p.ma

Abstract. In the last decade, several efforts have been achieved to integrate security in the Software Development Life-cycle (SDL). Analyzing software architecture in order to identify threats is an essential step in secure software development processes. However, performing this task manually can result in identifying false positives. It is thus time-consuming and error-prone. Therefore, there is a need for automated tool support to perform this task. Existing efforts are limited to specific, predefined security properties or threats that are checked either manually or using limited toolsets. In this paper, we present a general and constructive model-based approach for threat analysis. We employ domain-specific modeling language techniques to develop a set of modeling languages that enable the specification of the software architecture structure. We used the Object Constraint Language (OCL) for the purposes of precise specification and verification of security threats as properties of a modeled system. To validate our work, we explore a set of representative threats in the context of SCADA systems.

Keywords: Model-Based · OCL · Security · Threat Analysis · Software architecture

1 Introduction

Our society has become more dependent on software-intensive systems, such as Information and Communication Technologies (ICTs) systems, not only in safety-critical areas but also in areas such as finance, medical information management, and systems using web applications. The complexity of such systems during their design comes from the involvement of trans-disciplinary concerns. In addition, security experts, practitioners and researchers from different international organizations, associations, and academia have agreed that security should be treated in the early stages of the software and systems development life-cycle [8]. Otherwise, security vulnerabilities are more likely to be introduced in various stages and the cost of protecting them becomes increasingly more important. In this context, the use and application of security mechanisms through the life-cycle process would be easier if designers and developers had security guidelines during development. Architecture threat analysis is the process of identifying

J. Guiochet et al. (Eds.): SAFECOMP 2023, LNCS 14181, pp. 121–134, 2023.
https://doi.org/10.1007/978-3-031-40923-3_10

threats to an architecture. It is very useful when it comes to detecting threats at early stages. Reported vulnerabilities show that architecture design weaknesses represent half of the total vulnerabilities of a system. Several efforts have been done to assist threat identification [3]. However, the complexity of systems requires automated tool support.

This work is part of a more general process devoted to incremental pattern-based modeling and safety and security analysis for correct-by-construction systems design. In previous works, an approach and its tool support to support Security, Dependability and Resource Trade-offs using Pattern-based Development and Model-driven Engineering have been presented [5]. In this paper, a Model-Based Threat Analysis approach for software architecture and its tool support is introduced in order to allow automatic threat detection based on the Object Constraint Language (OCL). The remainder of the paper is organized as follows. Section 2 presents the main steps of the MBTA approach. In Sect. 3, the MDE framework supporting MBSPI is presented. The threat formalization process using OCL is explained step by step. Section 4, MBTA is assessed over a SCADA (Supervisory Control and Data Acquisition) system case study. Section 5 identifies related work tackling software architecture threat analysis. Finally, Sect. 6, concludes and sums up the contributions.

2 MBTA Approach

The approach depicted in Fig. 1 allows the analysis of software architectures in order to detect existing threats based on formalization. The first step consists of formalizing threats using OCL[1] from existing threat classification references **(step 0)**. Then, the software architecture model is passed to the analysis module **(step 1)** which outputs existing threats.

Specifying threats is based on experience in the security domain thus this activity should be done by security experts. Once formalized, these threats are stored in a knowledge base. Inputs are existing threat classification references: OWASP[2], STRIDE [6], Common Attack Pattern Enumeration and Classification (CAPEC)[3], Common Weakness Enumeration (CWE)[4]. These references describe informally a set of threats. Each threat has a signature. This signature specifies the conditions in which a threat can occur. Thus it defines the threats according to a certain scenario. However, the threats are described informally, and thus applying them manually is error-prone and time-consuming. The considered threats are discussed below. This is neither a comprehensive nor a complete list but two well-known categories which have been used in OWASP's top 10:

---

[1] https://www.omg.org/spec/OCL/.
[2] https://owasp.org/www-project-top-ten/.
[3] https://capec.mitre.org/.
[4] https://cwe.mitre.org/.

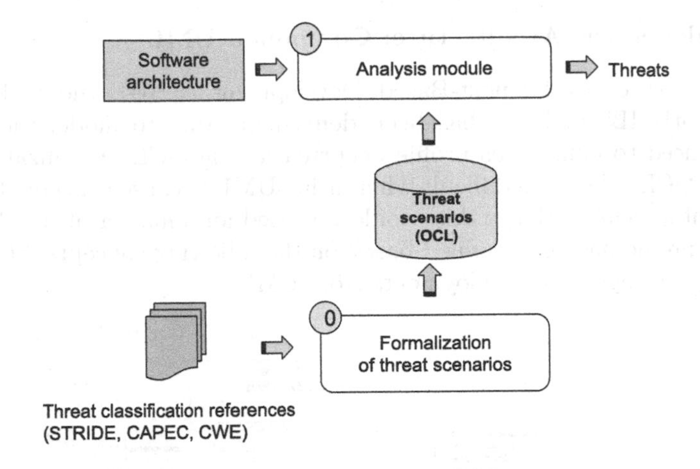

Fig. 1. Threat Analysis Process

- **Man-In-The-Middle (MITM)**: is responsible for relaying or altering messages between two parties. The signature of this threat is the lack and/or weakness of encryption and Authenticity mechanisms.
- **Injection**: is responsible for passing malicious inputs to gain higher privileges, alter data, or crash the system. The signature of this threat is the lack and/or weakness of input validation and the secure development of an application.

OCL is a formal language used to describe rules on UML models. We have used OCL to formalize the aforementioned threats as invariants. The analysis allows the detection of threats over the architecture. If an invariant is violated then the corresponding threat is relevant. In order to evaluate the formalized threats, the precision metric is used. It measures the soundness of the results. A high precision rate means that the detected threats contain more True Positives (TP) i.e., valid results than False Positives (FP) i.e., false results. It is computed as follows:

$$Precision = \frac{TP}{TP + FP} \tag{1}$$

3 Model-Driven Development

We now present the MDE framework supporting the previous approach and we detail its construction from the system architecture and security perspectives. For the system architecture aspects, we used a UML-Like[5] modeling language to describe software architecture using the component-port-connector fashion. The security perspective which consists of three iterations introduces additional architectural elements and uses OCL for the specification and analysis of the security threats.

[5] https://www.omg.org/spec/UML/.

3.1 Modeling the Architecture: ComponentUML

In the context of Component-Based Development (CBD), the UML profile "ComponentUML" in Fig. 2 has been defined in order to model the application. The need to define this profile occurred during OCL formalization using OCL. The OCL rules were difficult when using UML because concepts that were not relevant appeared. Hence this profile was used for a matter of simplification. The UML profile has been defined based on the following concepts: StructuredClassifiers, Messages and Deployments from UML.

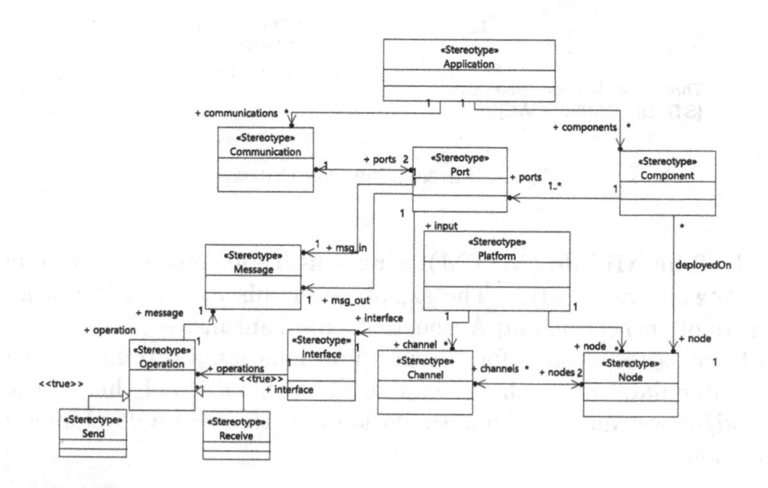

Fig. 2. UML profile for component-based software architectures

Working Example: Metamodel Instantiation. Figure 3 shows the software architecture of a three-tier web application[6]. The architecture consists of three component types: *Page*, *Webapp* and *Database*. Each component is associated with ports, interfaces, data types and messages, accordingly. For instance, a component *webpage* of type Page uses a Port Client Server for the communication with component *webapp* of type *Webapp*. For that, component *webapp* uses a port *Port Server Client*. The comments in blue show the different messages: *m1* to model the request sent to the application, *m2* to model the response from the application, *m3* to model the request sent to the database and m4 to model the response from the database. From the deployment perspective, the underlying platform consists of three nodes: *Browser* hosting *webpage*, *Server* (exposed to Internet) hosting *wepapp* and Back to host database. The software architecture model for the web application has been made intentionally not secure to test the OCL constraints. In fact, the model does not contain any sort of security

[6] https://www.ibm.com/topics/three-tier-architecture.

mechanisms: encryption and input validation. Hence it is vulnerable to injection and MITM attacks. The objective is to detect: one injection and one MITM threat.

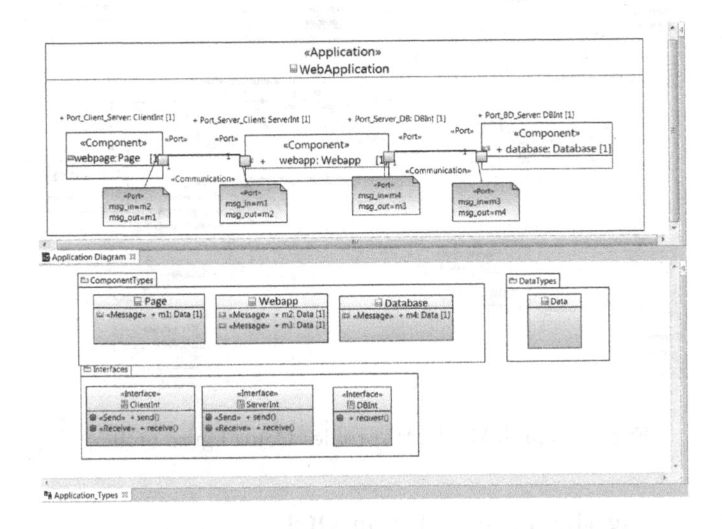

Fig. 3. Web application software architecture model and types (Color figure online)

3.2 Modeling the Security Solutions as Security Patterns

As introduced in Sect. 1, our work is part of a general approach to building secure software at high-level design stage using patterns (PBSE). We developed a UML profile called *SepmUML*, as depicted in Fig. 4 using UML notations (not all classes and attributes are shown on the diagram to avoid cluttering). *SepmUML* contains the necessary stereotypes for modeling a security pattern in UML environments (stereotypes in white). The solution of the security pattern is modeled using ComponentUML (stereotypes in grey). In addition pattern integration-related concepts (stereotypes in blue). The specification of the UML profile is out of scope this paper and is detailed in [5].

In the context of this work, during the formalization process, we considered the following security mechanisms:

- **Firewall**: This mechanism is responsible for input validation.
- **Encryptor**: encrypts transmitted messages using a key.
- **Decryptor**: decrypts received messages using a key.
- **Signer**: produces for each message a signature that guarantees the authenticity and integrity of the message. It is sent together with the message.
- **Verifier**: verifies the integrity and authenticity of the message via its accompanied signature.

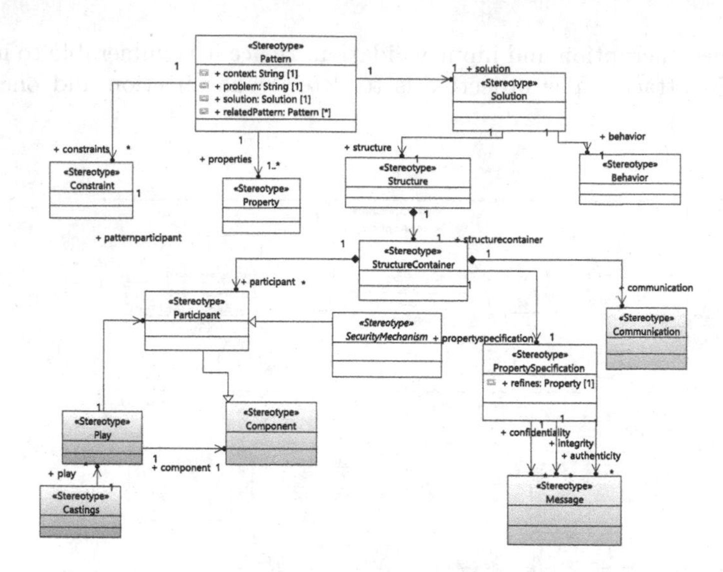

Fig. 4. SepmUML UML profile (Color figure online)

3.3 Formalizing the Threats Using OCL

The main objective of this work is to analyze software architectures allowing the detection of threats according to formalized threats. In this section, the integration-related detecting threats are described. As we shall see, this has required additional concepts in ComponentUML in Fig. 5. At each iteration, threats are formalized using OCL. The evaluation through the precision rate (TP and FP) is measured in the context of the working example from Sect. 3.1. *Iteration 1* starts with the initial ComponentUML. Man-In-The-Middle (MITM) and Injection threats are formalized. *Iteration 2* adds the concept of trust level to ComponentUML. Last but not least, *Iteration 3* adds the concept of port kind is added i.e., if the port is external (public) or internal (private).

Iteration 1. Man-In-The-Middle (MITM) and injection threats are formalized using OCL. MITM exploits the lack of encryption and integrity protection mechanisms. Injection threats exploit the lack of input validation. In Listing 1.1 and Listing 1.2 are given the OCL constraints of MITM and injection, respectively.

```
1  Context Application inv Man-In-The-Middle_v1
2  self.components->select(c1 |
3  self.components->exists(c2 |
4  not(c1.oclIsKindOf(PatternProfile::SecurityMechanism))
5  and
6  not(c2.oclIsKindOf(PatternProfile::SecurityMechanism))
7  and
8  /* if c1 and c2 are different*/
9  c1._'<>'(c2)
10 and
11 /* if c1 and c2 are deployed in different nodes*/
12 (c1.nodlater>'(c2.node) and c1.node.channels->exists(ch | c2.node.
       channels->includes(ch))
```

```
13 and
14 /* c1 and c2 communicate */
15 (c1.ports->exists(inp| c2.ports->exists(inpt2 | inpt2.communication
       = inp.communication)))
16  and
17
18 /* The security mechanisms exist: encryptor, decryptor, signer and
       verifier */
19 (self.components->select(enc |  self.components->exists(dec , mac1
       | self.components->exists(mac2|
20
21 enc.oclIsKindOf(PatternProfile::Encryptor) and  dec.oclIsKindOf(
       PatternProfile::Decryptor) and mac1.oclIsKindOf(PatternProfile
       ::Signer) and mac2.oclIsKindOf(PatternProfile::Verifier)
22 and
23 (enc.node = c1.node) and (dec.node = c2.node) and (mac1.node = c1.
       node) and (mac2.node = c2.node)
24 ))))))->size()
```

Listing 1.1. Man-In-The-Middle (MITM) threat formalized using OCL

The constraint in Listing 1.1 explores the application model via the stereotypes applied on them. For a given application, components are parsed and only those that are not security mechanisms are checked. For each two different components *c*1 and *c*2 deployed on different nodes and which can communicate. The constraint checks if the following security mechanisms (described in Sect. 3.2): "encryptor", "decryptor", "signer" and "verifier" exist and are deployed in the same nodes as *c*1 and *c*2. The second constraint is commented on Listing 1.2.

```
1 Context Application inv Injection_v1
2 self.components->select(c1 |
3 /* Firewall exists and is deployed on the same node as c1*/
4 self.components->select(firewall | firewall.oclIsKindOf(Firewall)
       and firewall.node = c1.node
5 ))->size()
```

Listing 1.2. Injection threat formalized using OCL

Results. Table 1 gives the number of threats, TPs, FPs. The actual version of the threats has a Precision of 60%. The precision rate indicates that 40% are FPs.

Table 1. Number of detected threats, TPs and FPs (iteration 1)

Threat Category	Detected	TP	FP
Man-In-The-Middle	2	1	1
Injection	3	2	1
Total	5	3	2

After investigating, three injection threats were detected for the three components: web page, web application and DBMS. The third one is an FP because an injection threat is more likely to happen when components are exposed. Hence, in iteration 2 a new concept is added in iteration 2: "port type".

Two MITM threats were detected: (1) between the browser and the web application and (2) between the web application and DBMS. The second one is an FP because a MITM is more likely to happen on "untrusted" zone. Hence, in iteration 3, the concept of "Trust Level" is added.

Iteration 2. In this iteration, the concept of trust level is added to ComponentUML in order to check if components are in a trusted network zone or not. Figure 5 shows ComponentUML model with the *TrustLevel* enumeration with two literals *trusted* and *untrusted.* The application model is modified and considers that DBMS node is in a "trusted" node while the web page and web application are in an "untrusted" node. In addition, the constraint does not only check for the existence of security mechanisms but that they are correctly used. This is done by verifying that the transmitted messages between two components are encrypted/unecrypted and signed/verified. The application model in Fig. 3 is modified. In this version the web page and application are considered in "untrusted" node while the DBMS is in an "untrusted" node. In addition, the OCL in Listing 1.1 constraint is modified. Listing 1.3 gives an extract of the OCL constraint. The lines that have already explained have been removed intentionally for matter of simplicity.

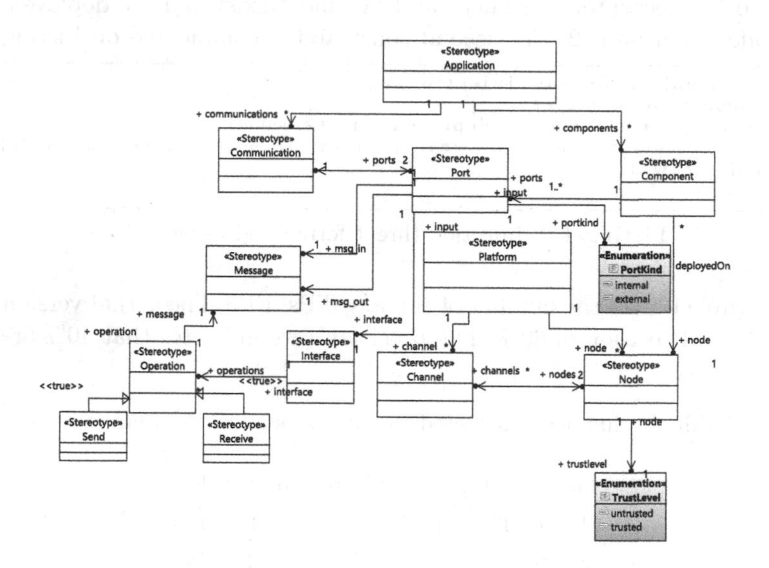

Fig. 5. Augmented ComponentUML model

```
1 Context Application inv Man-In-The-Middle_v2
2 self.components->select(c1 |
3 self.components->exists(c2 |
4 /* c1 or c2 are deployed in an untrusted node */
5 (c1.node.trustlevel= TrustLevel::untrusted or c2.node.trustlevel =
      TrustLevel::untrusted )
6 and
```

```
7 /* c1 and c2 two components communicating deployed in different
      nodes
8 [...]
9 and
10 /* The security mechanisms exist: encryptor, decryptor, signer and
      verifier and Mechanisms are connected to components */
11 [...]
12 /* The mechanisms are called correctly */
13 c1.ports->select(inp_c1_c2 | inp_c1_c2.msg_out._'<>'(null) and c2.
      ports->exists(inpt2 | inpt2.communication = inp_c1_c2.
      communication))->forAll(inp_c1_c2 | mac1.ports->exists(sign_in
      |c1.ports->exists(c1_inp| c1_inp.communication = sign_in.
      communication)
14  and
15  (enc.ports->exists(enc_in | c1.ports->exists(c1_signorEnc |
16 -- case 1: message flow encrypt and then sign
17      (c1_signorEnc.communication = enc_in.communication and
           sign_in.msg_out = inp_c1_c2.msg_out and sign_in.msg_in =
           enc_in.msg_out and enc_in.msg_in = c1_signorEnc.msg_out)
18      or
19 -- case 2: message flow sign and then encrypt
20      [...]
```

Listing 1.3. Man-In-The-Middle (MITM) threat version 2 formalized using OCL

The constraint explores the application model via the stereotypes applied on them. For each two different components c1 and c2, the constraint checks if they are deployed in an "untrusted" node (lines 5–7). It checks also if they are connected to the aforementioned security mechanisms (line 5–7). Then, it checks if they are calling the security mechanisms and that they are correctly used in message flows (lines 10–20). Two cases have been identified:

- The message sent from c1 to c2 is encrypted then signed
- The message sent from c1 to c2 is signed then encrypted

Results. Table 2 shows the results after checking the new version of the OCL constraints over the working example. The results show that three threats have been detected and are TPs and only one is an FP. Hence, the precision have increased to 75%.

Table 2. Number of detected threats, TPs and FPs (iteration 2)

Threat Category	Detected	TP	FP
Man-In-The-Middle	1	1	0
Injection	3	2	1
Total	4	3	1

Iteration 3. In this iteration, the concept of port kind is added. Figure 5 shows ComponentUML with the *PortKind* enumeration with two literals *external* (public ports) and *internal* (private ports). The application model in Fig. 3 is modified. In this version, the web page and application ports are considered "external" while the DBMS port is "internal". In addition, Listing 1.2 is modified.

Listing 1.4 gives an extract of the new version of the OCL constraint. The lines that have already been explained have been removed intentionally for a matter of simplicity. Only line 3 was kept and considers external ports.

```
1 Context Application inv Injection_v2
2 /*Publicly accessible port*/
3 self.components->select(c1 | c1.ports->exists(public_port | (
      public_port.portkind = PortKind::external)
4 and
5 /* Checks if Firewall exists and is connected to component c1 and
      message flow is correct*/
6 [...]
7 )->size()
```

Listing 1.4. Injection threat formalized using OCL

Results. As depicted in Table 3, the third version of the threats formalized with OCL has a precision of 100%. Of course, this is specific to the working example that has been presented which is a very simple example. In addition, the results are specific to the threats that have been considered.

Table 3. Number of detected threats, TPs and FPs (iteration 3)

Threat Category	Detected	TP	FP
Man-In-The-Middle	1	1	0
Injection v2	2	2	0
Total	3	3	0

4 Case Study: SCADA System

This section assesses the feasibility of the contributions of our work through the modeling and analysis of a SCADA (Supervisory Control And Data Acquisition) system. SCADA system applications are different from classical ITs (i.e., web applications) and have strong security requirements.

4.1 Description and Modeling

SCADA systems are meant to continuously control, monitor processes and acquire field information. In our experiment, we consider an adapted and simplified version of SCADA used in the context of smart grids [9]. In this context, the controlled process is power distribution. The control center consists of a control and a corporate network. The corporate network provides the operator with a Human-Machine Interface (HMI) that allows access to system data, SCADA servers, and databases that store operational and financial information. The SCADA server controls and gathers field information from geographically distributed substations or Remote Terminal Units (RTUs). The software components perform the following functions: (1) Perform control, (2) Poll Data, (3) System Start-up/shutdown, (4) Adjust Parameter Settings, (5) Log Field Data, (6) Archive Data, (7) Trigger Alarm, (8) Perform Trending: Select Parameters, Display Parameters, Zooming, Scrolling. Figure 6 depicts the software architecture model. In addition, ports, interfaces, data types, and transmitted messages are specified to provide a more detailed model of the application. The platform is also modeled to specify the relationship between components and nodes.

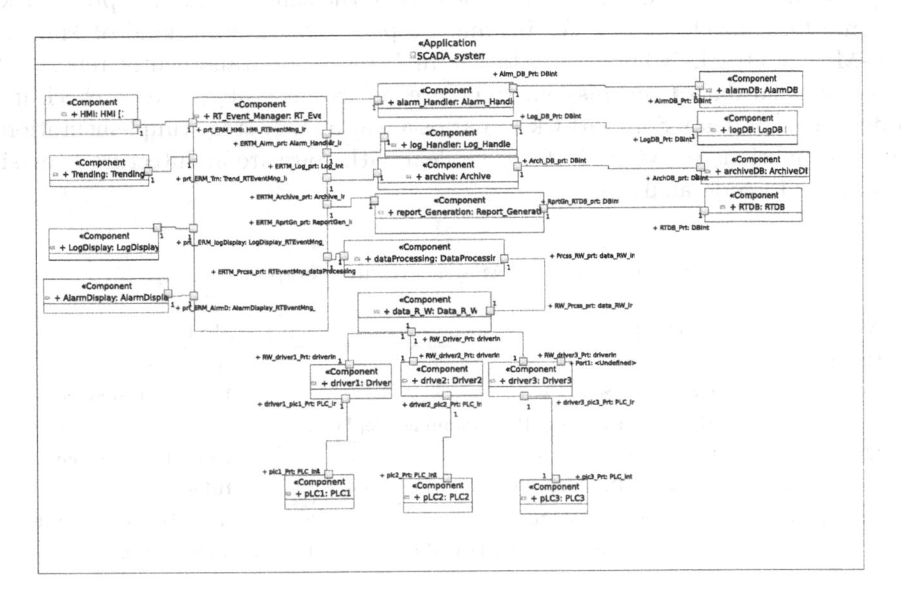

Fig. 6. SCADA software architecture model

4.2 Comparison of MBTA and ASTORIA

To assess MBTA, The obtained results are compared to the work of [9]. The latter proposes a framework named ASTORIA for attack scenario simulation for smart grid systems. The selection of the framework was motivated by the fact that ASTORIA is a simulation framework whereas ours is a formal verification-like framework. The ASTORIA team has simulated attack scenarios and evaluated their impact on the smart grid system to discover existing threats. In addition to the two threats presented previously, two more threats were formalized: Tampering and Denial of Service. Their formalization was omitted for simplification purposes. However, we give a brief explanation. *Denial of Service (DoS)* can make the system resources unavailable for authorized users. The signature of this threat is the lack or weakness of Firewall, Authentication, and Authorization mechanisms. *Tampering* is responsible for altering data at rest or in transit. The signature of this threat is the lack or weakness of Authenticity mechanisms.

Results. Table 4 presents the results obtained with the ASTORIA framework and "MBTA". For each asset, we conclude that all the detected threats are TPs. In addition, MBTA detected at the level of RTUs and communication new threats i.e., Tampering and Injection. FNs, i.e., threats that were not detected are due to different reasons. Some attack scenarios were simply not formalized or out of scope of our framework. For instance, Phishing is an attack scenario that attempts to obtain sensitive information such as credentials, and credit card details by using emails. This attack exploits social engineering which is out of scope of the study. Some attack scenarios are of the same kind or are pre-attacks of some formalized threats. For instance, replay attacks are a kind of Man-In-The-Middle attacks where the attacker maliciously or fraudulently repeats or delays a valid data transmission. Ping sweeps are generally used to check if a node is alive or dead. Some attack scenarios are at a lower stage (implementation) such as malicious software. In fact, we deal with software architecture analysis and not with code analysis.

Table 4. Threat Analysis results comparison

Assets	ASTORIA [9]	MBTA
Control Center	**Injection**	**Injection**
	Denial of Service	**Denial of Service**
	Malware, Phishing, Port scanning, Replay	
RTU	**Denial of Service**	**Denial of Service**
	Malware, Phishing, Port scanning, Replay	**Injection**
Communication	**Man-In-The-Middle**	**Man-In-The-Middle**
	Sniffing, Eavesdropping, Denial of Service, Replay	**Tampering**

Discussion. After analyzing the case study and the conducted assessment, some lacks in the current version of MBTA have been identified and are left for future work. Threat analysis can be generalized by replacing ComponentUML with OMG standards for Component-Based Development particularly UCM[7]. The second step is to construct a library of helpers to easily formalize threats. The specification of threats was done using OCL. OCL is a general language for constraining UML models. The goal is to enable security experts to contribute to the threat knowledge-base, who are not necessarily familiar with OCL and with less effort. In this context, we can inspect DSMLs for specifying these rules and then study mappings towards OCL.

5 Related Work

Security architecture assessment approaches can be categorized into two groups: scenario-based and property-based approaches. *Scenario-based Analysis.* focuses

[7] https://www.omg.org/spec/UCM.

on modeling security scenarios and then analyzing the architecture with regard to these scenarios. In literature, most of these works [1,2,7] have limitations in formalizing scenarios, in reusing and extending them, in automatizing the verification process and they also lack tool support. Recently Maidl et al. [4] have proposed a model-based threat modeling approach for Cyber-Physical Systems. It is based on a two-dimensional taxonomy that links system components and relevant attacks. The formalization language is OCL. The tool helps in prefiltering relevant attack actions and their documentation. In [7], the authors propose a framework for detecting architectural flaws in a code and introduce SCORIA as a formalization language. It starts by generating a graph describing a runtime architecture using static analysis. Then they assign security properties to the graph of objects. The constraints in this approach are highly dependent on the application and are not generic or reusable. The aim of "MBTA" is to foster reuse. In [2], the authors present a framework for detecting flaws in the code. The formalization language is OCL. The code is first transformed in STRIDE Data Flow Diagrams (DFDs) using static analysis. Then based on a 'best practice' repository where threat patterns are stored, an automatic check is performed to detect the threats and security measures that may be applied as annotations to DFDs to mitigate these threats. *Property-Based Analysis.* focuses on formalizing security properties to assess a software architecture. They defined a set of modularity properties used for analyzing the architecture. Table 5, compares "MBTA" to the aforementioned ones mainly: Almorsy et al. [1], Vanciu et al. [7] and Berger et al. [2] according to the following criteria: (C1) Foster reuse of the formalized threats, (C2) Verify that the architecture has the right security mechanisms, (C3) Verify that these security mechanisms are used correctly, and (C4) and Have a list of well-known threats.

Table 5. Positioning of the contribution with regards to other approaches

Approaches	(C1)	(C2)	(C3)	(C4)
MBTA	✓	✓	✓	✓
Maidl et al. [4]	✓	×	✓	✓
Almorsy et al. [1]	✓	✓	×	✓
Vanciu et al. [7]	×	✓	✓	✓
Berger et al. [2]	✓	✓	×	✓

6 Conclusion

In this paper, a model-based threat analysis for software architecture "MBTA" has been introduced. The contribution of this work is twofold. First, the approach enables detailed exploration of the software architecture. The formalized threats allow not only the verification of the existence of security mechanisms but also the verification of their correct usage. The second aspect is that the threats are reusable and extensible. OCL has been used to formalize Injection and Man-In-The-Middle threats. The formalization process has been explained through three iterations. For each iteration, the precision is evaluated. The formalized threats are not application dependent. They can be further extended if a threat exploits new vulnerabilities and weaknesses. The next step of this work consists of defining a correct-by-construction pattern-based security engineering process. It aims to provide the correct-by-construction integration of security patterns into an application while offering a certain degree of liberty to the designer using it. In order to be able to validate the integration, a formal specification of the pattern must be constructed, i.e., its properties, constraints, and related validation artifacts, as input to the pattern-based development process. Here, the concepts behind the formalized threats will be used and combined with patterns, to integrate security solutions in the application model and perform a security analysis within other types of threats.

References

1. Almorsy, M., Grundy, J., Ibrahim, A.S.: Automated software architecture security risk analysis using formalized signatures. In: Proceedings of the 2013 International Conference on Software Engineering, ICSE 2013, pp. 662–671. IEEE Press (2013)
2. Berger, B.J., Sohr, K., Koschke, R.: Extracting and analyzing the implemented security architecture of business applications. In: 2013 17th European Conference on Software Maintenance and Reengineering, pp. 285–294. IEEE (2013)
3. Fernandez, E.B., Yoshioka, N., Washizaki, H.: Modeling misuse patterns. In: 2009 International Conference on Availability, Reliability and Security, pp. 566–571 (2009)
4. Maidl, M., Münz, G., Seltzsam, S., Wagner, M., Wirtz, R., Heisel, M.: Model-based threat modeling for cyber-physical systems: a computer-aided approach. In: van Sinderen, M., Maciaszek, L.A., Fill, H.-G. (eds.) ICSOFT 2020. CCIS, vol. 1447, pp. 158–183. Springer, Cham (2021). https://doi.org/10.1007/978-3-030-83007-6_8
5. Motii, A., Hamid, B., Lanusse, A., Bruel, J.M.: Guiding the selection of security patterns for real-time systems. In: 21st International Conference on Engineering of Complex Computer Systems, ICECCS 2016, pp. 155–164. IEEE (2016)
6. Shostack, A.: Experiences threat modeling at Microsoft. In: Proceedings of the Workshop on Modeling Security, vol. 413, pp. 5:1–5:12. CEUR-WS.org (2008)
7. Vanciu, R., Abi-Antoun, M.: Finding architectural flaws using constraints. In: 2013 28th IEEE/ACM International Conference on Automated Software Engineering (ASE), pp. 334–344 (2013)
8. Weiss, M., Mouratidis, H.: Selecting security patterns that fulfill security requirements. In: 2008 16th IEEE International Requirements Engineering, RE 2008, September 2008, pp. 169–172 (2008)
9. Wermann, A.G., Bortolozzo, M.C., da Silva, E.G., Schaeffer-Filho, A., Gaspary, L.P., Barcellos, M.: ASTORIA: a framework for attack simulation and evaluation in smart grids. In: NOMS 2016–2016 IEEE/IFIP Network Operations and Management Symposium, April 2016, pp. 273–280 (2016)

Attribute Repair for Threat Prevention

Thorsten Tarrach[1], Masoud Ebrahimi[2], Sandra König[1], Christoph Schmittner[1], Roderick Bloem[2], and Dejan Ničković[1](✉)

[1] AIT Austrian Institute of Technology, Vienna, Austria
dejan.nickovic@ait.ac.at
[2] Graz University of Technology, Graz, Austria

Abstract. We propose a model-based procedure for preventing security threats using formal models. We encode system models and threats as satisfiability modulo theory (SMT) formulas. This model allows us to ask security questions as satisfiability queries. We formulate threat prevention as an optimization problem over the same formulas. The outcome of our threat prevention procedure is a suggestion of model attribute repair that eliminates threats. We implement our approach using the state-of-the-art Z3 SMT solver and interface it with the threat analysis tool THREATGET. We demonstrate the value of our procedure in two case studies from automotive and smart home domains.

1 Introduction

The proliferation of communication-based technologies requires engineers to have cybersecurity in mind when designing new applications. Historically, security decisions in the early stages of development have been made informally. The upcoming requirements regarding security compliance in soon-to-be-mandatory standards such as the ISO/SAE 21434 call for more principled security assessment of designs and the need for systematic reasoning about system security properties have resulted in threat modeling and analysis tools. One example of this new perspective is the Microsoft Threat Modelling Tool (MTMT) [8], developed as part of the Security Development Lifecycle. MTMT provides capabilities for visual system structure modelling. Another example is THREATGET [3,12], a threat analysis and risk management tool, originally developed in academia and following its success, commercialized and used today by leading world-wide companies in automotive and Internet-of-Things (IoT) domains. Threat modeling and analysis significantly reduces the difficulty of a security assessment, reducing it to accurate modeling of the systems and the security requirements.

Existing methods use ad-hoc methods to reason about the security of systems. As a result, it is not easy to extend such tools with model repair capabilities.

This project has received funding from the European Union's Horizon 2020 research and innovation programme under grant agreements No. 956123 (FOCETA), No. 871385 (TEACHING) and from the program "ICT of the Future" of the Austrian Research Promotion Agency (FFG) and the Austrian Ministry for Transport, Innovation and Technology under grant agreements No. 867558 (project TRUSTED).

J. Guiochet et al. (Eds.): SAFECOMP 2023, LNCS 14181, pp. 135–148, 2023.
https://doi.org/10.1007/978-3-031-40923-3_11

Although a trial-and-error method is always possible, it does not provide a systematic exploration of the space of possible prevention measures and leaves the question of optimizing the cost of the prevention to the designer's intuition. As a result, remedying a potential threat remains cumbersome and simple solutions may be missed, especially in presence of multiple interacting threats.

This paper proposes a procedure for preventing threats based on a formal model of the structure of the system and a logic-based language for specifying threats. The use of rigorous, formal languages to model the system and specify threats allows us to automate threat prevention. More specifically, we reduce the problem of checking presence of threats in the system model to a satisfiability modulo theory (SMT) check. A threat specification defines a class of potential threats and a witness of a system model that satisfies a threat specification defines a concrete threat in the model. This allows us to frame the problem of preventing concrete threats as an attribute parameter repair.

The attributes of system components define a large spectrum of security settings and, in presence of a threat, of possible preventive actions. This class of repairs enables simple and localized measures whose cost can easily be assessed by a designer. We formulate attribute repair as a weighted maximum satisfiability (MaxSAT) problem with a model of cost of individual changes to the system attributes. This formulation of the problem allows us to find changes in the model with *minimal* cost that result in removing as many threats as possible.

We introduce *threat logic* as a specification language to specify threats. We formalize the system model as a logic formula that consists of a conjunction of sub-formulas, called *assertions*, parameterized by attributes that specify security choices. The conjunction of the system model formula and a negated threat formula is satisfiable iff there are no threats in the system. We introduce clauses that change the specific instantiation of model attributes to a different value and associate a *weight* with each assertion. Then, the MaxSAT solution of this formula is the set of changes to system model attributes with minimum cost that ensure the absence of the threat. Given an incorrect system, we can choose the weights so that we compute the set of changes to system model attributes with the minimal cost to remove the existing threats from the model. To ensure that our method scales to industrial size models, we also define a heuristic that provides partial threat prevention by addressing repairable threats and explaining the reason why the others cannot be repaired. We believe that this method, even though partial and approximate in general, can compute near optimal repairs for many real-world problems.

We implemented the threat prevention method in the THREATGET tool and evaluated it on two case studies from the automotive and the IoT domains.

Motivating Example. We motivate this work with a smart home application from the IoT domain, depicted in Fig. 1. The smart home architecture consists of 7 *typed elements*: (1) a control system, (2) an IoT field gateway, (3) temperature and (4) motion sensors, (5) a firewall, (6) a web server and (7) a mobile phone. The elements are interconnected using *wired* and *wireless connectors.*

The elements and connectors have associated sets of *attributes* that describe their configuration. For instance, every connector has attributes Encryption, Authentication and Authorization. The attribute Encryption can be assigned the values No, Yes and Strong. We associate to each attribute a *cost* of changing the attribute value, reflecting our assessment of how difficult it is to implement the change. In this example, the temperature and the motion sensor communicate wirelessly with the gateway. If the motion sensor detects a movement, the user is notified by phone. It is possible to override the behavior, e.g., the heating can be turned on remotely in case of late arrival. The web server allows for access and information exchange from and to the smart home. The IoT sub-system protected by the firewall defines a *security boundary* called the IoT Device Zone. Communication should be confidential and encrypted outside the IoT Device, which is represented by the two associated *assets*.

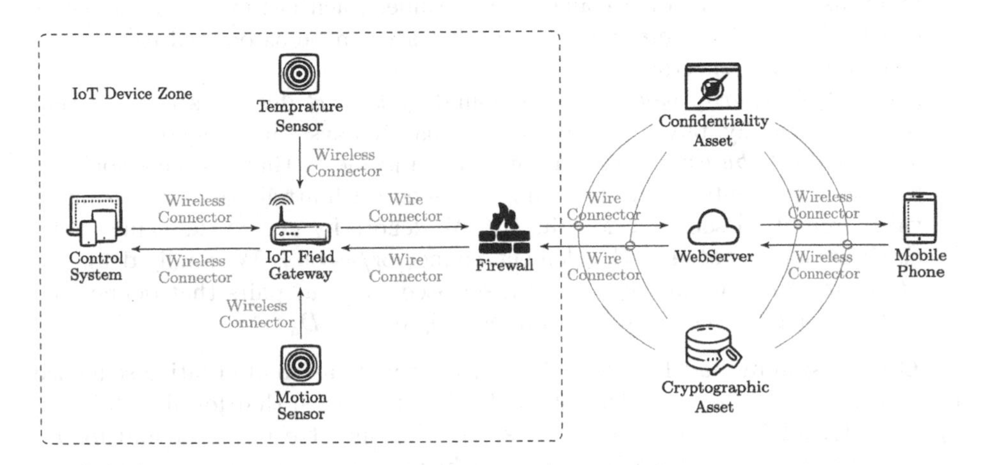

Fig. 1. Smart Home IoT model.

Threats in this smart home system are characterized by logical relations between elements, connectors and their attributes. Consider two potential threats that are applicable to this example: **Threat 1:** The web server enables data logging functionality without encrypting the data, and **Threat 2:** The mobile phone device is connected to the web server, without the web server enabling data logging. Assume that the web server has data logging enabled, but no data encryption, thus matching Threat 1. If we consider this threat in isolation, we can either repair it by turning off the data logging, or by implementing the data encryption on the web server. The first repair results in matching Threat 2. Only the second repair results in the removal of all security threats. Given two data encryption algorithms with costs c_1 and c_2, where $c_1 > c_2$, implementing the latter is the cost-optimal option. We see that an optimal preventive solution must consider simultaneous repair of multiple threats.

2 Threat Modelling

A threat model consists of two main components, a *system model* and a database of *threat rules*. A system model provides an architectural view of the system under investigation, representing relevant components, their properties, as well as relations and dependencies between them.

System Model. A system model M consists of:

- a set E of *elements*: an element $e \in E$ is a typed logical (software, database, etc.) or physical (ECUs, sensors, actuators, etc.) component.
- a set C of *connectors*: a connector $c \in C$ is a direct interaction between two elements, a *source* $\mathbf{s}(c) \in E$ and a *target* element $\mathbf{t}(c) \in E$.
- a set A of *security assets*: an asset $a \in A$ describes logical or physical object (such as an element or a connector) of value. Each element and connector can hold multiple assets. Similarly, each asset can be associated to multiple elements and connectors.
- a set B of *security boundaries*: a boundary $b \in B$ describes a separation between logically, physically, or legally separated system elements.
- a set $\mathfrak{A}$ of *attributes*: an attribute $\mathfrak{a} \in \mathfrak{A}$ is a property that can be associated to system elements, connectors and/or assets. Each attribute $\mathfrak{a}$ can assume a value from its associated domain $D_{\mathfrak{a}}$. We denote by $v(x, \mathfrak{a})$ the value of the attribute $\mathfrak{a}$ associated to the element/connector/asset x. We finally define an *attribute cost* mapping $w_{x,\mathfrak{a}}(v, v')$ associated to $(x, \mathfrak{a})$ pairs that defines the cost of changing the attribute value $v \in D_{\mathfrak{a}}$ to $v' \in D_{\mathfrak{a}}$.

Given a system model M, we define a *path* π in M as an alternating sequence $e_1, c_1, e_2, c_2 \cdots, c_{n-1}, e_n$ of elements and connectors, such that for all $1 \leq i \leq n$, $e_i \in E$, for all $1 \leq i < n$, $c_i \in C$, $\mathbf{s}(c_i) = e_i$, and $\mathbf{t}(c_i) = e_{i+1}$ and for all $1 \leq i < j \leq n$, $e_i \neq e_j$. We note that we define paths to be *acyclic*, since acyclic paths are sufficient to express all interesting security threats.

We use the notation elements(π) and connectors(π) to define the sets of all elements and of all connectors appearing in a path, respectively. The starting and the ending element in the path π are denoted by $e_{start}(\pi) = \mathbf{s}(c_1)$ and $e_{end}(\pi) = \mathbf{t}(c_{n-1})$, respectively. We denote by $P(M)$ the set of all paths in M.

Threat Logic. We provide an intuitive introduction of *threat logic* for specifying potential threats[1]. The syntax of threat logic is defined as follows:

$$\varphi := R(X \cup P) \mid \neg\varphi \mid \varphi_1 \vee \varphi_2 \mid \exists p.\varphi \mid \exists x.\varphi,$$

where $X = E \cup C \cup A \cup B$, $x \in X$, P is a set of path variables, $p \in P$, and $R(X \cup P)$ is a predicate. The predicate $R(X \cup P)$ is of the form:

[1] THREATGET uses its own syntax and semantics to express threats [3]. We use instead predicate logic to facilitate the encoding of the forthcoming algorithms into SMT formulas. Our implementation contains an automated translation from THREATGET syntax to threat logic.

1. $\mathtt{type}(x) = t$ - the type of $x \in X$ is t;
2. x $\mathtt{in}$ p - the element or the connector $x \in E \cup C$ is in the path $p \in P$;
3. $\mathtt{connector}(e, c)$ - the element $e \in E$ is either the source or the target of the connector $c \in C$;
4. $\mathtt{src}(c) = e$ - the source of the connector $c \in C$ is the element $e \in E$;
5. $\mathtt{tgt}(c) = e$ - the target of the connector $c \in C$ is the element $e \in E$;
6. $\mathtt{src}(p) = e$ - the source of the path $p \in P$ is the element $e \in E$;
7. $\mathtt{tgt}(p) = e$ - the target of the path $p \in P$ is the element $e \in E$;
8. $\mathtt{crosses}(c, b)$ - the connector $c \in C$ crosses the boundary $b \in B$;
9. $\mathtt{contained}(x, b)$ - the element or boundary $x \in E \cup B$ is contained in the boundary $b \in B$;
10. $\mathtt{holds}(x, a)$ - the element or the connector $x \in E \cup C$ holds the asset $a \in A$;
11. $\mathtt{val}(x, att) = v$ - the valuation of the attribute att associated to $x \in E \cup C \cup A$ is equal to v.

Example 1. Consider a requirement that there exists a path in the model such that all the elements in that path are of type Cloud. It is expressed with the threat logic formula: $\exists p.\forall e.(e \ \mathtt{in} \ p \implies \mathtt{type}(e) = \text{Cloud})$.

We define an *assignment* Π_M as a partial function that assigns element, connector, asset, security boundary and path variables to concrete elements, connectors, assets, security boundaries and paths from the system architecture model M. We denote by $\Pi_M[x \mapsto i]$ the item assignment in which x is mapped to i and otherwise identical to Π_M. Similarly, we denote by $\Pi_M[p \mapsto \pi]$ the path assignment in which p is mapped to π and otherwise identical to Π_M. The semantics of threat logic follow the usual definitions of predicate logic.

We say that a threat logic formula is *closed* when all occurrences of element, connector, asset and security boundary variables are in the scope of a quantifier. Any closed threat logic formula is a valid threat specification. Given a system model M and a closed threat logic formula φ, we say that M *witnesses* the threat φ, denoted by $M \models \varphi$ iff $\Pi_M \models \varphi$, where Π_M is an empty assignment.

From Threat Logic to First Order Logic (FOL). We interpret threat logic formulas over system models with a finite number of elements and connectors, and hence we can eliminate path quantifiers by enumerating the elements and connectors in the path. We thus obtain an equisatisfiable FOL formula that can be directly used by an SMT solver.

Example 2. We formalize the two threats described in Sect. 1:

Threat 1 $\exists e.(\mathtt{type}(e) = \text{WebServer} \wedge \mathtt{val}(e, \text{Data Logging}) = \text{Yes})$
$\wedge \mathtt{val}(e, \text{Data Encryption}) = \text{No})$

Threat 2 $\exists p, e_1, e_2.\mathtt{src}(p) = e_1 \wedge \mathtt{tgt}(p) = e_2 \wedge$
$\mathtt{type}(e_2) = \text{WebServer} \wedge \mathtt{type}(e_1) = \text{MobilePhone} \wedge$
$\mathtt{val}(e_2, \text{Data Logging}) \neq \text{Yes})$

3 Automated Threat Prevention

We now present our main contribution – a procedure to automatically repair a system model with one or more threats. We restrict our attention to the class of *attribute repairs* that consists in changing the model attribute values and show how to encode the problem using optimization modulo theories. We first present an exact algorithm for the minimal attribute repair using an SMT solver, and then propose a more scalable heuristic for partial repair.

3.1 Attribute Repair

In this work, repairing a model that has one or multiple threats consists in changing the attribute valuation function of the model. Not every model can be attribute repaired. For a threat model M we denote by $M[v' \backslash v]$ the threat model in which the attribute value assignment v is replaced by another assignment v'.

Definition 1 (Threat-repairable model). *Given a model M with an attribute valuation function v that witnesses a threat φ, $M \models \varphi$, we say that M is* attribute repairable *wrt φ iff there exists a v' such that $M[v' \backslash v] \not\models \varphi$.*

We specifically aim at finding the *optimal* repair, which has the minimal repair cost. To reason about this quantitative repair objective, Definition 2 specifies the *distance* $d(v, v')$ between two attribute valuation functions v and v' as the sum of altered attribute costs for attributes that differ in the two valuations.

Definition 2 (Attribute valuation distance). *Let M be a system model with attribute valuation v and attribute cost w. Let v' be another attribute valuation. The* distance $d(v, v')$ *between v and v' is defined as:*

$$d(v, v') = \sum_{x \in X, \mathfrak{a} \in \mathfrak{A}} w_{x,\mathfrak{a}}(v(x, \mathfrak{a}), v'(x, \mathfrak{a})) \text{ s.t. } v(x, \mathfrak{a}) \neq v'(x, \mathfrak{a}).$$

For instance, for the attribute 'Encryption' the cost of changing from 'None' to 'Weak' may be 20, but to change 'None' to 'Strong' may cost 30. A change from 'Weak' to 'Strong' could cost 15, but a change from 'Weak' to 'None' may only cost 1. Sensible cost functions will adhere to some restrictions (such as a variant of the triangle inequality) that we do not formalize here.

Definition 3 (Minimal attribute repair). *Let φ be a threat logic formula and M a system model such that $M \models \varphi$ and M is attribute repairable w.r.t. φ. The* minimal attribute repair *of M is another threat model $M[v' \backslash v]$ s.t.:*

$$v' \in \arg\min\{d(v, v'') | M[v'' \backslash v] \not\models \varphi\}$$

Other Notions of Minimal Repair. There are at least two other natural notions of minimal repair. In the first one, costs are associated with the attribute itself. This means that every change of the attribute carries the same cost. We can model this by assigning the same cost to all possible combinations of previous and new value for an attribute. Alternatively, engineers are often not interested in minimizing the overall real cost, but rather in minimizing the number of attributes that need to be repaired. We can model this restricted variant of the problem by associating the fixed cost of 1 each attribute in the model, thus effectively counting the number of individual attribute repairs. Both variants can be implemented in a straightforward manner in our framework.

3.2 Attribute Repair as Weighted MaxSMT

We encode the attribute repair problem (see Sect. 3.1) as a *weighted MaxSMT* problem, in which $\mathbf{F}$ represents a (hard) *assertion*, while $F_1, \ldots, F_m$ correspond to *soft assertions* and every soft assertion F_i has an associated cost $cost_i$.

Definition 4 (Weighted MaxSMT [1]**).** *Given an SMT formula* $\mathbf{F}$*, a set of SMT formulas* $F_1, \ldots, F_m$ *and a set of real-valued costs* $cost_1, \ldots, cost_m$*, the* weighted MaxSMT *problem consists in finding a subset* $K \subseteq M$ *of indices with* $M = \{1, \ldots, m\}$ *such that: (1)* $\mathbf{F} \wedge \bigwedge_{k \in K} F_k$ *is satisfiable, and (2) the total cost* $\sum_{i \in N \setminus K} cost_i$ *is minimized.*

We now sketch the encoding of the minimal attribute repair problem into weighted MaxSMT. We assume that we have a MaxSMT solver object, with the following functionality: *push*() - push new context to solver stack, *pop*() - pop context from solver stack, *add*(φ) - add new hard assertion φ, $add\_soft(\varphi, c)$ - add new soft assertion φ with weight c, *solve*() - check if formula is satisfiable, $max\_solver()$ - check if formula is max-satisfiable, and *model*() - generate and return a model witnessing satisfaction of a formula.

Given a system model M and a set of threat logic formulas $\Phi = \{\varphi_1, \ldots, \varphi_n\}$, we compute the MaxSMT formulas $\mathbf{F}$ that represents the hard assertion $\mathbf{F} = F_M \wedge \bigwedge_{j=1}^{n} \neg\varphi_j$ conjoins F_M that encodes the entire system model *except* its attribute valuations and costs with the negation of each threat logic formula φ_j. We also define one soft assertion $F_{x,\mathfrak{a},v}$ for each element x, attribute $\mathfrak{a}$ of x and possible value v of $\mathfrak{a}$, stating intuitively that $v(x, \mathfrak{a}) = v$. These soft attributes are mutually exclusive if they assert different values for the same attribute. We set up the costs of each $F_{x,\mathfrak{a},v}$ in such a way that asserting $F_{x,\mathfrak{a},v}$ leads to cost corresponding to changing the value of $\mathfrak{a}$ to v. (The exact value of the cost function can easily be computed by solving a linear system of equations.)

We use the weighted MaxSAT solver max solve($F_M \wedge \bigwedge_{j=1}^{n} \neg\varphi_j \wedge \bigwedge F_{x,\mathfrak{a},v}$) to obtain the satisfiability verdict and the optimization cost. Informally, the solver can return three possible verdicts:[2]

- sat verdict with total cost 0: the system model M does not contain a potential threat defined by any of the threat formulas φ_i,
- sat verdict with total cost k: the system model M contains a set of potential threats defined by a subset of threat formulas and can be repaired by changing

[2] We ignore here a fourth possible verdict unknown that can arise in practice and that happens if the solver is not able to reach a conclusion before it times out.

the values of model attributes with total cost k. The solver returns a model, which defines a possible repair, i.e., the altered attribute values that render the formula satisfiable,
- unsat verdict: the system model M is not attribute repairable with respect to at least one threat formula φ_i.

The encoding of the attribute repair into this MaxSMT problem provides an effective solution to the minimum attribute repair problem.

Theorem 1. *Let M be a system model and $\{\varphi_1, \dots, \varphi_n\}$ a set of closed threat logic formula. We have that* $\max\text{solve}(F_M \wedge \bigwedge_{i=1}^{n} \neg\varphi_i \wedge \bigwedge_{F \in \Psi} F)$ *provides the solution to the minimum attribute repair problem.*

3.3 Partial Repair of Unrepairable Models

The problem with the approach from Sect. 3.2 arises if there is a formula φ_i for which M is not attribute repairable. In that case, the entire problem is unsatisfiable, even if other threats could be repaired. This outcome, although correct, is not of particular value to the security engineer. Ideally, the objective is to repair attributes for threats that can be repaired and explain the others.

We observe that attribute-unrepairable threats have a particular form and correspond to formulas without constraints on attribute valuations. An inductive visit of the formula allows a syntactic check has_attr(φ) whether a threat formula φ has any constraint on attribute valuations. The MaxSMT algorithm from Sect. 3.2 can be adapted to compute a partial repair of M with respect to a subset of repairable threat formulas. The procedure removes all threat logic formulas that are satisfied by the model and that are known to be unrepairable, before computing MaxSMT.

The partial repair procedure implies that the MaxSAT applied to the subset of (potentially) repairable threat formulas corresponds to the minimum attribute repair restricted to that subset of threat formulas.

Corollary 1. *Let M be a system model, $\Phi = \{\varphi_1, \dots, \varphi_n\}$ a set of closed threat logic formulas and $G \subseteq \Phi$ a subset of repairable threats, i.e. for all $\varphi \in G$, $M \not\models \varphi$ or* has_attr(φ) *is true. We have that* $\max\text{solve}((F_M \wedge \bigwedge_{\varphi \in G} \neg\varphi) \wedge \bigwedge_{F \in \Psi} F)$ *provides the solution to the minimum attribute repair problem restricted to the set G of threat formulas.*

Explaining Unrepairable Threats: The partial repair method is useful in the presence of threats that cannot be addressed by attribute repair only. The SMT solver can be used to provide in addition an explanation of why a threat cannot be repaired – the solver assigns values to variables in threat logic formulas that witness its satisfaction (i.e. the presence of that threat). This witness explains exactly why that formula is satisfied and locates in the system model the one set of items that are responsible for that verdict. The threat may match at multiple locations of the model, which could be discovered by multiple invocations of the solver (while excluding the previously found sets). Note that the MaxSMT will repair all occurrences though, because the threat is negated there and the negated existential quantifier becomes a forall quantifier.

Algorithm 1 Approximate partial repair: HPartialRepair

```
   Input  : M, {φ_1, ..., φ_n}, F_M, Ψ
   Output: Repair status and cost, set of repaired and non-repaired threats,
           repaired set of attributes
1  nothreat ← ∅; repairable ← ∅; totalcost ← 0 ;
2  solver ← SMTMaxSolver() ;
3  solver.add(F_M) ;
4  for φ ∈ {φ_1, ..., φ_n} do
5      solver.push() ;
6      for F ∈ Ψ do
7          solver.add(F) ;
8      solver.add(T(φ)) ;
9      status ← solver.solve() ;
10     solver.pop() ;
11     if status = unsat then
12         nothreat ← nothreat ∪ {φ} ;
13     else if status = sat then
14         if has attr(φ) then
15             solver.push() ;
16             for F ∈ Ψ do
17                 solver.add soft(F, cost(F)) ;
18             solver.add(¬T(φ)) ;
19             status ← solver.max solve() ;
20             if status = sat then
21                 m = solver.model() ;
22                 Ψ̂, c ← repair(m, Ψ) ;
23                 solver.pop() ;
24                 solver.push() ;
25                 solver.add(⋁_{φ' ∈ repairable ∪ nothreat} T(φ')) ;
26                 for F ∈ Ψ̂ do
27                     solver.add(F) ;
28                 status ← solver.solve() ;
29                 if status = unsat then
30                     repairable ← repairable ∪ {φ} ;
31                     Ψ ← Ψ̂ ;
32                     totalcost ← totalcost + c ;
33             solver.pop() ;
34 return status, totalcost, repairable, nothreat, Ψ
```

We propose a heuristic procedure for partial repair in Algorithm 1. The proposed procedure works for the important subset of system models in which all the (item, attribute) pairs are assigned the same attribute cost. For every threat logic formula φ, the procedure first checks if that threat is present in the system model using the SMT solver (lines 5–10). If the threat is absent, it is added to the set of formulas that do not represent any threat (lines 11–12). Otherwise, the procedure attempts to compute a repair for that particular threat (lines 13–33). It first checks whether the threat logic formula refers to any attribute valuations (line 14). If not, the formula is unrepairable. Otherwise (lines 15–19), the formula is added as a hard assertion, and the set of attribute valuations are added as soft assertions are added as soft assertions to the solver, and the MaxSMT solver is invoked. If the solver gives unsat verdict, it means that the model cannot be repaired to satisfy the threat logic formula. Otherwise (lines 20–32), we use the model witnessing the satisfaction of the threat logic formula to compute the partial repair for that formula (line 22). The outcome of the repair method is a the repaired set of attribute assertions and the number of assertions that needed to be altered. The procedure checks that this partial repair is consistent with the previous repairs (i.e. that it does not lead to violation of other previously processed threat logic formulas), and the repair is accepted only upon passing this last consistency check (lines 24–32).

4 Implementation and Case Studies

We did a prototype implementation of the proposed methods in Java and integrated it to THREATGET. We used THREATGET's threat database from automotive and IoT domains originating from multiple sources, including security-related standards and previously discovered threats. The tool imports system models as JSON files and threat descriptions as THREATGET rules, translates both to FOL SMT formulas and uses Z3 as the MaxSMT solver. The MaxSMT solver's results are used to compute the repair suggestions. While THREATGET is a proprietary software (with a free academic license), our threat repair extension is distributed under the BSD-3 license. We assume the default attribute change cost 1 and allow the user to change it using a CSV file. We apply our tool to two case studies from two domains: the smart home IoT application introduced as our motivating example in Sect. 1, and the vehicular telematic gateway.

Smart Home IoT Application. This case study was introduced in Sect. 1. In this section, we report on the experimental results obtained by applying our threat repair approach. The model has been analysed against 169 IoT-related threat descriptions given in the form of threat log formulas. We applied both the full MaxSMT optimization procedure and its heuristic variant.

Both the model and the database of IoT threat formulas are publicly available. Table 1 summarizes the outcomes. To accurately report the number of repairable formulas we implement Algorithm 1 without line 31. The cost reflects the number of attributes (per item) changed in the model because we set the cost per attribute to 1 in our experiments. We can observe that the heuristic procedure was able to repair 27 out of 36 found threats in less than 50 s.

We illustrate the repair process on two threats from the database of IoT security threats. We first consider the threat with the title "Attacker can deny the malicious act and remove the attack foot prints leading to repudiation issues".

Table 1. Results of attribute repair applied to Smart Home IoT case study.

verdict	SAT	# formulas w/t threat	133
total # formulas	169	total cost	77
# repairable formulas	27	time (s)	46.7
# unrepairable formulas	9		

This threat is formalized using the threat logic formula

$$\exists e.\ \texttt{type}(e) = \text{Firewall} \wedge (v(e, \text{Activity Logging}) \neq \text{Yes} \vee v(e, \text{Activity Logging}) = \text{Missing}).$$

The SMT solver finds that the model satisfies the above formula and hence has a threat. The witness shows that one element of type 'Firewall' has the 'Activity Logging' attribute set to 'undefined', thus explaining the threat. The proposed repair consists in implementing the activity logging functionality. This repair was found to be consistent with the other ones and is reported as part of the overall repair suggestion. The second threat has the title "Spoofing IP" and is reported as an irreparable threat. It is formalized using the threat logic formula

$$\exists c.\exists e_1.\exists e_2\ \texttt{type}(c) = \text{Internet Connection} \wedge \texttt{src}(c) = e_1 \wedge \texttt{tgt}(c) = e_2.$$

This threat cannot be repaired by changing the model attributes. On the contrary, this threat formula states that any connection to the internet constitutes a potential IP spoofing threat.

Vehicular Telematic Gateway. This study is based on an industrial-strength model of a vehicular telematic gateway (VTG) [9]. The VTG connect internal elements of the vehicle with external services. it offers vehicle configuration and entertainment and navigation to the user. An item is a term introduced by ISO 26262, describing a system or combination of systems, enabling a function on the vehicle level. The industrial company that devised the THREATGET model develops such systems for usage by different vehicle manufacturer so that the item is developed based on assumptions about the vehicle that need validation.

We consider a telematic ECU that offers remote connectivity for the on-board network to support various remote services including data acquisition, remote control, maintenance, and over-the-air (OTA) software update. It also provides a human-machine interface (HMI) for navigation, configuration, and multimedia control. HMI represents the central element in a vehicle, connecting the control system to the human operator and the backend. It has cellular and wireless local area network (WLAN) interfaces for wireless connectivity. For local connectivity, it includes a USB port for software updates and application provision. The telematics system is also connected to other onboard ECUs.

For our analysis we considered two variants of time-triggered control. A simplified model contains 15 elements, 24 transitions and 5 assets, while the full model has 25 elements, 48 transitions and 5 assets. The presence of two models reflects the iterative design process in which the high-level simplified model was refined into the complete model based on the previous analysis.

The results of the attribute repair are presented in Table 2. We see that the tool is able to scale to large industrial models. It analyses 95 threat formulas in 497s, repairing 19 out of 42 threats with the cost 57. The same set of rules were repaired in 127s on the simplified model. The main complexity in repairing large models comes from the threat formulas that contain quantification over path. This is not surprising because each such formula corresponds to a bounded model checking (reachability) problem. To confirm this observation, we analysed the full model without formulas with quantification over paths. The analysis of 82 such formulas was done in less than 118s, resulting in the repair of 18 out of 39 threats.

Table 2. Results of attribute repair for two vehicular telematic gateways.

	with flow		w/o flow
	simple	full	full
verdict	SAT	SAT	SAT
total # formulas	95	95	82
# repairable formulas	15	19	18
# unrepairable formulas	25	23	21
# formulas w/t threat	55	53	43
total cost	29	57	57
time (s)	127	497	118

One repaired threat was *Spoofing Sensors by External Effects.* It was present because a CAN interface is connected to an ECU (the secondary CPU) and they both hold an asset (the Communication Interface). It represents the possibility that the assets could be attacked to send incorrect data to vehicle sensors (i.e., radar signals). That could lead to giving incorrect decisions based on the tampered input signal, affect the safe operation of the vehicle, or impact on usual vehicle functionalities. Our tool suggests to this threat by implementing an input validation on the Secondary CPU, which enables the CPU to detect false data.

5 Related Work

Threat Modeling and Analysis has received increasing interest in the recent years, both in academia and industry. A plethora of commercial and open-source threat modeling tools have been developed, including THREATGET [3,12], Microsoft Threat Modeling Tool [8], ThreatModeler [15], OWASP Threat Dragon [7] and pytm [17], Foreseeti [6], Security Compass SD Elements [14] and Tutamen Threat Model Automator [16]. These tools can be divided into three categories: (1) manual approaches based on excel sheets or questionnaires [16], (2) graphical modelling approaches *without* an underlying formal model [7,8,14,15], and (3) model-based system engineering tools *with* an underlying formal model [3,6,12,17]. The first class of tools does not admit automated threat analysis and any threat prevention measure must be manually identified and selected. Several tools from the second and third class [3,6,8,12] provide some limited form of hard-coded measures that are associated to individual threats or assets, without considering threat inter-dependencies nor mitigation costs and are not able to compute a global and consistent set of threat prevention measures. Hence our threat prevention approach could be integrated to theses classes of tools.

Optimization Modulo Theories (OMT) combine SMT solvers with optimization procedures [2,4,5,10,13]. to find solutions optimizing some objective function Parameter synthesis using SMT solvers does not require optimization objectives in general. Bloem et al. [11] synthesized parameter values ensuring safe behavior of cyber-physical systems through solving an $\exists\forall$ SMT formula. In this work, users specify safe states in terms of state and parameter values; then, the synthesizer attempts to compute correct parameter values conforming to an invariant template such that for all possible inputs all reachable states are safe.

6 Discussion and Future Work

We presented a framework that enables automated threat prevention by repairing security-related system attributes. Although widely applicable, attribute-value repair is not enough to cover all interesting preventive procedures. For example, protecting safety-critical components connected to a Controller Area Network (CAN) bus in a vehicle cannot be done just by encrypting messages. In fact, encryption is not part of the CAN protocol. The preventive measure would require separating trusted (safety-critical) part of the system from the untrusted one (entertainment system, etc.) with a firewall, a measure that is beyond the attribute-value repair. Despite a few similar examples, attribute repair remains a suitable repair strategy for the majority of threats present in common architectures. Intuitively, this is the case because the attributes document the counter-measures taken against common classes of threats, e.g., authentication as a counter-measure against escalation of privilege. The more general *model repair* problem can addresses the limitations illustrated by the CAN example by enabling addition and removal of elements in the model. Unrestricted alteration of models would lead to trivial and uninteresting repairs, e.g. it suffices to disconnect all elements in the model from each other to disable the vast majority of threats. Hence, model repair requires a restriction of repair operations and even just identifying a set of useful repair operations is a challenging task. The more general model repair is a separate research problem that differs in several key aspects from the attribute-value repair and we plan to tackle it as future work.

References

1. Bjørner, N., Phan, A.: νz - maximal satisfaction with Z3. In: 6th International Symposium on Symbolic Computation in Software Science, SCSS 2014, Gammarth, La Marsa, Tunisia, December 2014, pp. 1–9. EasyChair (2014)
2. Bjørner, N., Phan, A., Fleckenstein, L.: νZ - an optimizing SMT solver. In: Baier, C., Tinelli, C. (eds.) TACAS 2015. Lecture Notes in Computer Science, vol. 9035, pp. 194–199. Springer, London, UK (2015)
3. Christl, K., Tarrach, T.: The analysis approach of ThreatGet. CoRR, abs/2107.09986 (2021)
4. Cimatti, A., Griggio, A., Schaafsma, B.J., Sebastiani, R.: A modular approach to MaxSAT modulo theories. In: Järvisalo, M., Van Gelder, A. (eds.) SAT 2013. LNCS, vol. 7962, pp. 150–165. Springer, Heidelberg (2013). https://doi.org/10.1007/978-3-642-39071-5_12
5. Dillig, I., Dillig, T., McMillan, K.L., Aiken, A.: Minimum satisfying assignments for SMT. In: Madhusudan, P., Seshia, S.A. (eds.) CAV 2012. LNCS, vol. 7358, pp. 394–409. Springer, Heidelberg (2012). https://doi.org/10.1007/978-3-642-31424-7_30
6. Foreseeti AB. Foreseeti (2020). Online. Accessed 29 Nov 2020
7. Goodwin, M., Gadsden, J.: OWASP threat dragon (2020). Online. Accessed 29 Nov 2020
8. McRee, R.: Microsoft threat modeling tool 2014: identify & mitigate. ISSA J. **39**, 42 (2014)
9. Mürling, M.W.: Security by design: new "THREATGET" tool tests cyber security in vehicles and systems (2021). Online Article
10. Nieuwenhuis, R., Oliveras, A.: On SAT modulo theories and optimization problems. In: Biere, A., Gomes, C.P. (eds.) SAT 2006. LNCS, vol. 4121, pp. 156–169. Springer, Heidelberg (2006). https://doi.org/10.1007/11814948_18
11. Riener, H., Könighofer, R., Fey, G., Bloem, R.: SMT-based CPS parameter synthesis. In: Frehse, G., Althoff, M. (eds.) 3rd International Workshop on Applied Verification for Continuous and Hybrid Systems, ARCH@CPSWeek 2016, Volume 43 of EPiC Series in Computing, Vienna, Austria, pp. 126–133. EasyChair (2016)
12. El Sadany, M., Schmittner, C., Kastner, W.: Assuring compliance with protection profiles with ThreatGet. In: Romanovsky, A., Troubitsyna, E., Gashi, I., Schoitsch, E., Bitsch, F. (eds.) SAFECOMP 2019. LNCS, vol. 11699, pp. 62–73. Springer, Cham (2019). https://doi.org/10.1007/978-3-030-26250-1_5
13. Sebastiani, R., Trentin, P.: OptiMathSAT: a tool for optimization modulo theories. J. Autom. Reason. **64**(3), 423–460 (2018). https://doi.org/10.1007/s10817-018-09508-6
14. Security Compass Ltd.: Security compass SD elements (2020). Accessed 29 Nov 2020
15. ThreatModeler Software, Inc.: ThreatModeler (2020). Online. Accessed 29 Nov 2020
16. Tutamantic Ltd.: Tutamen threat model automator (2020). Online. Accessed 29 Nov 2020
17. Was, J., Avhad, P., Coles, M., Ozmore, N., Shambhuni, R., Tarandach, I.: OWASP pytm (2020). Online. Accessed 29 Nov 2020

Safety of Autonomous Driving

Probabilistic Spatial Relations for Monitoring Behavior of Road Users

Lennart Siefke[1,2(✉)], Volker Sommer[1], Murat Can Baylan[1], and Lars Grunske[2]

[1] Berliner Hochschule für Technik, Luxemburger Straße 10, 13353 Berlin, Germany
lennart.siefke@informatik.hu-berlin.de, volker.sommer@bht-berlin.de
[2] Humboldt-Universität zu Berlin, Unter den Linden 6, 10099 Berlin, Germany
grunske@informatik.hu-berlin.de

Abstract. Safe autonomous driving requires monitoring the movement of other road users to detect potential problems as early as possible. Road users and regions of interest can be modeled as time-dependent areas. However, current algorithms for monitoring spatial relations between such areas do not consider uncertainty. Thus, they are not able to cope with sensor inaccuracy and errors in trajectory prediction, which can lead to false verdicts regarding possible collisions. In this paper, spatial relations between regions are generalized using a probabilistic approach to treat uncertainties. This makes verdicts of monitored movements more reliable, especially when the positions or boundaries of spatial regions are not known precisely. Therefore, this allows monitoring inherently uncertain spatial relations, like collision estimations or insight into regions of interest using range sensors. The applicability of the presented probabilistic spatial relations is demonstrated by monitoring a potentially hazardous turn maneuver simulated with the Open Urban Driving Simulator CARLA.

Keywords: Runtime Monitoring · Spatial Relations · Uncertainty · Road Traffic

1 Introduction

Monitoring behavior of road users during runtime is an important step to the goal of safe autonomous driving. Potential problems could be detected early, which allows vehicles to adapt their behavior. However, uncertainties in localization, perception, and prediction are a major issue when evaluating correctness of behavior. Nevertheless, such uncertainties need to be considered during monitoring to obtain reliable verdicts.

Formal languages are suitable to specify and monitor behavior during runtime. Commonly, formal languages based on temporal logic are used in the domain of software or hardware systems [18]. In contrast to such systems, autonomous vehicles are cyber-physical systems moving through space. To specify behavior of movements, spatio-temporal logic can be applied [20]. However, literature does not provide spatial relations between areas in which locations and boundaries are not known precisely.

J. Guiochet et al. (Eds.): SAFECOMP 2023, LNCS 14181, pp. 151–164, 2023.
https://doi.org/10.1007/978-3-031-40923-3_12

This work generalizes spatial relations operating on two-dimensional geometric objects using probabilistic evaluations. The contributions of this paper are:

- Introducing probabilistic geometric objects, which allow modeling uncertainties caused by localization and trajectory prediction (Subsect. 3.1),
- Generalizing spatial relations between areas using probabilistic geometric objects (Subsects. 3.2 and 3.3),
- Applying probabilistic spatial relations in general (Sect. 4) and in road traffic scenarios to monitor collision probability (Sect. 5).

2 Related Work

Runtime monitoring is a technique to observe, whether a system behaves according to its specifications [11]. The specifications are evaluated automatically using current observations. Therefore, specifications need to be written using machine-readable formal languages. Temporal logic is a common foundation for such specification languages, as time-dependent states are common in a lot of systems [18]. Signal Temporal Logic (STL) allows specifying real-valued properties on metric time [21].

In addition to time-dependant states, autonomous mobile systems carry space-dependent states, such as location or range sensors. To reason about such spatial properties, spatial logic can be used [2]. Clementini et al. [8,9] introduced spatio-topological relations between points, lines, and areas. Examples of these spatio-topological relations are, among others, *overlaps*, *disjoints*, *meet*, *covers*, and *covered_ by*. The geometric objects are defined by three sets of points: interior, boundary, and exterior points. Each point in space is uniquely associated to one of those sets regarding one geometric object. Therefore, those relations are not able to cope with uncertainties in the objects and map the verdict to truth values. These relations were extended by Vazirgiannis to quantify how much those relations are satisfied, e.g., A overlaps B significantly [24]. To specify behavior of road users on multi-lane roads (e.g., merging), Hilscher et al. [15] developed multi-lane spatial logic to reason about occupancy of lane segments. Runtime monitoring of spatio-temporal properties in cyber-physical systems with support for uncertainties was proposed by Visconti et al. [25]. Their work introduces spatio-temporal reach and escape logic, which supports evaluation of distance relationships, but does not support evaluation of spatio-topological relationships between regions. Thus, spatio-topological relations between regions with uncertain positions and boundaries are still missing in literature.

Uncertainty is a broad term in context of cyber-physical systems [26]. In the context of possible crashes in the nearby future, uncertainty can be understood as a probability of occurrence. On the other hand, uncertainty can be understood as errors in sensor measurements or lack of knowledge about the behavior of other road users for example. With regard to uncertainties in traffic scenarios, localization [22] and trajectory prediction [19] for road users require probabilistic estimation by nature. In this paper, we use the terms path and trajectory as described by Kant [16]: A path is a geometric object; a trajectory contains additional temporal information as well. Predicting future positions and their arrival times can be useful for further navigation. Trajectory prediction is an active field of research. The input of prediction algorithms usually depends on the method, whereas the output can be modeled as a ordered set of time-dependant poses $X_t = \{X_1, X_2, ..., X_{T_{obs}}\}$ [19]. A pose describes the position and the orientation of an object in two- or three-dimensional space. In a 2D-plane, poses can be modeled as $X = [x, y, \theta]$ with θ representing the orientation. Note that the predicted poses describe points whereas the shape of the spatial objects is ignored. Modeling road users as point-shaped during trajectory prediction causes algorithms to become less complex. Predictions are inherently uncertain, as future movements cannot be known perfectly in advance. To quantize the uncertainty of a predicted pose, the pose can be modeled as a random variable using a probability distribution. The approaches for trajectory prediction in literature can be broadly categorized, whether they are based on neural networks [7] or dynamic Bayesian networks [12, 17]. Additionally, many approaches are specialized to concrete scenarios, like highways [3] or dense traffic in urban areas [6].

Approaches for estimating collision probabilities usually do not build up on predicted trajectories, as with such trajectory predictions vehicles are considered as point-shaped usually, whereas the shapes of vehicles are crucial for estimating collision probabilities. Althoff et al. [4,5] developed stochastic reachable sets, which are probability distributions over all possible regions the mobile system can reach from its current position. For example, the stochastic reachable sets are constrained by the maximum acceleration, deceleration, and driving style. To estimate the collision probability of two vehicles, each possible future position is checked for overlap with the position of the other car at the same future time. If they do overlap, the probabilities for the vehicles being located at the same time and location are multiplied. The sum of all those probabilities is the estimation of the collision probability. Hermann and Schroven [13, 14] extend this approach by using more information to build the stochastic reachable sets.

Predicted severity of upcoming collisions is of great interest as it allows vehicles to change steering just before an unavoidable crash with the goal of minimizing the severity. Parseh et al. [23] partitioned cars into two-dimensional zones and use these to rate severity of unavoidable crashes. They assume positions and trajectories of the vehicles to be known without uncertainty. Their approach tells the severity of the upcoming crash, but not the probability of this crash to actually occur with this severity. Composing predictions of severity with estimating whether the crash will actually occur would enable the vehicle to make more sophisticated decisions during motion planning.

3 Probabilistic Spatial Relations

The spatial relations which are enriched with probabilistic concepts in this paper are overlaps, disjoints, covers, and covered_by, as defined in [9] and visualized in Fig. 1. Classic spatial relations operate on regions consisting of points, which belong to either the regions' interior, boundary, or exterior. To define spatial relations considering uncertainty, the uncertainty must be modeled within the spatial regions.

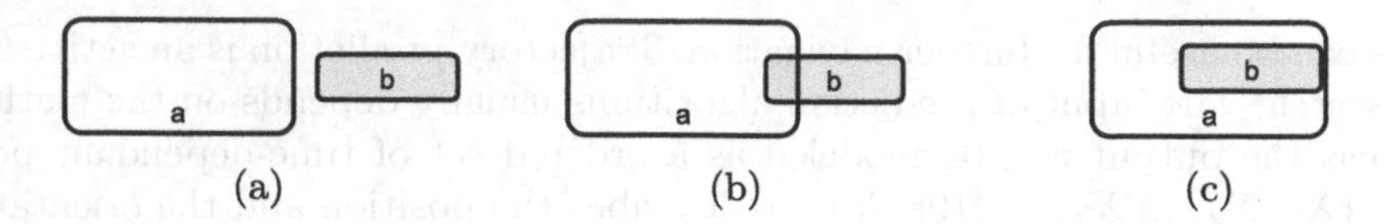

Fig. 1. Spatio-topological relations: *disjoints*(a, b) in (a), *overlaps*(a, b) in (b), *covers* (a, b) or *covered_by*(b, a) in (c)

3.1 Probabilistic Geometric Objects

We define probabilistic geometric objects (PGOs) to evaluate spatial relations between regions and support modeling of uncertainties in poses or boundaries of those regions. A PGO is a model of a region, with each point of the PGO is mapped to a value in the interval $[0, 1] \in \mathbb{R}$. A value of 1 indicates the point certainly belongs to the region; a value of 0 indicates the point to certainly not belong to the region. Therefore, a two-dimensional PGO is defined as follows:

$$PGO : (x \times y) \rightarrow [0, 1] \in \mathbb{R} \tag{1}$$

PGOs are not probability functions, as the integral can be not equal to 1, when the area of the region is not equal to 1. The area is defined as the set of all coordinates (x, y), where $PGO(x, y) > 0$:

$$area(PGO) = \{(x, y) \in \mathbb{R} \times \mathbb{R} \mid PGO(x, y) > 0\} \tag{2}$$

PGOs are useful to model the boundary of areas spawned by distance sensors with the uncertainty described in the sensors' data sheet, for example.

Note that there are more spatial relations defined on regions, such as *a meets b*, which is true if and only if there is an overlap on their boundaries but nowhere else. To evaluate overlaps on boundaries, a definition for the boundary of a PGO is required. Since this is an ambiguous question it is thus omitted from this paper and left for future work.

3.2 Probabilistic Disjoints and Overlaps

Given two PGOs a and b, which are located in the same coordinate system, the probability of $disjoints(a, b)$ and $overlaps(a, b)$ can be estimated. The result of those relations is a probabilistic value, therefore we can note:

$$disjoints(a, b) \rightarrow [0, 1] \in \mathbb{R} \tag{3}$$

$$overlaps(a, b) \rightarrow [0, 1] \in \mathbb{R} \tag{4}$$

$Overlaps(a, b)$ is the probability of the PGOs to occupy the same region. The size of this region is not considered, $overlaps(a, b)$ evaluates whether two regions overlap in at least one point. We know that $overlaps(a, b)$ is the converse probability of $disjoints(a, b)$ and vice versa, therefore:

$$overlaps(a, b) = 1 - disjoints(a, b) \tag{5}$$

To calculate, whether a single point (x, y) is occupied by each object, the product of each PGO at the according coordinate is used:

$$disjoints_{point}(a, b, (x, y)) = 1 - a(x, y) \cdot b(x, y) \tag{6}$$

$$overlaps_{point}(a, b, (x, y)) = a(x, y) \cdot b(x, y) \tag{7}$$

Building upon this, we extend the formulas to evaluate the probability of $overlaps(a, b)$ and $disjoints(a, b)$ in any points in the two-dimensional plane:

$$disjoints(a, b) = \min_{(x,y) \in \mathbb{R} \times \mathbb{R}} 1 - a(x, y)b(x, y) \tag{8}$$

$$overlaps(a, b) = \max_{(x,y) \in \mathbb{R} \times \mathbb{R}} a(x, y)b(x, y) \tag{9}$$

Correctness of Eq. 8 is supported using the following special cases. First, when a and b overlap at any point where both PGOs have a probability of 1, then $disjoint(a, b) = 0$. Second, when a and b do not overlap in any point, $disjoint(a, b)$ is expected to be 1.
Example. $Disjoints(a, b)$ is 0 if there is a point (x, y) exists where $a(x, y) = b(x, y) = 1$. The disjoint probability for this is calculated using Eq. 6: $disjoint_{point} = 1 - 1 \cdot 1 = 0$. Inside Eq. 8, the minimum disjoint probability found in any point is set to the disjoint probability of the two regions. As the disjoint probability per point cannot be lower than 0, the disjoint probability of two regions equals 0 when they share a probability of 1 in at least one coordinate.
Example. $Disjoints(a, b)$ is 1 if there is no overlapping area, e.g., there is no point (x, y) where $a(x, y) > 0 \wedge b(x, y) > 0$. In this case, the product in Eq. 8 is always 0 and the formula evaluates to 1.

3.3 Probabilistic Covers and CoveredBy

To describe whether a spatial region is located completely inside another region, the relations covers and covered_by are used. The expressions $covers(a, b)$ and $covered\_by(a, b)$ are read as "a covers/is covered_by b", the order is important as the relations are not commutative. When a covers b, b is covered by a, therefore:

$$covers(a, b) = covered\_by(b, a) \tag{10}$$

$covers(a, b)$ is true, if and only if a and b overlap and no point of b is located in the exterior of a. To consider uncertainties when evaluating, the probability of a and b to overlap is multiplied with the converse probability of encountering b outside the area of a. This way, the probability of $covers(a, b)$ increases when there is an overlapping, but decreases when b covers area outside of a. To calculate the probability of b occupying a specific area, we need to adjust the boundaries of the max operation. The boundaries for encountering b outside a is set as $area(b) \backslash area(a)$. Formally, this operation is represented in Eq. 11.

The probability of $covered\_by(a, b)$ is calculated similarly as in Eq. 12. The factor $overlaps(a, b)$ stays, but the other factor is replaced with the converse probability of encountering a outside of b.

$$a\ covers\ b = \{\max_{(x,y)\in a\cup b} a(x,y)b(x,y)\} \cdot \{1 - \max_{(x,y)\in b\setminus a} b(x,y)\} \tag{11}$$

$$a\ covered\_by\ b = \{\max_{(x,y)\in a\cup b} a(x,y)b(x,y)\} \cdot \{1 - \max_{(x,y)\in a\setminus b} a(x,y)\} \tag{12}$$

Finally, the spatial relations and their probabilistic generalizations are collated in Table 1.

Table 1. Spatial relations and their generalized probabilistic spatial relations.

Operator	Boolean Condition	Probabilistic Condition
overlaps	$a \cap b \neq \emptyset$	$\max_{(x,y)\in\mathbb{R}^2} a(x,y)b(x,y)$
disjoints	$a \cap b = \emptyset$	$\min_{(x,y)\in\mathbb{R}^2} 1 - a(x,y)b(x,y)$
covers	$a \cap b = b$	$\{\max_{(x,y)\in a\cup b} a(x,y)b(x,y)\} \cdot \{1 - \max_{(x,y)\in b\setminus a} b(x,y)\}$
covered_by	$a \cap b = a$	$\{\max_{(x,y)\in a\cup b} a(x,y)b(x,y)\} \cdot \{1 - \max_{(x,y)\in a\setminus b} a(x,y)\}$

3.4 Example

As an example, we define two rectangular regions a with (3×2)m and b with (1×1)m. The uncertainty of the positions of both objects is modeled using a 2D normal distribution parameterized with $\sigma_x = \sigma_y = 0.1$. This uncertainty in the poses leads to the following situation: The area of points which certainly belong to the objects, i.e., $PGO(x,y) = 1$, is smaller than the ground-truth area of the objects. However, the objects now have a non-sharp boundary of points which belong to the object with a probability of $0 < PGO(x,y) < 1$. Because of this boundary, the PGOs define a greater area than the ground-truth area of the region. First, the PGOs are located next to each other without overlapping. The smaller object b will be moved along the x-axis until it is located at the other side of a. This movement is visualized with the arrow in Fig. 2a.

For every distance b is moved, the relations $overlaps(a,b)$, $disjoints(a,b)$, $covers(a,b)$ and $covered\_by(a,b)$ are evaluated. The obtained probabilities in relation to the distance PGO b is moved are visualized in Fig. 2b. We see that $overlaps(a,b)$ increases immediately as the boundaries start overlapping. As expected with Eq. 5, $disjoints(a,b)$ and $overlaps(a,b)$ are converse to each other, which leads to the probabilities being mirrored at $p = 0.5$. $covers(a,b)$ increases when the left boundary of a overlaps with the left boundary of b. The probability of $covered\_by(a,b)$ is always 0, which is caused by the area of a being much larger than the area of b.

4 Building Probabilistic Geometric Objects

In this section, we first describe how to model uncertainty caused by localization errors into probabilistic geometric objects (PGOs). Building on this, uncertainties caused by predicted trajectories are modeled into PGOs. From now on, the two-dimensional space is discretized to allow efficient processing of PGOs and probabilistic spatial relations using computers.

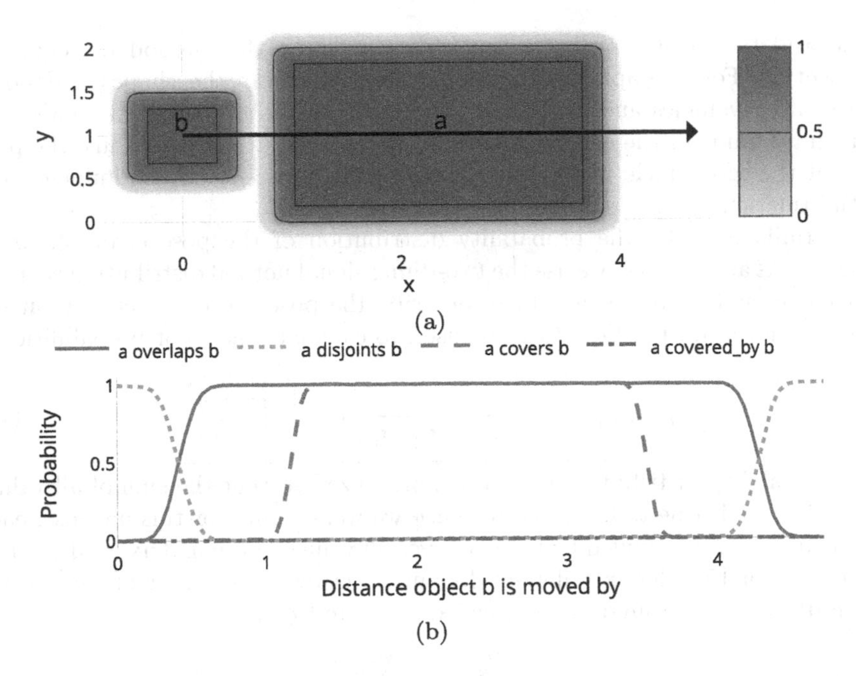

Fig. 2. The small PGO b is moved through the bigger PGO a, as visualized with the arrow in (a). The color indicates the probability to meet a point of the PGO at the coordinate. The probabilities of the spatial relations depending on the distance b is moved are visualized in (b).

4.1 Model Uncertainties Caused by Pose Errors

Knowledge about poses of road users is subject to uncertainty, which is usually described by probability distributions. The pose of the ego vehicle can be obtained using sensors and localization algorithms, whereas poses of other road users can be obtained using sensors and tracking algorithms for example. As measurements of sensors are subject to uncertainty, the poses of road users are also subject to uncertainty. In simplified cases, the unimodal normal distribution is suitable to model pose errors. However, our algorithms work with any discrete probability distributions. This makes the approach composable with any localization- or tracking algorithm, providing poses as random variables using a probability distributions.

First, the assumptions we made are explained. We assume the vehicles to be equipped with sensors to measure and estimate the positions of themselves and other road users. Road users can be considered as two-dimensional geometric objects given by an orthographic projection from above. One method to obtain shape of other vehicles is to assume the vehicles as symmetric when two sides are obtained using sensor data. We operate in the local coordinate frame of the object. In the local coordinate frame, the center of the object is the coordinate origin and the orientation of the coordinate frame is equal to the orientation of the object. Uncertainties in the transformation between a global coordinate frame and the local coordinate frame do not affect the method presented in this section. For example, when the ego vehicle tracks the shapes and poses of two other vehicles and evaluates a spatial relation between their PGOs, the coordinate frame of the ego vehicle would be appropriate to use. Thus, the pose error of the ego vehicle in the global coordinate frame would not influence the spatial relation.

To build a PGO, the probability distribution of the pose error $p(x, y)$ is required. As an example, we use the two-dimensional normal distribution without covariance, and therefore $\rho = 0$, to describe the pose error. Depending on the standard deviation [1], Eq. 13 can be used to calculate the error probabilities.

$$p(x, y|\rho = 0) = \frac{1}{2\pi \cdot \sigma_x \cdot \sigma_y} \cdot e^{-\frac{1}{2}\left(\frac{x^2}{\sigma_x^2} + \frac{y^2}{\sigma_y^2}\right)} \tag{13}$$

This probability distribution has to be normalized so that the sum of all values equals 1. This is due to the discrete space we work with. For this purpose, each cell of the matrix is divided by the sum over all values in the matrix, as described in Eq. 14 and 15. The axis-aligned distances between the center of the normal distribution to the boundaries are called x_{max} and y_{max}.

$$norm = \sum_{x=-x_{max}}^{x_{max}} \sum_{y=-y_{max}}^{y_{max}} p(x, y) \tag{14}$$

$$p(x, y) = \frac{p(x, y)}{norm} \tag{15}$$

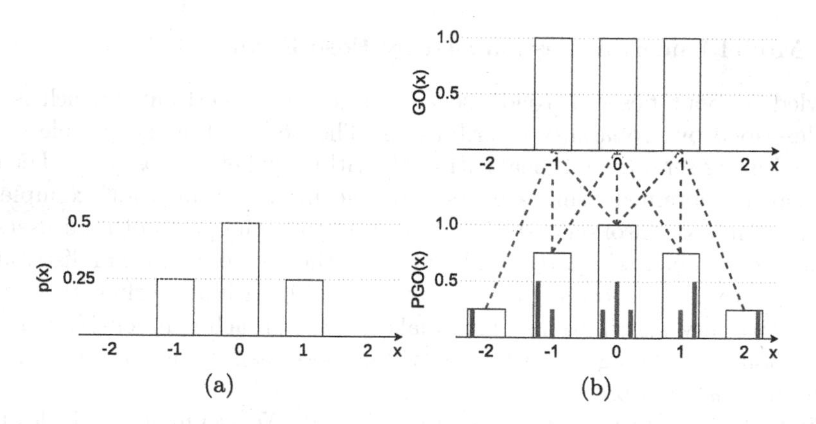

Fig. 3. Pose error $p(x)$ described by a discrete probability function (a) is modeled into a probabilistic geometric object using convolution (b). For convenience, the one-dimensional space is used in this figure.

Using the road users shape and the pose error, we can build a two-dimensional $PGO(x, y)$, which probability values state whether the road user occupies the according coordinate values. An important assumption is that the whole geometric object is subject to the same pose error, otherwise the geometric object will be distorted with the following method. To model the uncertainty in the pose, the rasterized shape of the road user and the discrete probability distribution describing the pose error are required. When the shape is known, each point in the plane has a probability of either 0 or 1 to belong to the geometric object $GO(x, y)$. Figure 3a shows the pose error, Fig. 3b shows how the pose error affects the PGO. For convenience, the one-dimensional space is used in the figure. The probability value of each raster cell belonging to the geometric object is multiplied with each probability value of the pose error, whereas the products are shifted spatially by an offset. The offset is the distance of the according raster cell of the pose error to its center. Then, the resulting $PGO(x, y)$ is obtained by summing up all probability values which overlap at the same coordinate. This sum at each coordinate expresses the probability to meet an arbitrary point of the geometric object at this coordinate. Equation 16 realizes this method.

$$PGO(x, y) = \sum_{\widetilde{x}=-x_{max}}^{x_{max}} \sum_{\widetilde{y}=-y_{max}}^{y_{max}} p(\widetilde{x}, \widetilde{y}) \cdot GO(x - \widetilde{x}, y - \widetilde{y}) \tag{16}$$

This mathematical operation is known as two-dimensional convolution. To be concise, the pose error $p(x, y)$ is convoluted with the shape of the road user $PGO(x, y)$ to model the uncertainty into $PGO(x, y)$, as described in Eq. 17.

$$PGO(x, y) = p(x, y) ** GO(x, y) \tag{17}$$

The ground truth size of the area modeled by a PGO can be obtained as follows. Using Fubini's theorem, we see that the sum over a PGO will not change after convolution with a probability function as the integral over a probability function equals 1 and $\int (f * g)(x)dx = (\int f(x)dx)(\int g(x)dx)$. This integral, which is a sum in discrete space, over a rasterized shape of a region multiplied with the area each discrete cell occupies, equals the area of the shape:

$$\sum_{(x,y)} PGO(x, y) \cdot cell\_area = area\_geometric\_object \tag{18}$$

Figure 4 shows the 2D convolution of $PGO_{shape}(x, y)$ with the normal distribution $p(x, y)$ describing the pose error. The pose error is modeled asymmetrical in this example with $\sigma_x = 0.1$ and $\sigma_y = 0.02$. The resulting PGO consists of all raster cells, which contain a part of the geometric object with a probability above zero. In this example the two-dimensional shape of a road user served as the geometric object. However, the method works for any geometric area which is known with subject to a pose error.

4.2 Model Uncertainties Caused by Trajectory Predictions

Modeling errors in predicted trajectories is similar to modeling pose errors caused by localization or object tracking. However, the convolution needs to be applied multiple times and the coordinate transformation to the predicted poses will be

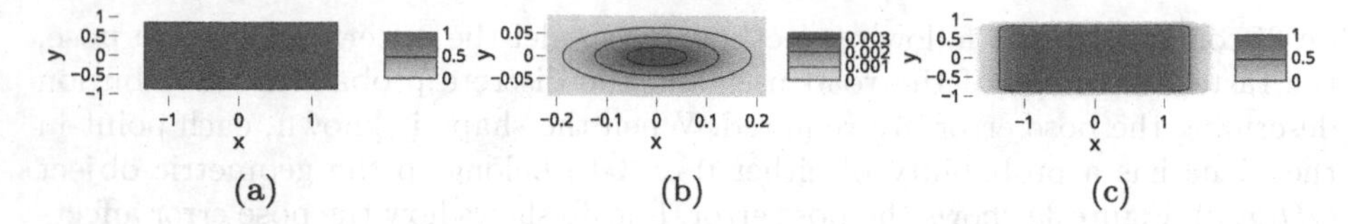

Fig. 4. Creating a probabilistic geometric object to model an error in localization. The shape of the road user is shown in (a). The distribution of the pose error is shown in (b). The convolution of the shape with the pose error shows the area (c), where the road user can be encountered.

considered now. The vehicle follows a trajectory, which is spatially discretized using n poses. Each pose carries the estimated timestamped pose and its probability distribution $p(x, y)$. The approach is composable with any external trajectory prediction, providing a probability function to describe the uncertainty of the predicted pose. A PGO is modeled for each pose of the trajectory. This is done by recursively applying the convolution on the PGOs of neighbored poses with a discrete probability function, $p(x, y)$ as shown in Eq. 19. For the first predicted pose, the current PGO will be used.

$$PGO_{i+1}(x, y) = PGO_i(x, y) ** p(x, y) \tag{19}$$

Figure 5 presents the effect of the recursive algorithm. The dashed line represents the predicted trajectory; the dots represent the poses. The x- and y-coordinates are local coordinates. Beginning with $i = 0$, the PGO already comprises a certain pose error. For $i = 1$ and $i = 2$, the inflation of the PGOs due to the recursive convolution is visible. To evaluate spatial relations, PGOs need to be transformed into the same coordinate transformations. The coordinate transformations into a global coordinate system are given by the poses of the trajectory. This approach assumes the uncertainty of poses to increase or stay constant in successive timesteps in the future. In the case of predicted trajectories which contain a pose with lower uncertainty than the previous pose, a new probability distribution needs to be initialized to convolute with the shape of the according road user.

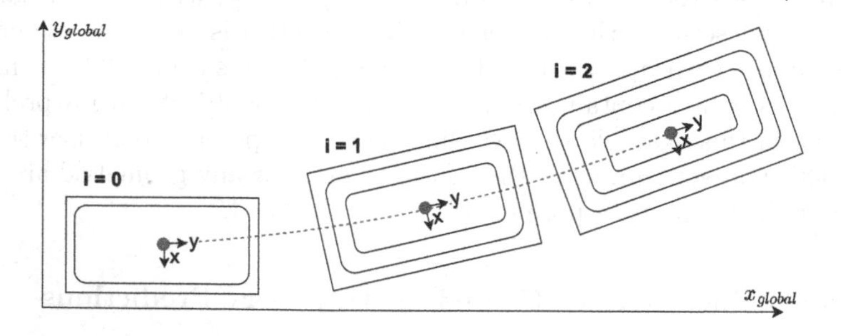

Fig. 5. Probabilistic geometric objects built for each pose of a predicted discrete trajectory. Poses farther away have a greater prediction error, which is visible by means of the inflated PGOs.

5 Application in Road Traffic

To demonstrate applicability of the proposed methods, the $overlaps(a, b)$ operator is used to monitor collision probabilities in road traffic. Using CARLA, a simulator for autonomous driving systems [10], we investigate the situation shown in Fig. 6. Two vehicles are moving toward an intersection, we call the blue vehicle 'Blue' and the red vehicle 'Red'. Red wants to turn left, Blue drives straight. According to the road traffic regulations, Blue has the right of way on the intersection and Red must wait. However, if Blue is slow or far away, Red can turn before blue. This scenario is simulated repeatedly, with Blue having a different speed in each simulation and constant configuration otherwise. The speed of Blue is set to values in the interval [15, 35] km/h, Red has a constant speed of 30 km/h.

For each vehicle, a discrete trajectory will be predicted at every point in time. The predicted trajectories are based on the planned trajectories and assume the vehicles to stay in their lanes. Uncertainty is assumed in the moment the vehicles will reach predicted poses, as perfectly constant speed cannot be assumed in reality. To consider these assumptions, the poses of the trajectory are modeled using a normal distribution. The standard deviation oriented along the direction of motion increases by 0.75 per meter, and 0 orthogonal to the direction of motion. Neighbored poses have a distance of 0.5 m to each other. The length of the predicted trajectories is 9 m, which is approximately the braking distance for a car driving with 30 km/h ignoring reaction time. The vehicles strictly follow their planned trajectory, even if an unacceptable collision probability is predicted, as we are interested in monitoring and not interested in control.

To estimate the risk of collision in this turn maneuver, we monitor and estimate the collision probability using the approach presented in Sect. 4.2 and the

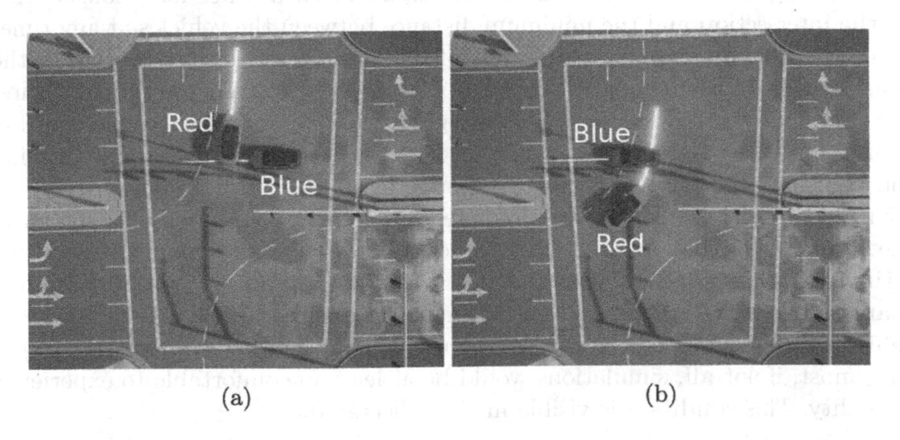

Fig. 6. Monitoring behavior at an intersection benefits from estimating collision probability. The red car wants to turn, either before (a) or after the blue car passes the intersection (b). Blue has a speed of 20.3 km/h in (a) and a speed of 30.1 km/h in (b). Planned trajectories are visualized in green. (Color figure online)

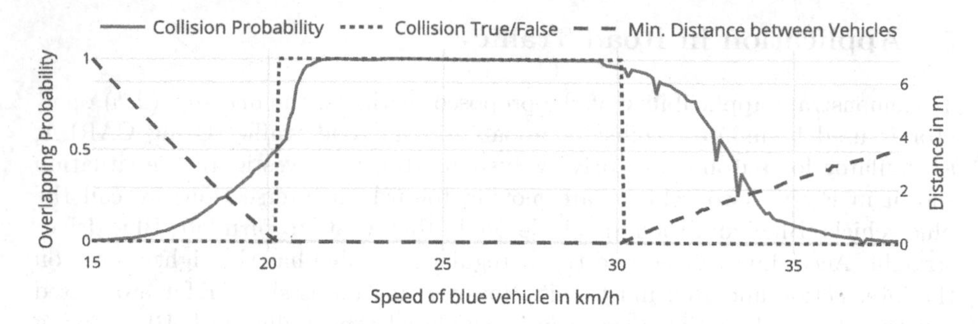

Fig. 7. Estimated collision probabilities using probabilistic $overlaps(a, b)$ operator in relation to the speed of the blue vehicle. Whether a collision happens is presented with the dotted line. The dashed line shows the minimum distance between the vehicles at any point in time per simulation. (Color figure online)

probabilistic overlaps-operator for $i = [0.5, 1, ..., 9]$:

$$overlaps_i(PGO_{red,i}, PGO_{blue,i}) \tag{20}$$

This spatio-temporal relation is evaluated every $16.\overline{6}$ ms for each pose in the predicted trajectory. Note that a turn is not necessarily safe when no collision happens: The vehicles could pass each other very closely. The maximum overlapping probabilities predicted at any future point in time per simulation are visualized in Fig. 7. With speed of Blue inside the interval $[20.3, 30.1]$ km/h, a crash happens. The severity of the crash is not accounted here though the crashes at both ends of the interval are probably less severe than the other crashes. Interestingly, the collected data shows the collision estimations being asymmetric. This is caused by the spatial arrangement on the intersection. In the case of Red turning after Blue, the vehicles are closer to each other for a longer time on the intersection and the minimum distance between the vehicles at any timepoint tends to be shorter. The PGOs of both vehicles therefore overlap in the nearby future more frequently. As the prediction error is smaller in nearby future, the probability values of the PGOs are greater. As a consequence, the collision estimation tends to be higher when Red turns after Blue. The graph showing the collision probabilities is not very smooth, which is due to the distance of 0.5 m between the poses in the predicted trajectory. Using a smaller distance, the graph would be smoother and the collision estimations more correct. The estimated collision probability at $v_{Blue} = 20.3$ km/h and $v_{Blue} = 30.1$ are less than 1. Certain collisions are not estimated as such, which is plausible as the vehicles could still avoid the collision by replanning their behavior shortly. In the end, most, if not all, simulations would be at least uncomfortable to experience in reality. This condition is visible in the collected data.

6 Conclusion

The presented probabilistic spatial relations allow for the consideration of uncertainties in poses and boundaries of regions. Therefore, evaluation of such spatial relations is not a binary decision anymore and becomes more reliable. A use case is monitoring the behavior of road users, which are permanently subject to relationships between themselves and other spatial objects. Those spatial objects might be other road users or regions of interest on the road. Formal specifications of a road users' behavior are now able to express the acceptable risk of spatial relations between objects to occur. Further work is needed to validate the usage of probabilistic geometric relations alongside state-of-the-art methods from localization, tracking and trajectory prediction to estimate collision probabilities in realistic scenarios.

Acknowledgements. This research was funded by the Berlin Institute for Applied Research (IFAF Berlin) within the CARS-project.

References

1. Abramowitz, M., Stegun, I.A.: Handbook of Mathematical Functions with Formulas, Graphs, and Mathematical Tables, vol. 55. US Government Printing Office (1948)
2. Aiello, M., Pratt-Hartmann, I., van Benthem, J. (eds.): Handbook of Spatial Logics. Springer, Netherlands (2007). https://doi.org/10.1007/978-1-4020-5587-4
3. Altché, F., de La Fortelle, A.: An LSTM network for highway trajectory prediction. In: 2017 IEEE 20th International Conference on Intelligent Transportation Systems (ITSC), pp. 353–359 (Oct 2017). https://doi.org/10.1109/ITSC.2017.8317913
4. Althoff, M., Stursberg, O., Buss, M.: Stochastic reachable sets of interacting traffic participants. In: 2008 IEEE Intelligent Vehicles Symposium, June 2008, pp. 1086–1092 (2008). https://doi.org/10.1109/IVS.2008.4621131
5. Althoff, M., Stursberg, O., Buss, M.: Model-based probabilistic collision detection in autonomous driving. IEEE Trans. Intell. Transp. Syst. **10**(2), 299–310 (2009). https://doi.org/10.1109/TITS.2009.2018966
6. Chandra, R., Bhattacharya, U., Roncal, C., Bera, A., Manocha, D.: RobustTP: end-to-end trajectory prediction for heterogeneous road-agents in dense traffic with noisy sensor inputs. In: Proceedings of the 3rd ACM Computer Science in Cars Symposium, CSCS 2019, October 2019, pp. 1–9. Association for Computing Machinery, New York (2019). https://doi.org/10.1145/3359999.3360495
7. Chandra, R., et al.: Forecasting trajectory and behavior of road-agents using spectral clustering in graph-LSTMs, August 2020
8. Clementini, E., Di Felice, P., van Oosterom, P.: A small set of formal topological relationships suitable for end-user interaction. In: Abel, D., Chin Ooi, B. (eds.) SSD 1993. LNCS, vol. 692, pp. 277–295. Springer, Heidelberg (1993). https://doi.org/10.1007/3-540-56869-7_16
9. Clementini, E., Sharma, J., Egenhofer, M.J.: Modelling topological spatial relations: strategies for query processing. Comput. Graph. **18**(6), 815–822 (1994). https://doi.org/10.1016/0097-8493(94)90007-8
10. Dosovitskiy, A., Ros, G., Codevilla, F., Lopez, A., Koltun, V.: CARLA: an open urban driving simulator. In: Proceedings of the 1st Annual Conference on Robot Learning, October 2017, pp. 1–16. PMLR (2017)
11. Havelund, K., Ro, G.: Java PathExplorer - a runtime verification tool. In: International Space Conference, p. 8 (2001)

12. Hermes, C., Wohler, C., Schenk, K., Kummert, F.: Long-term vehicle motion prediction. In: 2009 IEEE Intelligent Vehicles Symposium, June 2009, pp. 652–657 (2009). https://doi.org/10.1109/IVS.2009.5164354
13. Herrmann, S.: Kollisionswarnung im urbanen Straßenverkehr auf Basis einer probabilistischen Situationsanalyse. Logos Berlin, Berlin, July 2013
14. Herrmann, S., Schroven, F.: Situation analysis for driver assistance systems at urban intersections. In: 2012 IEEE International Conference on Vehicular Electronics and Safety, ICVES 2012, July 2012, pp. 151–156 (2012). https://doi.org/10.1109/ICVES.2012.6294295
15. Hilscher, M., Linker, S., Olderog, E.-R., Ravn, A.P.: An abstract model for proving safety of multi-lane traffic manoeuvres. In: Qin, S., Qiu, Z. (eds.) ICFEM 2011. LNCS, vol. 6991, pp. 404–419. Springer, Heidelberg (2011). https://doi.org/10.1007/978-3-642-24559-6_28
16. Kant, K., Zucker, S.W.: Toward efficient trajectory planning: the path-velocity decomposition. Int. J. Robot. Res. **5**(3), 72–89 (1986). https://doi.org/10.1177/027836498600500304
17. Lambert, A., Gruyer, D., Saint Pierre, G.: A fast Monte Carlo algorithm for collision probability estimation. In: Robotics and Vision 2008 10th International Conference on Control, Automation, December 2008, pp. 406–411 (2008). https://doi.org/10.1109/ICARCV.2008.4795553
18. Lamport, L.: Specifying Systems: The TLA+ Language and Tools for Hardware and Software Engineers. Addison-Wesley Longman Publishing, USA (2002)
19. Leon, F., Gavrilescu, M.: A review of tracking and trajectory prediction methods for autonomous driving. Mathematics **9**(6), 660 (2021). https://doi.org/10.3390/math9060660
20. Loreti, M., Bortolussi, L., Bartocci, E., Nenzi, L.: A logic for monitoring dynamic networks of spatially-distributed cyber-physical systems. Log. Meth. Comput. Sci. **18** (2022)
21. Maler, O., Nickovic, D.: Monitoring temporal properties of continuous signals. In: Lakhnech, Y., Yovine, S. (eds.) FORMATS/FTRTFT -2004. LNCS, vol. 3253, pp. 152–166. Springer, Heidelberg (2004). https://doi.org/10.1007/978-3-540-30206-3_12
22. Panigrahi, P.K., Bisoy, S.K.: Localization strategies for autonomous mobile robots: a review. J. King Saud Univ. Comput. Inform. Sci. **34**(8, Part B), 6019–6039 (2022). https://doi.org/10.1016/j.jksuci.2021.02.015
23. Parseh, M., Asplund, F., Svensson, L., Sinz, W., Tomasch, E., Törngren, M.: A data-driven method towards minimizing collision severity for highly automated vehicles. IEEE Trans. Intell. Veh. **6**(4), 723–735 (2021). https://doi.org/10.1109/TIV.2021.3061907
24. Vazirgiannis, M.: Uncertainty handling in spatial relationships. In: Proceedings of the 2000 ACM Symposium on Applied Computing, SAC 2000, Como, Italy , pp. 494–500. ACM Press(2000). https://doi.org/10.1145/335603.335928
25. Visconti, E., Bartocci, E., Loreti, M., Nenzi, L.: Online monitoring of spatio-temporal properties for imprecise signals. arXiv:2109.08081 [cs], September 2021
26. Zhang, M., Selic, B., Ali, S., Yue, T., Okariz, O., Norgren, R.: Understanding uncertainty in cyber-physical systems: a conceptual model. In: Wąsowski, A., Lönn, H. (eds.) ECMFA 2016. LNCS, vol. 9764, pp. 247–264. Springer, Cham (2016). https://doi.org/10.1007/978-3-319-42061-5_16

Concept and Metamodel to Support Cross-Domain Safety Analysis for ODD Expansion of Autonomous Systems

Jan Reich[1(✉)], Daniel Hillen[1], Joshua Frey[1], Nishanth Laxman[1], Takehito Ogata[2], Donato Di Paola[2], Satoshi Otsuka[3], and Natsumi Watanabe[3]

[1] Fraunhofer Institute for Experimental Software Engineering IESE, Kaiserslautern, Germany
{jan.reich,daniel.hillen,joshua.frey,nishanth.laxman}@iese.fraunhofer.de

[2] European Research and Development Centre, Hitachi Europe, London, UK
{takehito.ogata,donato.di-paola}@hitachi-eu.com

[3] Research and Development Group, Hitachi, Ltd., Omika City, Japan
{satoshi.otsuka.hk,natsumi.watanabe.jv}@hitachi.com

Abstract. Automated driving systems (ADS) can improve efficiency in logistics and last-mile delivery, but a major challenge is ensuring safety for operational design domain (ODD) expansion or cross-domain deployment. Various ontologies and formats exist for modeling and representing the operational environment. However, their structuring schemes are not suitable for safety engineering activities, as the safety-relevant aspects of the environment differ from those relevant for other purposes, e.g., simulation scenario representation. This paper addresses the problem of effectively supporting safety engineers in performing environmental safety analyses considering cross-domain aspects and the impact of environmental changes. We contribute a concept for modeling and comparing operational design domains as well as algorithms for semi-automatically analyzing change impact. The approach is model-based, integrated within the Digital Dependability Identity (DDI) framework, and has been evaluated qualitatively for ADS cross-deployment from a logistic warehouse to urban environments. The evaluation suggests that the approach is a suitable starting point for explicitly linking ontological domain modeling with safety engineering. It also helps safety engineers to think about ODDs in a structured way, performing change impact analyses regarding specification gaps, and enabling cross-domain learning.

Keywords: SOTIF · safety assurance · autonomous system · autonomous vehicle

1 Introduction

Context. Automated driving systems (ADS) promise economic and societal benefits in various use cases such as smart logistics warehouses, last-mile delivery, or passenger transport services. The lack of methods for systematically analyzing the impact of a given operational environment on ADS safety is a significant challenge to enabling

J. Guiochet et al. (Eds.): SAFECOMP 2023, LNCS 14181, pp. 165–178, 2023.
https://doi.org/10.1007/978-3-031-40923-3_13

the benefits [1]. To manage operational domain complexity, manufacturers develop their systems for a constrained environment, the operational design domain (ODD). The ODD specifies "operating conditions under which a given driving automation system or feature thereof is specifically designed to function" [2]. Thus, the ODD specifies the context for engineering, validating and arguing the system's safety.

In practice, manufacturers start with small controllable ODDs and gradually extend them to more complex situation classes in line with increasing technology maturity and experience [3]. In contrast, Tier 1 suppliers developing ADS technology are interested in efficiently performing cross-domain deployment [4]. For instance, ADS technology can be used both in smart logistics warehouses and urban environments to realize different use cases. Although there are many differences regarding the environments, there are many common concepts with similar impacts on safety, too. To realize the two use cases, both OEM and suppliers need efficient methods for analyzing the impact of an extended or changed ODD on safety.

SotA Deficiencies. Analyzing the impact of ODD changes on safety requires two components: a representation of the ODD and a way to compare multiple ODD representations regarding their impact on safety artifacts. Rich ADS environment ontologies like the 6-layer model [5], BSI PAS 1883 [6] or the ADS operational world model ontology [7] have been proposed to inform ODD specification. In addition, ASAM OpenODD [8] specifies a format capable of representing a defined ODD. The main focus of these approaches is to enable ADS verification and validation, i.e. to derive validation scenarios covering the ODD. However, the relationship to other important safety engineering artifacts such as hazard analysis and risk assessment (HARA) or safety requirements is not considered. In addition, analyzing the impact of extended or changed ODD representations on these artifacts is still an open problem. These gaps lead to the necessity to research how to support safety engineers in terms of a) linking ODD representations to safety engineering artifacts and b) performing change impact analyses for changed ODDs. Without systematic model-based approaches, it will be hardly possible to manage changes in complex ODDs with both acceptable efficiency and adequate confidence in the safety analyses.

Contribution. In this paper, we propose a model-based method to address the research gaps. To that end, we introduce a metamodel that supports explicit traceability between safety artifacts and their dependencies on an ODD representation, independent of the environment ontology used. Cross-domain safety analysis is enabled through hierarchization and domain abstraction concepts. Based on the metamodel, an algorithm is defined that enables the identification of safety artifacts and associated context elements needing attention by safety engineers. The method has been qualitatively evaluated with industry experts from Hitachi in a use case, where an ADS safety monitor assured for a logistics warehouse intersection is to be cross-deployed to an urban intersection.

Paper Structure. Section 2 briefly summarizes the related work surrounding existing models and methods for ODD specification and analysis. Section 3 introduces the concepts to enable environmental safety analysis. In Sect. 4, the formalization of the concepts in a metamodel and a change impact analysis algorithm are explained. Section 5 describes the results of the industrial case study along with a critical discussion of the results. Section 6 concludes the paper and highlights future research.

2 Related Work

A common approach to modeling the operational environment is the use of ontologies for structuring and instantiating the model. BSI PAS 1883 [6] defines a standardized minimal set of categories and elements that should be defined in the ODD for autonomous vehicles. Another environment structuring approach has been proposed by Scholtes et al. with the 6-layer model. A more detailed ontology is provided by Czarnecki [7] which is structured like the 6-layer model but has five layers. Furthermore, a list of elements is provided for each layer. These environment models describe the automotive environment and focus on structuring the elements for verification and validation, but they are not tailored toward safety engineering activities.

To enable practical use, ontologies and environment models must be formalized. Two large initiatives are working on defining standardized technical formats for modeling the environment. OpenODD [8] defines a language for specifying an ODD via allowed or restricted operating conditions on top of the elements of an assumed ontology. The OpenXOntology [9] project defines a standardized ontology for scenarios in the automotive domain. For OpenODD, an initial proposal is publicly available and OpenXOntology is still under development. While these formats enable engineers to model the operational domain and its ODD subset, there is no support for explicitly linking environment elements to other safety artifacts.

In Kemmanns' [10] dissertation, a method is proposed for creating environment models to support model-based hazard analysis and risk assessment in the context of ISO 26262. A metamodel and ontologies for structuring the environment are proposed in the OASIS framework (Ontology-based Analysis of Situation Influences on Safety). The concepts by Kemmann are a foundation for the work presented in this paper. While he focuses on HARA and thus on scenarios in the context of ISO 26262, our concept is more generic as it is not restricted to functional safety artifacts only.

The approaches described above are related to safety-driven environment modeling and do not explicitly consider ODD expansion scenarios. Gyllenhammar [11] proposed an approach for the continuous development and management of ODDs where use cases are connected to operating conditions. When extending the environment, new use cases arise that are linked to operating conditions. Only if all operating conditions are located within the ODD can the AV drive safely within the use case. Therefore, a change impact analysis could identify whether new use cases are located within the ODD so that operation is safe. This approach can be implemented with our concept by linking use case-specific artifacts to environmental elements.

SotA Gap Summary. A variety of research provides support for environment modeling of verification and validation activities, such as describing scenarios for simulation. However, these models are not yet sufficiently connected and tailored for safety activities. While creating environment ontologies is an important contribution, only their formal connection to safety engineering processes will solve the broader challenge of safe ODD change management. These gaps lead to the necessity to research how to support safety engineers in terms of a) linking ODD representations to safety engineering artifacts and b) performing change impact analyses for changed ODDs.

3 Concept

3.1 Method Overview

This section describes the methodological big picture of the steps required to engineer a safety-driven environment model suitable for analyzing the change impact on safety engineering artifacts (see Fig. 1).

The assumed safety engineering process for ADS consists of common activities such as hazard analysis and risk assessment (HARA), the derivation of safety requirements realizing the safety goals, and the allocation of the requirements to architectural elements. It is especially challenging for ADS that the safety artifacts have complex dependencies to elements of the ODD. For instance, safety requirements are dependent on the presence of particular situations containing various context element combinations out of a large set of possibilities. Thus, managing the possible situation space and its impact on safety engineering gets top priority. Operational domain ontologies describing the relevant concepts of the particular mobility domain are an important input to the ODD modeling process performed by domain experts. Unfortunately, an ontology's adequacy is highly dependent on the goals of its usage. Existing ontologies decompose the world into concepts relevant to enable scenario-based validation. In contrast, the goal of the method described in this paper is to bring environment models closer to a representation that is more suitable for safety engineering.

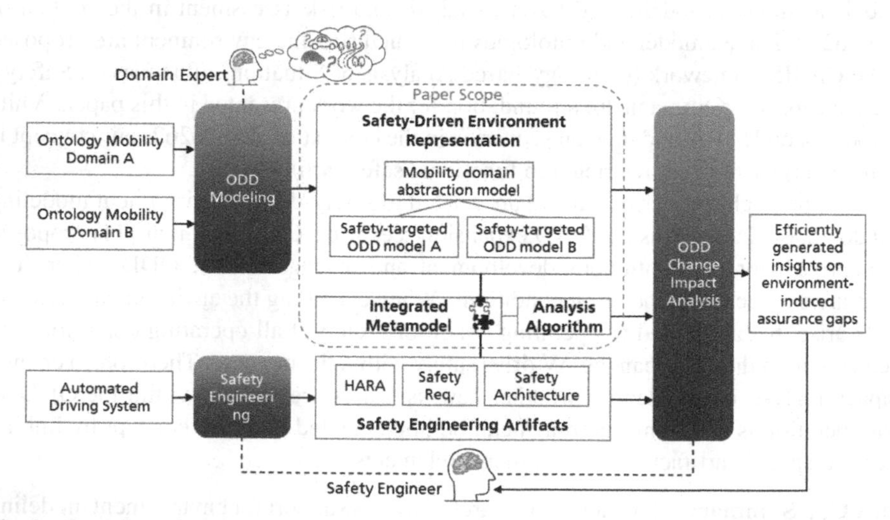

Fig. 1. Overview of method for creating & analyzing safety-driven environment representations

The key contribution to achieving this is an integrated metamodel that supports expressing the dependencies between safety engineering artifacts and the safety-driven environment. Since ODD extensions or changes between multiple mobility domains are a practical challenge, concepts are provided to support creating a mobility domain abstraction model, which is then refined into more concrete safety-targeted ODD models. For instance, a priority signal is an abstract concept relevant to many concrete mobility domains. In logistics warehouses, orange blinking lights realize priority signaling, while in urban environments, there are pedestrian or vehicle traffic lights. If there is a safety requirement for handling the right of way, the environmental dependencies can be leveraged to analyze relevance. Thus, the domain abstraction model is a key element to enable conceptual similarity comparison of context elements when ODDs are changed. To support safety engineers in performing this analysis, an analysis algorithm is defined based on the introduced concepts. The result is a set of efficiently generated insights on environment-induced assurance gaps, e.g., in the shape of context elements, for which no safety requirements have been specified yet. Section 3.2 introduces the required modeling concepts with an example. In this paper, we focus only on the modeling aspects. A method for enhancing environment models, which are then formally specified through the metamodel, will be presented in a future publication.

3.2 Concepts for Safety-Driven Environment Representation

Figure 2 visualizes the three conceptual features used to create a safety-driven environment representation that can be referenced, from safety engineering artifacts. The concepts build upon each other logically, i.e., concept 2 assumes the presence of concept 1, and concept 3 assumes the presence of concepts 1 and 2.

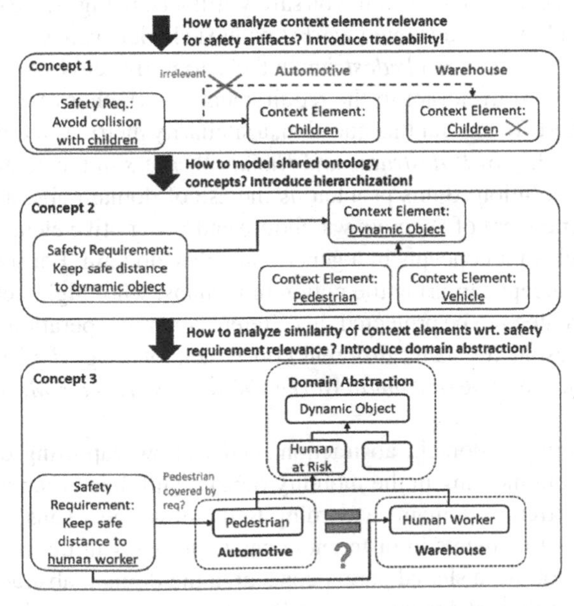

Fig. 2. Illustration of concepts for safety-driven environment representation

Concept 1: Linking Safety Engineering Artifacts to Context Elements they Refer to is the Basis for Analyzing the Safety Impact of a Changing ODD. The first step is to explicitly express the dependencies between safety artifacts such as safety requirements and the context elements they refer to. On the one hand, this enables navigation to all referenced context elements of any safety requirement. This is useful when specifying the relevance or irrelevance of context elements in different domains. Children appear in automotive environments and ADS therefore need to consider them. In warehouses, process measures usually ensure that children do not get access and thus no special requirements are needed for ADS. On the other hand, explicit context elements enable providing an additional specification of context elements by adding attributes.

Concept 2: Hierarchization Enables Representing Shared Ontological Concepts. Existing environment ontologies are structured hierarchically. This allows engineers to group ontology elements with similar attributes and according to conceptual dimensions relevant for achieving particular inference goals. An additional value of hierarchization in the safety engineering context is that safety artifacts can be bound to parent context elements, thereby subsuming child context elements. For instance, in Fig. 2, a high-level safety requirement demands keeping a safe distance from dynamic objects. The hierarchical structure of the environment model enables algorithmic reasoning that the relevant concrete dynamic objects are pedestrians and vehicles. Note that the hierarchization concept can be applied separately for multiple semantical domains, e.g. for an automotive ODD or a warehouse ODD.

Concept 3: Shared Domain Abstraction Enables Similarity Specification. Concept 3 builds on concepts 1 and 2 to enable operationalizing a model structure supporting the analysis of the impact of different ODDs on safety artifacts. In Fig. 2, the starting point is a safety requirement bound to a *Human Worker* context element in a warehouse ODD. The question to answer is whether a *Pedestrian* in the automotive environment is a) covered by the requirement, b) irrelevant to the requirement, or whether c) a new requirement needs to be specified. To answer this question, a similarity metric is required that enables comparing the equality of *Pedestrians* and *Human Workers* in the context of the safety requirement. Our solution to this problem is the use of domain abstraction to create a model that is independent of concrete warehouse and automotive elements but specifies common safety-relevant concepts in a generic mobility domain. For instance, a *human at risk* is such a concept capturing the notion of humans, who might get endangered by ADS and therefore need to be protected. In different concrete operational environments, this concept is present for many different concretizations, e.g., *Pedestrian, Bicyclist, Police Officer* in automotive or *Human Worker, Warehouse Leader, Maintenance Person* in warehouses.

The concepts in the domain abstraction model allow capturing essential aspects relevant to safety engineering in the mobility domain and thus grouping concrete context elements in different domains according to these aspects. Thus, to determine the similarity of context elements in different concrete domains in the context of a safety requirement, we look for a shared parent concept in the domain abstraction. While this criterion may be sufficient for classifying *humans at risk*, it may be not so obvious for other safety-relevant abstract concepts like *vulnerability* or *predictability.* There, the equivalence is dependent on the concretely considered safety artifact. Therefore, a modeling approach needs to provide the possibility to explicitly specify the presence or absence of equivalence for two context elements, despite them being grouped under a similar abstract concept. This acknowledges the common theme that safety engineering is often case-dependent and (un-)setting context element equivalence with a documented rationale allows tracing and analyzing the decision at any later point in time.

In summary, the three introduced concepts enable us to conceptually bridge the gap between environment ontologies created by domain experts for different mobility domains and the artifacts created by safety engineers during the safety engineering process for an ADS. To make the concepts practically usable by safety engineers and consumable by an algorithm, a formal metamodel and algorithm are introduced in Sect. 4.

4 Concept Formalization

4.1 Metamodel

Figure 3 shows the metamodel with the elements and relationships necessary to formally represent the introduced concepts.

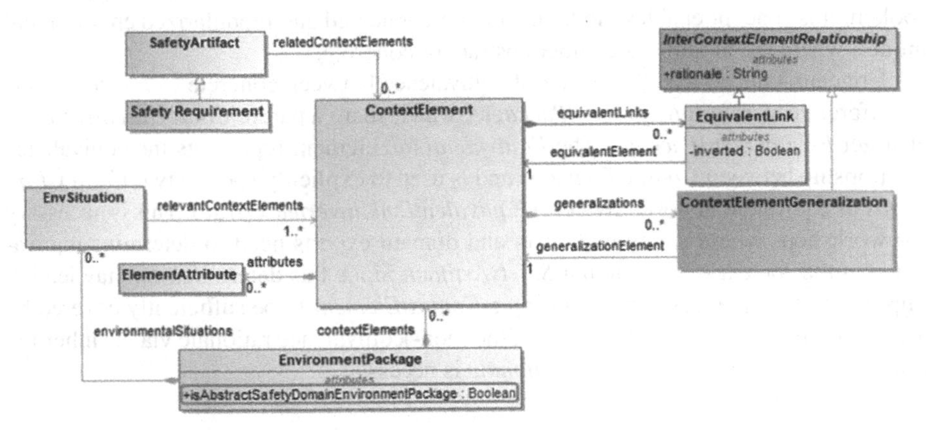

Fig. 3. Metamodel for integrating environment models with safety artifacts

Context Representation. The fundamental basis for making environmental elements referencable is a representation of context elements, realized by the **orange** elements and relationships. An *EnvironmentPackage* is the container of *ContextElements*, which can be further specified by *ElementAttributes. EnvSituations* capture a set of *Context Elements* representing higher-level situations.

Safety Artifact Reference. Concept 1 of Sect. 3.2 is realized by the **blue** elements and relationships. There exist many *SafetyArtifacts* in a safety engineering lifecycle, therefore concrete artifacts like *SafetyRequirements* inherit from *SafetyArtifact*. This is the point where the expressiveness of the Open Dependability Exchange (ODE) metamodel for extensively capturing and relating artifacts of the safety engineering lifecycle can be realized (see Sect. 5 for details). The *SafetyArtifact.relatedContextElements* relationship enables binding *SafetyArtifacts* to *ContextElements*.

Hierarchization Concept. Concept 2 of Sect. 3.2 is realized by the **red** elements and relationships. To represent hierarchization, the *ContextElementGeneralization* element is used. The element represents a relationship between two *ContextElements*. It inherits a rationale attribute from *InterContextElementRelationship*, which is used to give a rationale for why the two *ContextElements* are in a hierarchy relationship. This is required for documentation and process traceability purposes. Navigating from a child *Context Element* to its parent is possible via *ContextElement.generalizations*. The multiplicity of this relationship makes it possible to relate a child *Context Element* to multiple parents. This is required because a child context element needs to be related to multiple parents in the context of the safety-driven domain abstraction concept 3.

Domain Abstraction Concept. Concept 3 of Sect. 3.2 is realized by the **green** elements, relationships, and attributes. Domain abstraction implies the presence of concrete *EnvironmentPackages* and an abstract *EnvironmentPackage*. This is modeled by the *EnvironmentPackage.isAbstractSafetyDomainEnvironmentpackage* attribute and enables modular specification of the domain abstraction model and its reuse in different projects. *ContextElements* from concrete *EnvironmentPackages* can be linked via *ContextElementGeneralization* to *ContextElements* in abstract *EnvironmentPackages*. Figuratively spoken, this concept enables modeling an interconnected and modularized environment model, where the modules are either abstract or concrete.

Concept 3 introduced the notion of equivalence between concrete *ContextElements* in different concrete *EnvironmentPackages*, which share a parent *ContextElement* in an abstract *EnvironmentPackage*. The *EquivalentLink* element represents the equivalence relationship between *ContextElements* and is used to explicitly specify two *ContextElements* as equivalent or not equivalent (*EquivalentLink.inverted* = *true*). This is necessary for workshops where safety engineers and domain experts need to determine equivalence in the context of a particular *SafetyArtifact*. Since this determination may lead to important consequences, e.g., requiring a *ContextElement* to be sufficiently covered by a *SafetyRequirement*, documentation of the (non-)equivalence rationale via the inherited *InterContextElementRelationship.rationale* is necessary.

4.2 Analysis Algorithm

The change impact analysis can be supported by algorithms that are defined on the metamodel. There are two main steps, (i) preparation and (ii) analysis. The preparation step (i) requires as input the environment model of the old and the new domain as well as the abstract environment model. The algorithm should output lists of context elements that are exclusive to (a) the old domain and (b) the new domain, as well as (c) a list of candidates containing pairs of inter-domain siblings, i.e., two elements of a different domain having the same parent in the abstract model. The algorithm first retrieves all elements and then decides based on the attached generalization link and its environment package whether it is exclusive to one domain. This is the case if the lowest-level abstract parent element has no child from a different domain. As an intermediate step, the algorithm creates a mapping between the lowest-level abstract parent element and a list of connected concrete context elements. Then it can filter for elements of different domains and the same parent. Additionally, the algorithm collects all safety artifacts that have a reference to the context element by iterating over the artifacts and gathering those that are referenced to the specific context element. For elements in list (a), the algorithm will further retrieve the referenced safety artifacts. These elements require no further action because they are irrelevant when deploying the system in the new domain. Related safety requirements therefore do not need to be fulfilled. Elements in list (b) are not relevant in the old domain, so new safety artifacts must be created. List (c) is the input for another expert analysis. Each pair of siblings is enriched with safety artifacts that reference the element from the old domain. Safety engineers must actively determine whether the siblings are equivalent between domains in the context of the safety artifact. The enhanced model instance then serves as input for the next step.

The analysis algorithm (ii) retrieves all elements that have an equivalent link attached. It can provide a list of elements that are (d) equivalent between domains and (e) explicitly non-equivalent even though they have the same parent. For elements in list (d), no new safety artifacts need to be created; the existing ones can be reused. Last, safety engineers know that no reusable safety artifacts exist for elements of list (e). For those elements, new artifacts must be created.

The metamodel is an enabler for script-based change impact analysis. The described algorithms support engineers in analyzing the environment and identifying reuse potential. The algorithm output therefore does not generate artifacts but proposes elements and artifacts so that engineers do not need to collect the relevant information manually.

5 Case Study

Case Study Introduction. The case study for demonstrating and evaluating the developed concepts targets a use case, where an ADS system is assumed to be assured for straight crossing of an intersection in a logistics warehouse environment (Fig. 4). This means that safety artifacts such as safety requirements exist and consider the relevant context elements of the warehouse ODD. The ODD is to be expanded such that the ADS can be safely deployed to an intersection in an automotive urban environment. Thus, the safety engineers need to determine the change impact of the ODD expansion in terms of the need to modify existing safety requirements and/or specify new requirements for automotive-specific context elements, and the warehouse-specific requirements that are irrelevant for the automotive ODD and thus do not require revalidation.

Evaluation Method. To apply the method described in Sects. 3 and 4, we instantiated both domain-specific ODD models with domain experts from Hitachi in separate workshops for each domain. The experts used a mobility domain abstraction model, which had been created previously inspired by the PEGASUS 6-layer model, as input and concretized the abstract concepts for each concrete ODD. For the warehouse domain, example safety requirements were derived and linked to the respective context elements of the warehouse ODD model. Afterwards, context elements from automotive and warehouse with shared abstract concepts were algorithmically determined and analyzed in the context of the particular requirements regarding their equivalence. Then equivalent links were created for elements that are similar in both domains. These processes resulted in the target models, which were subsequently analyzed utilizing the change impact algorithm. The safety engineers are then provided with the algorithm results showing safety requirements and context elements relevant to required safety modifications for the ODD expansion. The results were evaluated qualitatively by industry experts regarding their benefits and drawbacks compared to a non-model-based approach as a baseline. Given the early stage of the research conducted, an empirical evaluation was deemed inappropriate due to the cost-benefit imbalance.

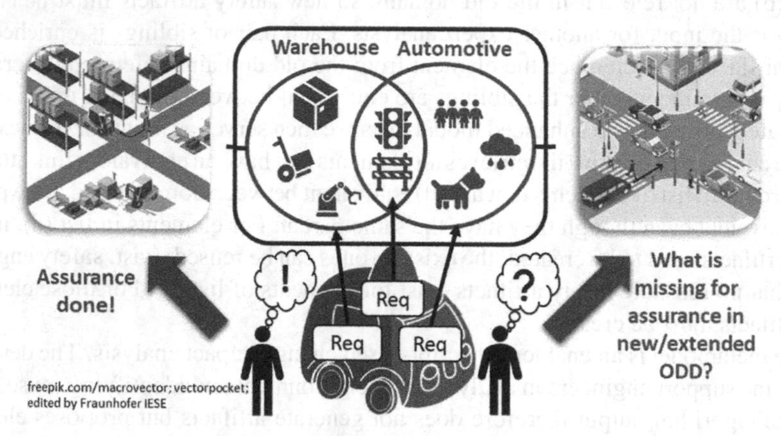

Fig. 4. Case study context: assuring expansion of an ODD from warehouse to automotive

Implementation. The proposed model-based method was realized technically and integrated with the Digital Dependability Identity (DDI) framework [12]. The framework contains the Open Dependability Exchange (ODE) metamodel with integrated capabilities to express the artifacts of the safety engineering lifecycle. In addition, a scripting engine based on the Eclipse Epsilon framework [13] enables executing algorithms on technical DDI instances conforming to the ODE. Our environment representation extends the ODE metamodel so that we can link context elements to specific safety artifacts. The change impact algorithm was realized in the Epsilon Object Language (EOL). Note that, although the case study focuses on safety requirements, the change impact analysis can be easily extended to analyze environmental dependencies of other safety artifacts, e.g., from HARA, due to the ODE's expressiveness.

5.1 Creation and Analysis of Safety-Driven Environment Representation

Figure 5 shows an excerpt of the models created within the case study. For the sake of clarity, it only contains the case study elements needed to illustrate the working principle of the approach. These include the domain abstraction model (top), two domain-specific safety-targeted ODD models (distinguished by color – purple: warehouse, blue: automotive), and safety requirements referencing the context elements. The domain abstraction model is realized as *EnvironmentPackage* with *isAbstractSafetyDomainEnvironmentPackage* set to true. The domain-specific models are also *EnvironmentPackages*, but non-abstract. All elements in the *EnvironmentPackages* are *ContextElements* hierarchized by *ContextElementGeneralization,* and in the bottom layer are *Safety Requirements.* In the following, four scenarios are described to highlight the features of the models.

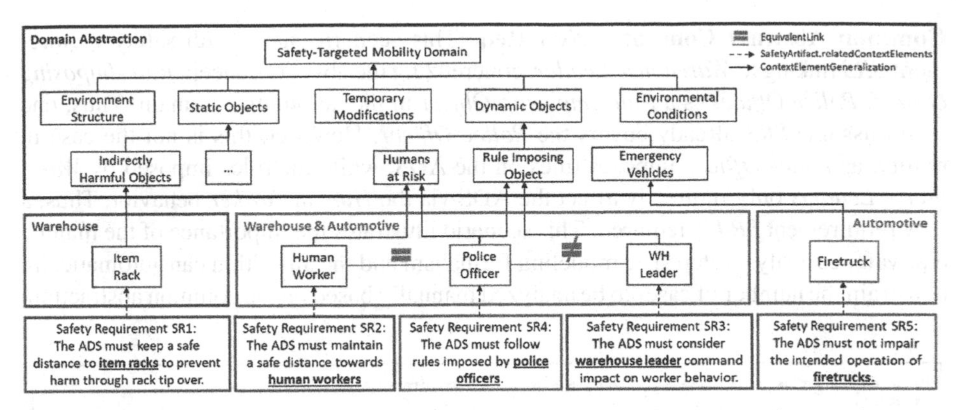

Fig. 5. Excerpt of the safety-targeted ODD models with implications on safety requirements (Color figure online)

Irrelevant Requirements. Safety Requirement *SR1* refers to the context element *Item Rack*, which is part of the warehouse package and a concretization of the abstract concept *Indirectly Harmful Objects*. If the domain experts would find a context element in the automotive domain concretizing this concept, there would be another child element for *Indirectly Harmful Objects*. Since this is not the case, *SR1* is deemed irrelevant for the automotive domain, so no revalidation, and validation effort is saved. This case illustrates the general comparison concept: The potential similarity of two concrete context elements is indicated via the abstract concept of a common parent.

New Requirements. The context element *Firetruck* in the automotive package is a concretization of the abstract concept *Emergency Vehicles*. For this abstract concept, no concrete concept exists in the warehouse package. The conclusion is that a firetruck was not envisioned to exist in warehouses and consequentially, no safety requirement existed for its assured deployment in the warehouse ODD. Thus, a new safety requirement *SR5* needs to be specified to address *Firetruck*. Note that the safety requirement would not be generated automatically, but the lack of consideration of *Firetruck* would be reported to the safety engineer.

Common Abstract Concept – Covered by Req. The next two scenarios deal with abstract concepts with concretizations in both domains. Safety requirement *SR2* demands a safe distance from *Human Workers* in the warehouse. The relevant abstract concept within *SR2* is that *Human Workers* can be endangered by ADS and are thus *Humans at Risk*. In the automotive domain, a *Police Officer* is equally deemed to be a *Human at Risk*, expressed by the *ContextElementGeneralization* relationship. Merely using the common parent concept as a sufficient criterion to indicate equality may be hazardous. Therefore, an extra step has been introduced in that the safety engineers and domain experts discuss the concrete case in the context of the concrete safety requirement and actively set the *EquivalentLink,* documenting of the rationale within the link's *rationale* attribute. After equality is determined, coverage of *SR2* for *Police Officer* can be assumed and no new requirement is needed.

Common Abstract Concept – New Req. This scenario starts with safety requirement *SR3* linking to *Warehouse Leader*, concretizing the abstract concept *Rule-Imposing Object*. *Police Officer* is a *Rule-Imposing Object* in the automotive domain. Thus, one could assume *SR3* already covers the *Police Officer*. However, this is not the case in reality, as *Police Officer* imposes rules on the ADS, while the rules imposed by *Warehouse Leaders* only indirectly affect the ADS via the *Human Worker* behavior. Thus, a new requirement *SR4* is required. This scenario illustrates the importance of the manual equivalence analysis step. The modeling formalism and the algorithm can automatically constrain the number of cases to be analyzed manually based on the common abstraction.

```
New elements exclusive for domain: Automotive - New safety requirements must be derived:
Firetruck
...
================================================================================================
New elements of domain: Automotive with equivalent elements in domain: Warehouse - Requirements can be reused:
<abstract> Humans at Risk:
    Police Officer = Human Worker -> Rational: Both are vulnerable humans that are in danger
        |-> SR2 - The ADS must maintain a safe distance towards Human Worker
        |-> Proposal: SR2.1 - The ADS must maintain a safe distance towards Police Officer
...
================================================================================================
New elements of domain: Automotive that are nonequivalent to elements in domain: Warehouse - Safety requirement
must be derived:
<abstract> Rule Imposing Object:
    Police Officer != WH Leader -> Rational: Police officer imposes rules on the ADS and WH leader to workers
        |-> SR3 - The ADS must consider warehouse leader command impact on worker behavior
...
================================================================================================
Elements exclusive for domain: Warehouse - No action required:
Item Rack
        |-> The ADS must keep a safe distance to item racks to prevent harm through rack tip over
...
```

Fig. 6. Snapshot of the output of the automated model-based change impact analysis algorithm

Figure 6 shows the output of the script execution realizing the change impact analysis algorithm based on the introduced metamodel. The script output consists of exactly the context elements deemed relevant or irrelevant in the illustrative scenarios in Fig. 5. The output is structured by the specific call to action for the engineers.

5.2 Discussion

The case study demonstrates the capability of the presented approach to filter and categorize relevant ODD model parts to efficiently support safety engineers in performing ODD change impact analyses. The baseline approach against which this paper's approach was evaluated was a non-model-based approach that uses the BSI PAS 1883 ontology to identify the impact of ODD changes on requirements. The evaluation feedback revealed that especially the domain abstraction model and the introduction of safety-motivated intermediate abstract concepts like *Indirectly Harmful Object, Humans at risk, rule-imposing objects,* or *emergency vehicles* improved the quality of the domain-specific ODD models. The clear separation of abstract safety-driven mobility concepts and domain-specific concepts facilitates reuse and even learning across domains. For instance, when a new abstract concept is formulated based on a new concrete domain, domain experts can reflect on whether this abstract concept might be concretizable in other concrete domains and thus reveal gaps. The algorithm's output was deemed appropriate for enabling safety engineers to identify concrete next steps (identification of new requirement, context element equivalence analysis, documentation of coverage rationale or requirement irrelevance).

One issue that emerged in the case study was the termination criterion for context concretization, i.e., when to stop differentiating a particular concept. The conclusion was that this is highly dependent on which safety artifact is concerned: While for behavioral requirements, the predictability or vulnerability aspects regarding humans are relevant, appearance aspects are more relevant for perception requirements. This means that the modeling formalism should be extended in the future to condition context element concretization by aspects particularly relevant to the questions appearing in different safety engineering processes. This will enable the definition of improved methodological support to create safety-targeted ODD models in which high confidence in completeness can be argued.

The introduced approach provides the means to make existing domain ontologies more safety-targeted. However, the suitability of intermediate safety-driven abstract concepts to provide a complete decomposition of ontological aspects has not been evaluated. An evaluation of a real-world system is planned in the upcoming iteration.

6 Conclusion and Future Work

For safety engineers, it is a challenge to integrate existing ADS environment ontologies into ODD models and systematically analyze the impact of extended or changed ODDs on safety engineering artifacts. Therefore, we explored the conceptual needs for an ODD representation that is suitable for integration into safety engineering processes. The result is a formalized metamodel that provides means to model (a) links between safety engineering artifacts and context elements; (b) hierarchical context element structures to express shared ontological concepts; and (c) abstract domain structures to make similar context elements in different concrete ODDs like warehouses or urban environments comparable through abstraction.

In addition, we introduced an algorithm that leverages the introduced formalisms to identify safety artifacts and their linked context elements requiring attention by safety engineers in an extended ODD specification. The benefit of using our approach is that manual reanalysis can be avoided in those cases where context elements in the extended ODD share similar concepts with context elements in the previous ODD. The industrial evaluation demonstrated that the proposed model-based approach is a suitable starting point for systematically integrating environment modeling and analysis activities in an overall safety engineering process. In addition, using the mobility domain abstraction model makes it easier to create environment models for new concrete ODDs and even support cross-domain learning. Thanks to the independence of concrete environment ontologies, safety engineers can instantiate the approach with existing ontologies such as [5, 7, 14], then run the algorithms to analyze the impact on safety changes. In parallel to exploring suitable formalisms for the representation and analysis of ODDs, a concrete model for safety-driven ADS domain abstraction has been created, which can be used to accelerate the creation of ODD models for new mobility environments by refinement. This model will be the subject of another publication.

The case study revealed an important open issue: Since every safety engineering task needs a different perspective on the ODD, it is necessary to support safety engineers in constructively creating task-specific environment models that are sufficiently complete. We believe this can be achieved by adding task-specific guiding questions that prompt domain experts to cognitively think about the ODD from the perspective of a particular safety engineering task. In future work, we want to research what such methodological guidance for instantiating the models presented in this paper might look like.

Acknowledgment. Part of this work for writing the article has been funded by the project Layers of Protection Architecture for Autonomous Systems (LOPAAS) funded by the Fraunhofer Gesellschaft.

References

1. Burton, S., Hawkins, R.: Assuring the safety of highly automated driving, state-of-the-art and research perspectives, Assuring Autonomy International Programme, Report (2020)
2. SAE International, SAE J3016:2021 Taxonomy and Definitions for Terms Related to Driving Automation Systems for On-Road Motor Vehicles. https://doi.org/10.4271/J3016_202104
3. Templeton, B.: Will it be hard or easy for self-driving cars to expand their territory? (2021). https://www.forbes.com/sites/bradtempleton/2021/03/30/will-it-be-hard-or-easy-for-self-driving-cars-to-expand-their-territory/?sh=3ea1a8996fc4 Accessed: 20/02/2023
4. Reich, J., et al.: Engineering dynamic risk and capability models to improve cooperation efficiency between human workers and autonomous mobile robots in shared spaces. In: Seguin, C., Zeller, M., Prosvirnova, T. (eds.) IMBSA 2022. Lecture Notes in Computer Science, vol. 13525, pp.251 Springer, Cham (2022). https://doi.org/10.1007/978-3-031-15842-1_17
5. Scholtes, M., et al.: 6-layer model for a structured description and categorization of urban traffic and environment. IEEE Access **9**, 59131–59147 (2021). https://doi.org/10.1109/ACCESS.2021.3072739
6. British Standards Institution (BSI), PAS 1883:2020: Operational Design Domain (ODD) taxonomy for an automated driving system (ADS) - Specification
7. Czarnecki, K.: Operational design domain for automated driving systems - taxonomy of basic terms (2018). https://doi.org/10.13140/RG.2.2.18037.88803
8. ASAM OpenODD Project. https://www.asam.net/standards/detail/openodd/. Accessed 20 Feb 2023
9. ASAM OpenXOntology Project. https://www.asam.net/project-detail/asam-openxontology/. Accessed 20 Feb 2023
10. Kemmann, S.: SAHARA-a structured approach for hazard analysis and risk assessments. Dissertation. Technical University Kaiserslautern, Germany (2015)
11. Gyllenhammar, M., et al.: Towards an operational design domain that supports the safety argumentation of an automated driving system. In: 10th European Congress on Embedded Real Time Software and Systems (ERTS), TOULOUSE, France (2020). ⟨hal-02456077⟩
12. DEIS Consortium: Dependability Engineering Innovation for Cyber-Physical Systems Project Dissemination. http://www.deis-project.eu/dissemination/. Accessed 20 Feb 2023
13. Eclipse Epsilon Framework. https://www.eclipse.org/epsilon/. Accessed 20 Feb 2023
14. Westhofen, L., Neurohr, C., Butz, M., Scholtes, M., Schuldes, M.: Using ontologies for the formalization and recognition of criticality for automated driving. IEEE Open J. Intell. Transp. Syst. **3**, 1 (2022). https://doi.org/10.1109/OJITS.2022.3187247

Security Engineering

Pattern-Based Information Flow Control for Safety-Critical On-Chip Systems

Tobias Dörr[(✉)], Florian Schade, and Jürgen Becker

Karlsruhe Institute of Technology (KIT), Karlsruhe, Germany
{tobias.doerr,florian.schade,juergen.becker}@kit.edu

Abstract. Implementing safety-critical systems on heterogeneous multicore platforms is a challenging task. Especially in mixed-criticality scenarios, it requires fine-grained control over potentially feasible information flows. This paper presents a design-time procedure to achieve such control. Following the X-by-Construction (XbC) paradigm, it performs an automatic configuration of on-chip isolation units and employs a lattice-based integrity propagation mechanism to support developers in the creation of sufficiently isolated system implementations. The procedure is integrated into the XANDAR toolchain for safety-critical system design and illustrated using an automotive example scenario based on the Zynq UltraScale+ MPSoC platform.

Keywords: X-by-Construction · On-chip isolation · Information flow tracking · Multiprocessor system-on-chip · Safety pattern

1 Introduction

Modern embedded systems are increasingly characterized by the consolidation of functions on shared hardware. In the automotive domain, for example, the ongoing shift to zonal architectures is associated with various benefits [2].

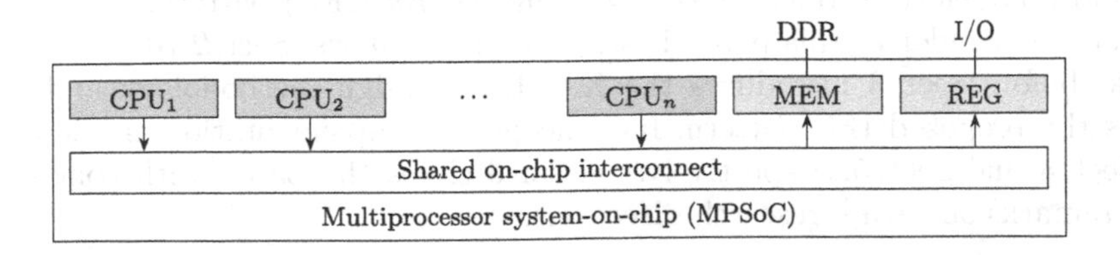

Fig. 1. Block schematic of a heterogeneous MPSoC platform.

Heterogeneous multiprocessor system-on-chip (MPSoC) platforms, such as the one shown in Fig. 1, are usually equipped with an on-chip interconnect that gives every core cluster ($\mathrm{CPU}_1, \ldots, \mathrm{CPU}_n$) access to shared on-chip resources.

J. Guiochet et al. (Eds.): SAFECOMP 2023, LNCS 14181, pp. 181–195, 2023.
https://doi.org/10.1007/978-3-031-40923-3_14

We consider these resources to be random access memory (MEM) and memory-mapped registers (REG), which represent units such as input/output (I/O) controllers. MPSoCs offer benefits in terms of cost and area efficiency [8] but come with various safety-related challenges. A core cluster with write access to the global address space, for example, can interfere with any other application on the chip. In mixed-criticality systems that run less trusted runtime environments such as a general-purpose operating system, this is a crucial issue.

Therefore, the on-chip interconnect of modern MPSoCs is often equipped with runtime-configurable isolation units to control transactions based on, e.g., their source and their destination. In the following, such a unit will be referred to as *access protection unit* (APU). One example of an APU is the XMPU of the Zynq UltraScale+ MPSoC [13]. APUs can contribute to the required on-chip isolation, but they are often difficult to configure [6] and must be combined with other techniques to address the full set of isolation requirements.

This paper is based on the XANDAR toolchain for safety-critical system design [12] and tackles the described isolation issue from a safety point of view. As part of this, it presents three main contributions:

1. A lattice-based integrity framework that allows designers to describe accepted information flows in a safety-critical on-chip system.
2. A strategy to auto-generate APU configurations for heterogeneous MPSoCs from a target deployment performed by the XANDAR toolchain.
3. A design-time technique to verify that certain entities in the APU-protected system receive inputs with a sufficient integrity level.

They address the impact that information flow due to random and systematic faults has on safety-critical on-chip components, e.g., due to an erroneous write to a system-level control register. Timing interferences are beyond the scope of the paper, but state-of-the-art approaches to achieve timing isolation, such as [17], are conceptually compatible with our contributions.

The proposed concept follows the X-by-Construction (XbC) paradigm [3] to generate implementations that meet certain safety requirements by construction. It is integrated into the *XbC pattern library* of the XANDAR toolchain and will therefore be referred to as *information flow control* (IFC) pattern.

The remainder of this paper is structured as follows: Sect. 2 reviews related work, before Sect. 3 introduces the XANDAR design methodology and motivates the proposed IFC pattern. Its concept and implementation are described in Sect. 4 and Sect. 5, respectively. Section 6 closes the paper with conclusions and remarks on future research directions.

2 Related Work

Dynamic information flow tracking is an established technique to detect and prevent prohibited information flows during the runtime of a program [19]. Following a similar goal, the constructive strategy presented in [18] facilitates the creation of programs that are guaranteed to meet certain confidentiality

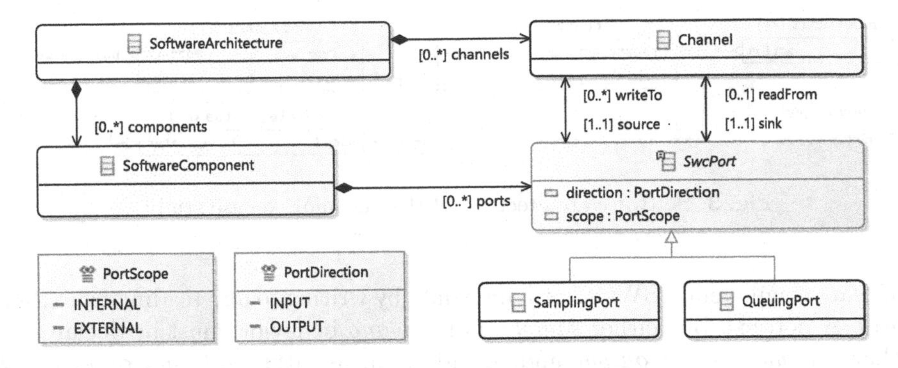

Fig. 2. Excerpt of the XANDAR software architecture metamodel.

requirements. The lattice-based methodology from [16] generalizes this strategy to achieve information flow control for both confidentiality and integrity by construction, i.e., without the need for post-hoc information flow analyses. While these approaches focus on information flows within programs, the presented pattern is concerned with information flows at the on-chip integration level.

Information flow tracking can also be applied at the gate level to reason about interactions at the granularity of individual bits [10,14] and has been extended to support a lattice-based treatment of information flows [9].

The model-based methodology presented in [15] generates system-on-chip designs in which applications of different criticality levels are isolated by dedicated hardware wrappers. It does not leverage runtime-configurable isolation units and, therefore, needs to be applied during the hardware design phase.

The model-based frameworks proposed in [1] and [4] target component-based embedded systems. While they focus on the security aspect, our work targets safety-critical systems and, therefore, considers random and systematic faults of hardware units or software stacks explicitly.

The proposed concept is similar to the methodology we describe in [6], which also performs an automatic APU configuration. In comparison to this work, the IFC pattern tracks information flow at the granularity of individual ports and has explicit support for mixed-criticality systems.

3 Background and Motivation

This section gives an overview of the XANDAR design methodology, presents the target deployment strategy that this work is built upon, and uses an automotive case study to motivate the IFC pattern in detail.

3.1 The XANDAR Design Methodology

To develop an embedded system using the XANDAR toolchain, a designer first describes the envisaged software architecture using the metamodel in Fig. 2.

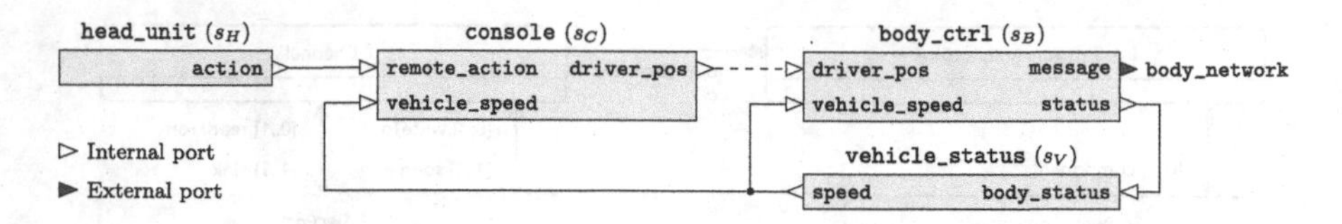

Fig. 3. Software architecture of the automotive case study.

Software components (SWCs) exchange data by writing to and reading from their ports. A port exhibits either *sampling* or *queuing* behavior, must be declared as either an *input* or an *output* and, in addition, as either *internal* or *external.* A channel represents a sender-receiver relationship between an internal output and an internal input port. External ports are used for the interaction with the environment, e.g., by reading a sensor value from an I/O controller.

Given code for each SWC, this architecture is then automatically deployed to a possibly distributed network of target platforms. In this work, we consider a particular deployment strategy of the XANDAR toolchain:

Definition 1 (deployment strategy). *The deployment strategy maps a software architecture to a platform with core clusters* $\{CPU_1, \ldots, CPU_n\}$ *such that:*

1. *Every core cluster* CPU_i *with* $i \in \{1, \ldots, n\}$ *executes a runtime environment* RTE_i *such as a bare-metal hypervisor.*
2. *Each specified SWC is executed by exactly one runtime environment.*
3. *To each* RTE_i *with* $i \in \{1, \ldots, n\}$*, a dedicated memory region* M_i *and a dedicated set of memory-mapped registers* R_i *is assigned.*
4. *For each pair of runtime environments* $\{RTE_i, RTE_j\}$ *with* $i, j \in \{1, \ldots, n\}$ *and* $i \neq j$*, a dedicated memory region* $M_{i \leftrightarrow j}$ *is assigned to this pair if and only if* RTE_i *and* RTE_j *communicate to implement a channel.*
5. *Every runtime environment* RTE_i *with* $i \in \{1, \ldots, n\}$ *implements time and space partitioning for its SWCs. It forwards selected portions of* M_i *and* R_i *for exclusive use to a SWC, but all other resources assigned to* RTE_i *are managed by itself, e.g., to implement channels between SWC ports.*

The partitioning and management capabilities given in Item 5 can be subject to systematic and random faults. They are trusted only to a certain extent, which we will later quantify using a provided integrity level.

3.2 Automotive Case Study

Consider the software architecture visualized in Fig. 3, where rectangles represent SWCs, triangles represent ports, and solid or dashed lines represent channels. The head unit (s_H) and the console controller (s_C) implement convenience applications that allow users to interact with the vehicle, e.g., via their smartphone. The body controller (s_B) uses its external `message` output to control features such as exterior lights or the position of the driver's seat, which have a potential safety impact. Therefore, the `message` output of s_B must be protected, most importantly, from failures of the convenience applications. In the remainder of this work, we further assume that the vehicle status controller (s_V) is sufficiently trusted to interact with the safety-critical features.

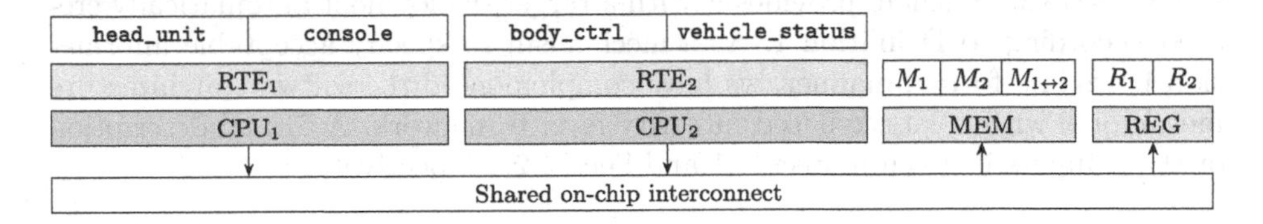

Fig. 4. Sample deployment performed by the XANDAR toolchain.

For the Zynq UltraScale+ MPSoC, which is equipped with $n = 2$ core clusters, assume that the deployment strategy from Definition 1 returned the target deployment shown in Fig. 4. Here, the non-critical applications are implemented on RTE_1, while the safety-critical body controller runs on RTE_2. Furthermore, suppose that RTE_1 is expected to fail in an unpredictable manner at any time, while RTE_2 is trusted.

Example 1 If the dashed channel in Fig. 3 is not implemented, then there is no specified information flow path from RTE_1 to the body controller. Nevertheless, without a suitable APU configuration in place, failures of RTE_1 can cause the contents of M_2 to become corrupted. Since this is the memory region RTE_2 stores its kernel-space data in, such events might cause the safety-critical body controller application to fail. This issue can be solved by using the APU to grant each core cluster access to only those resources that are allocated to it.

Example 2 If the channel visualized using a dashed line in Fig. 3 is implemented, RTE_2 reads from $M_{1\leftrightarrow 2}$ and might forward potentially erroneous data to the `driver_pos` port of s_B, which requests the body controller to trigger an adjustment of the driver's seat. Due to the safety-related nature of this feature, the body controller code is expected to ensure that potentially erroneous inputs are processed safely. This can be achieved, for example, by triggering seat adjustments only if the vehicle is standing still.

Such software-based protection mechanisms complementing the APU protection are an integral part of this paper. We refer to them as *flow barriers* and will formally define them at a later point in time.

Example 3 If register set R_1 contains a system-level control register that allows RTE_1 to trigger a reset of the entire MPSoC, a failure of RTE_1 can again have a significant effect on the body controller. The allocation of such registers to an RTE is therefore a critical aspect that the IFC pattern must consider.

In the following, we will use the scenario outlined in Example 2 (with the channel from s_C to s_B and the flow barrier in place) as a running example.

4 Concept and Input Modeling

The presented IFC pattern is integrated into the XANDAR toolchain as visualized in Fig. 5. One of its inputs is the software architecture model that the toolchain user defines according to the metamodel described above. Furthermore, the XANDAR toolchain provides it with a target deployment automatically created according to Definition 1. As a mechanism to specify acceptable information flows in a formal manner, we have complemented the software architecture metamodel with a lattice-based integrity level framework. A formal description of these inputs is given in Sect. 4.1 and Sect. 4.2, respectively.

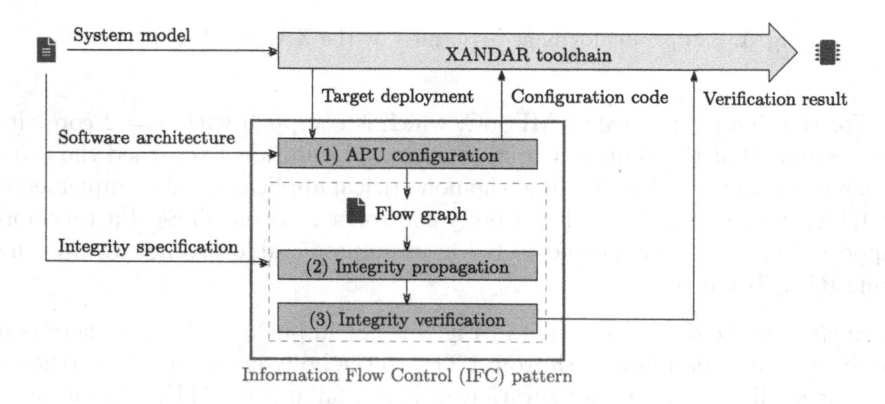

Fig. 5. Process steps and toolchain integration of the IFC pattern.

The pattern itself performs the three steps shown in Fig. 5. It uses knowledge of the software architecture and the associated target deployment to generate an APU configuration for the considered MPSoC (1), which is fed back to the XANDAR toolchain in the form of C code. Following this, it creates a *flow graph* capturing relevant information flow paths that remain possible with the APU configuration in place and, using provided integrity levels annotated by the designer, determines the least trusted source that is able to affect potentially safety-relevant system portions (2). Finally, this knowledge about the least trusted source is compared to the integrity requirements of internal and external SWC ports (3). The result of this check is then reported back to the XANDAR toolchain as a binary success value.

4.1 Formalization of Inputs from XANDAR

For the purposes of this paper, we formalize a software architecture that conforms to the metamodel excerpt in Fig. 2 as follows:

Definition 2 (software architecture). *A software architecture is given as*

$$(S, P, C, \varphi_0, \varphi_s, \varphi_d),$$

where S is a set of software components, P is a set of ports, and the relation $C \subseteq P \times P$ describes channels. $\varphi_0 \colon P \rightarrow S$ maps every port to its SWC. The function $\varphi_s \colon P \rightarrow \{\text{INT}, \text{EXT}\}$ specifies whether a port is an internal or an external one, while $\varphi_d \colon P \rightarrow \{\text{IN}, \text{OUT}\}$ indicates the direction of a port.

The pattern handles both sampling and queuing ports in a uniform manner. Therefore, the formal definition does not distinguish between them. Furthermore, the XANDAR toolchain defines and enforces additional constraints on how channels can be used, e.g., that $\varphi_s(p_1) = \varphi_s(p_2) = \text{INT}$ holds for every $(p_1, p_2) \in C$. A full list of these constraints is skipped for the sake of brevity.

Based on the software architecture (and further inputs beyond the scope of this paper), the pattern receives a deployment that we formalize as follows:

Definition 3 (target deployment). *A target deployment is given by*

$$(\Omega, \lambda, \Psi_{M1}, \Psi_{M2}, \Psi_R, \mu, \mu'),$$

where $\Omega = \{RTE_1, \ldots, RTE_n\}$ denotes the set of deployed runtime environments and $\lambda \colon S \rightarrow \Omega$ maps every SWC to the RTE it is executed on. Ψ_{M1}, Ψ_{M2}, and Ψ_R are sets of utilized on-chip resources.

Ψ_{M1} and Ψ_R contain resources managed by a particular RTE or forwarded to a specific SWC. In the Ψ_{M1} case, these resources are portions of the memory; in the Ψ_R case, they are memory-mapped registers. For elements from these two sets, the mapping is given by $\mu \colon (\Psi_{M1} \cup \Psi_R) \rightarrow (\Omega \cup S)$.

Ψ_{M2} contains memory portions allocated to a pair of RTEs for the purposes of inter-RTE communication. The mapping from such a resource to its RTE pair is given by the function $\mu' \colon \Psi_{M2} \rightarrow \{\{\omega_1, \omega_2\} \colon \omega_1, \omega_2 \in \Omega\}$.

4.2 Lattice-Based Integrity Framework

Inspired by the lattice-based information flow model proposed by Denning [5] and our previous work in [7], the integrity framework facilitates the specification of accepted information flows in a software architecture. Therefore, it expects the designer to specify an *integrity lattice* defined as follows:

Definition 4 (integrity lattice). *An integrity lattice is a bounded lower semilattice of integrity levels $(L, \leq)$, i.e., a partially ordered set L such that (1) every subset $\{x, y\} \subseteq L$ has a greatest lower bound and (2) the lattice has a greatest element. They will be referred to as $x \wedge y$ and $\top$, respectively.*

The fundamental idea of the framework is to let the designer annotate provided integrity levels to the MPSoC itself, every runtime environment, every output port, and values read from external input ports. Furthermore, it allows required integrity levels to be specified for values written to the environment and internal input ports of SWCs in the system.

Example 4. For the set of integrity levels $L = \{0, 1, 2, 3\}$ following the usual order $\leq$ of $\mathbb{N}_0$, the greatest elements is $\top = 3$. It corresponds to the highest integrity in the system. A provided integrity of 3 can therefore be assigned to an output port written by SWC code that meets the requirements of the highest ASIL rating [11] in an automotive system.

In the flow graph, all these entities will later be represented by two classes of vertices: integrity providers and integrity receivers.

Definition 5 (integrity provider). *The set of integrity providers is given by*

$$V_P = V_I \cup V_K \cup P_{\text{OUT}},$$

where a vertex in V_I represents the environment read by an external port. V_K contains $n+1$ vertices representing every RTE and the MPSoC itself, and the set $P_{\text{OUT}} = \{p \in P \colon \varphi_d(p) = \text{OUT}\}$ contains all SWC output ports.

Definition 6 (integrity receiver). *The set of integrity receivers is given by*

$$V_R = V_O \cup P_{\text{IN}},$$

where a vertex in V_O represents the environment written by an external port and the set $P_{\text{IN}} = \{p \in P \colon \varphi_d(p) = \text{IN}\}$ refers to all SWC input ports.

Before the formal definition of the integrity specification itself can be given, we must formalize isolation capabilities potentially integrated into trusted SWC code, such as the safety check discussed in Example 2.

Definition 7 (flow barrier). *A flow barrier between an input $p_1 \in P$ and an output $p_2 \in P$ of a SWC is a mechanism integrated into the SWC code to ensure that potentially faulty values received via p_1 do not have a safety-related effect on p_2. Since it is implemented by the SWC code writing to p_2, it automatically inherits the provided integrity specified for p_2.*

Flow barriers are an opt-in mechanism that should be used sparingly and require a careful safety analysis of relevant SWC code. With this definition, we can describe the full set of inputs to be provided by the designer:

Definition 8 (integrity specification). *The integrity specification is a tuple (B, ℓ_P, ℓ_R), where $B \subseteq P \times P$ (with $\varphi_0(p_1) = \varphi_0(p_2)$ for all $(p_1, p_2) \in B$) describes flow barriers, the function $\ell_P \colon V_P \rightarrow L$ maps every integrity provider to the integrity level it provides, and the partial function $\ell_R \colon V_R \rightharpoonup L$ can be used to map an integrity receiver to the integrity level it requires.*

```
1  ifc_pattern: {
2      lattice: "demo_lattice",
3      flow_barriers: [
4          { swc: "body_ctrl", output: "message", input: "driver_pos" },
5      ],
6      provided_integrity: [
7          { provider: "mpsoc", level: 3 },
8          { provider: "port", name: "vehicle_status.speed", level: 1 },
9          { provider: "port", name: "body_ctrl.message", level: 3 },
10         // 5 entries hidden for brevity...
11     ],
12     required_integrity: [
13         { receiver: "env", port: "body_ctrl.message", level: 1 },
14     ],
15 }
```

Listing 1. Sample integrity specification using the JSON5 notation of XANDAR

As part of this work, a textual JSON5 notation to express integrity specifications was developed and integrated it into XANDAR toolchain, which is built on Kotlin for the Java Virtual Machine (JVM). The notation is currently limited to numerical integrity lattices such as the one presented in Example 4.

Example 5. The excerpt in Listing 1 shows a possible integrity specification for the running example. It is based on an integrity lattice with $L = \{0, 1, 2, 3\}$, which is referenced as `demo_lattice` in line 2 of the specification. Based on this lattice, in line 13, the required integrity level of data `body_ctrl` writes to the environment via its `message` port is set to $\ell_P = 1$, for instance.

5 Pattern Implementation

Based on the input set described in Sect. 4, which includes an integrity specification such as the one shown in Listing 1, the IFC pattern performs the three steps covered in the following subsections.

5.1 APU Configuration

As an integral part of the isolation strategy, the pattern automatically generates an APU configuration that is as prohibitive as possible but allows every RTE to access the resources assigned to it. In general, this is a platform-specific procedure that must be implemented for every supported MPSoC.

Example 6. In the deployment shown in Fig. 4, the set of memory-mapped registers allocated to RTE_2 was visualized as R_2. If, for example, this set contains a timer that RTE_2 needs to schedule SWCs and an I/O controller that one of its SWCs uses to read from a sensor, the APU must be configured accordingly. The pattern receives these requirements towards the APU configuration as part of Ψ_R and μ from the target deployment.

As a proof of concept, this work considers the APU configuration procedure for the Zynq UltraScale+ MPSoC [13]. On this platform, the aforementioned XMPU and a unit referred to as the XPPU complement each other, where the XMPU focuses on on-chip and external DDR memory, while the XPPU protects units such as timers, I/O controllers and system-level control registers. The address information encoded into each $x \in (\Psi_{M1} \cup \Psi_{M2})$ is therefore used to configure a corresponding region in the XMPU. The core cluster used to execute the software that μ or μ' point to is then given full (read/write) access to this region. Entries in Ψ_R are associated with a unique identifier (such as `ttc0` for a timer or `uart1` for a UART controller) rather than with address boundaries. Each identifier corresponds to a fixed aperture register of the XPPU. For all $x \in \Psi_R$, this aperture register is used to give the core cluster pointed to by $\mu(x)$ full access to the resource. For all remaining XMPU regions and XPPU apertures, read and write access by any master is disallowed.

The result of this design-time procedure is C code executable on any core cluster of the MPSoC; it writes to the XMPU and the XPPU to set up the automatically generated protection. The task of integrating this code into the system implementation is finally delegated to the XANDAR toolchain.

Note that the integrity specification is not required to generate the APU configuration. This step depends solely on the availability of the target deployment from Definition 1 and, implicitly, knowledge of the SWCs.

5.2 Integrity Propagation

The goal of this step is to determine the effective integrity of every $v \in V$, where $V = V_P \cup V_R$ is referred to as the set of integrity vertices.

Definition 9 (effective integrity). *The effective integrity of $v \in V$ is the greatest $\ell' \in L$ that fulfills the following condition: for every $u \in V_P$ with a potentially safety-related impact on v, $\ell' \leq \ell_P(u)$ holds.*

The result of the integrity propagation will therefore be a function $\ell' : V \to L$. Its determination is performed based on the following graph:

Definition 10 (flow graph). *In the flow graph $G = (V, E)$ of integrity vertices, a directed edge $(q, r) \in E$ represents a direct information flow path limiting the effective integrity of r to $\ell'(q)$, i.e., the effective integrity of q.*

The procedure to generate G from a software architecture and a target deployment is given in Algorithm 1, which will be explained in the following.

Algorithm 1. Creation of the flow graph $G = (V, E)$

1: $V \leftarrow V_P \cup V_R$, $E \leftarrow C$ ▷ Vertices and explicit channels
2: $E \leftarrow E \cup \{(k_0, k(\omega)) : \omega \in \Omega\}$ ▷ MPSoC to all RTEs
3: $E \leftarrow E \cup \{(k(\omega), p) : p \in P, \omega = \lambda(\varphi_O(p))\}$ ▷ RTE to all of its ports
4: $E \leftarrow E \cup \{(q, r) \in P_{\text{IN}} \times P_{\text{OUT}} : \varphi_O(q) = \varphi_O(r)\} \setminus B$ ▷ SWC-internal flows
5: $E \leftarrow E \cup \{(v, e(v)) : v \in V_I\} \cup \{(e(v), v) : v \in V_O\}$ ▷ External inputs/outputs
6: **for all** $x \in \Psi_R$ **do** ▷ Implicit paths via registers
7: $\quad \Delta \leftarrow \text{ReachableUnits}(x)$
8: $\quad$ **if** $|\Delta| \geq 0 \wedge \mu(x) \notin \Omega$ **then return** null
9: $\quad E \leftarrow E \cup \{(k(\mu(x)), \delta) : \delta \in \Delta\}$
10: **return** (V, E)

Since every channel facilitates information flow that can have a safety-related impact on the receiving input port, edges from C are transferred to E in line 1. This assignment makes use of the premise that by default, every runtime environment manages reads from and writes to SWC ports in a correct manner.

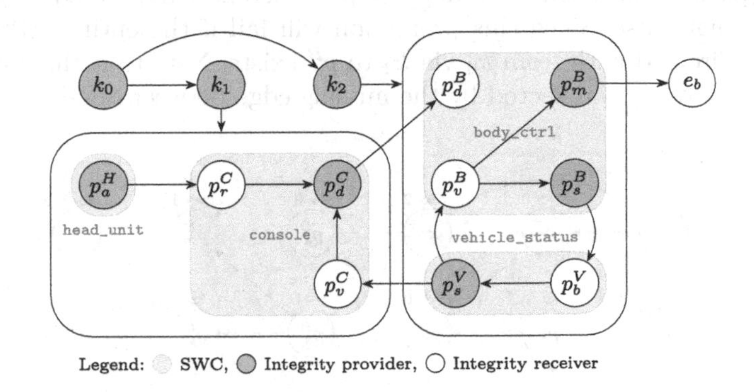

Fig. 6. Simplified flow graph G for the running example.

The edges added in lines 2 and 3 are related to $V_K = \{k_0, k_1, \ldots, k_n\}$, where k_0 represents the MPSoC and k_i with $i \in \{1, \ldots, n\}$ represents RTE$_i$. The helper function $k : \Omega \rightarrow V_K$ maps every RTE to its associated vertex in V_K. Since the MPSoC's provided integrity limits the effective integrity of every RTE, edges are added from k_0 to k_1 through k_n in line 2. Afterwards, for each $\omega \in \Omega$, an edge from this RTE to all ports of its executed SWCs is added. This represents the fact that whenever the RTE fails, it is no longer justified to trust, e.g., its scheduling of SWCs or management of SWC ports.

In line 4, every input-output pair of a SWC is connected unless a corresponding flow barrier is in place. The assignment in line 5 connects every external SWC port to the V_I or V_O vertex that represents the data read from or written to the environment. Here, the helper function $e : \{p \in P : \varphi_s(p) = \text{EXT}\} \rightarrow (V_I \cup V_O)$ maps each of these ports to its associated environment vertex.

Finally, the loop in line 6 creates edges to capture implicit information flows enabled by the assignment of registers to RTEs. Therefore, it makes use of the platform-specific ReachableUnits function, which returns a list of RTEs that are implicitly affected by writes to a given register. Since the current version of the flow graph does not model SWCs itself, registers with such side effects can be assigned to RTEs only. Line 8 ensures that this constraint is met.

Example 7. The graph in Fig. 6 is a simplified version of G for the running example. It is simplified in the sense that edges from an RTE vertex to its ports are not drawn individually; they are replaced by one edge from an RTE vertex to a wrapper around all ports mapped to RTE. The notation used to label a port makes use of a superscript letter to indicate its SWC and a subscript letter to indicate the name of the port within the letter. For example, the `speed` port of the `vehicle_status` is represented by the p_s^V vertex. Data written to the environment via p_m^B is represented by $e_b \in V_O$. It is important to understand that this graph reflects, among other things, the protection provided by the APU. This protection prevents RTE_1 from writing to the memory region that RTE_2 uses to implement the channel from p_s^V to p_v^B. Therefore, a directed edge from k_1 to p_v^B does not exist. Since this protection will fail if the entire MPSoC fails, however, a directed path from k_0 via k_2 to p_v^B exists. Note that the flow barrier between p_d^B and p_m^B is reflected by the missing edge between these vertices.

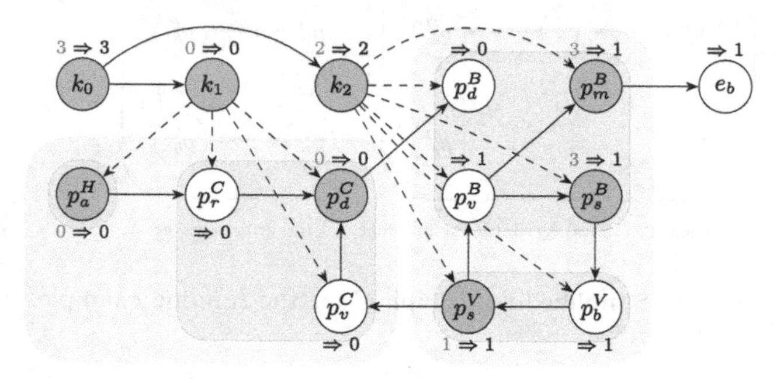

Fig. 7. Flow graph with provided and effective integrity levels.

The integrity propagation algorithm we apply to the flow graph has been adapted from the work in [7] and is shown in Algorithm 2.

Algorithm 2. Integrity propagation via flow graph edges

1: **for all** $v \in V$ **do** $\ell'(v) \leftarrow$ **if** $v \in V_P$ **then** $\ell_P(v)$ **else** $\top$
2: $Q \leftarrow$ Queue(V_P) ▷ Queue of vertices in V
3: **while** NotEmpty(Q) **do**
4: $\quad u \leftarrow$ Dequeue(Q) ▷ Get the next vertex to handle
5: $\quad$ **for** $v \in V: (u, v) \in E$ **do**
6: $\quad\quad \ell'_0 \leftarrow \ell'(v)$
7: $\quad\quad \ell'(v) \leftarrow \ell'(u) \land \ell'(v)$ ▷ Propagate its ℓ' to successor v
8: $\quad\quad$ **if** $\ell'(v) \neq \ell'_0$ **then** Enqueue(v) ▷ Enqueue v if its ℓ' value changed

First, the algorithm initializes the ℓ' function to provided integrity levels for vertices from V_P and to the greatest integrity level, $\top \in L$, for vertices from V_R. Using a queue, ℓ' values are then iteratively propagated through the flow graph until a steady state is reached.

Example 8. A possible result of the integrity propagation step for the running example is given in Fig. 7. Edges connecting an RTE vertex to its ports are now drawn using dashed arrows to set them off visually. Every $v \in V_P$ is labeled with the provided and the effective integrity level using the notation "$\ell_P(v) \Rightarrow \ell'(v)$". Analogously, every $v \in V_R$ is labeled with the notation "$\Rightarrow \ell'(v)$".

From the depicted ℓ' values, the effective integrity available to the body network, for example, can be obtained as $\ell'(e_b) = 1$. Various platform-specific details, such as the fact that the APU prohibits RTE_1 from accessing the memory regions assigned to RTE_2, contribute to this specific value.

5.3 Integrity Verification

Finally, the integrity verification step of the IFC pattern determines whether or not the specified system (with the APU configuration in place) meets its integrity requirements. Therefore, the pattern considers every integrity receiver $v \in V_R$, i.e., every SWC input port or environment writer. Whenever an integrity requirement is specified for such a vertex of the flow graph, the underlying integrity lattice is consulted to ensure that the effective integrity (propagated to the vertex using Algorithm 2) is not lower than the required level. By requiring all these checks to succeed, the pattern derives the following binary result:

Definition 11. *The* ***verification result*** *is a value* $\sigma \in \{\text{TRUE}, \text{FALSE}\}$*, which is* TRUE *if and only if the following condition is fulfilled: for each* $v \in V_R$ *such that* $\ell_R(v)$ *is defined,* $\ell_R(v) \leq \ell'(v)$ *holds.*

Example 9. For the effective integrity levels shown in Fig. 7, an integrity specification that contains $\ell_R(e_b) = 1$ as its only integrity requirement will lead to a positive verification result ($\sigma = \text{TRUE}$). If, however, the integrity required to write to the body network is raised to $\ell_R(e_b) = 2$, a negative verification result ($\sigma = \text{FALSE}$) is reported back to the XANDAR toolchain.

If inputs provided to the IFC pattern are accurate and the generated APU configuration is applied, a positive verification result guarantees the fulfilment of all specified integrity requirements. Therefore, the pattern reduces the challenging problem of implementing a logical isolation for software on complex MPSoC platforms to the more manageable task of defining flow barriers as well as provided and required integrity levels.

6 Conclusions and Future Work

To achieve a fine-grained on-chip isolation on heterogeneous MPSoCs, capabilities of APUs often need to be combined with safety mechanisms in application software and detailed knowledge about implicit information flow paths within the underlying platform, e.g., via a system-level control register.

The presented IFC pattern automates the APU configuration process and ensures that feasible information flows in the resulting system are acceptable from an integrity point of view. While timing interferences and confidentiality considerations are beyond the scope of this paper, we used an automotive example scenario to show that the pattern is successfully able to protect a safety-critical software component from failures caused by, e.g., a less trusted runtime environment running on the same multicore platform.

A promising direction for future work is to make the pattern applicable to a broader set of deployment strategies, e.g., by lifting the restriction to target deployments that map a SWC directly and statically to the RTE of a core cluster. Furthermore, a tool-supported integration with functional safety standards such as ISO 26262 [11] is another topic for future research.

Acknowledgments. This work is part of a project that has received funding from the European Union's Horizon 2020 research and innovation programme under grant agreement No 957210.

References

1. Abdellatif, T., Rouis, N., Saïdane, W., Jarboui, T.: Enforcing the security of component-based embedded systems with information flow control. In: 2010 International Conference on Wireless and Ubiquitous Systems (2010)
2. Bandur, V., Selim, G., Pantelic, V., Lawford, M.: Making the case for centralized automotive E/E architectures. IEEE Trans. Veh. Technol. **70**(2) (2021)
3. ter Beek, M.H., Cleophas, L., Schaefer, I., Watson, B.W.: X-by-construction. In: Margaria, T., Steffen, B. (eds.) ISoLA 2018. LNCS, vol. 11244. Springer, Cham (2018). https://doi.org/10.1007/978-3-030-03418-4_21
4. Ben Said, N., Abdellatif, T., Bensalem, S., Bozga, M.: Model-driven information flow security for component-based systems. In: Bensalem, S., Lakhneck, Y., Legay, A. (eds.) ETAPS 2014. LNCS, vol. 8415, pp. 1–20. Springer, Heidelberg (2014). https://doi.org/10.1007/978-3-642-54848-2_1
5. Denning, D.E.: A lattice model of secure information flow. Commun. ACM **19**(5), 236–243 (1976)
6. Dörr, T., Sandmann, T., Becker, J.: Model-based configuration of access protection units for multicore processors in embedded systems. Microprocess. Microsyst. **87**, 104377 (2021)
7. Dörr, T., Sandmann, T., Mohr, H., Becker, J.: Employing the concept of multi-level security to generate access protection configurations for automotive on-board networks. In: 2021 24th Euromicro Conference on Digital System Design (2021)
8. Hassan, M.: Heterogeneous MPSoCs for mixed-criticality systems: challenges and opportunities. IEEE Design Test **35**(4), 47–55 (2018)
9. Hu, W., Mu, D., Oberg, J., et al.: Gate-level information flow tracking for security lattices. ACM Trans. Des. Autom. Electron. Syst. **20**(1), 1–25 (2014)
10. Hu, W., Oberg, J., Irturk, A., et al.: Theoretical fundamentals of gate level information flow tracking. IEEE Trans. Comput.-Aided Design Integr. Circ. Syst. **30**(8), 1128–1140 (2011)

11. ISO 26262-1:2018: Road vehicles—Functional safety—Part 1: Vocabulary (2018)
12. Masing, L., Dörr, T., Schade, F., et al.: XANDAR: exploiting the X-by-construction paradigm in model-based development of safety-critical systems. In: 2022 Design, Automation & Test in Europe Conference & Exhibition (2022)
13. McNeil, S., Schillinger, P., Kolarkar, A., et al.: Isolation methods in Zynq UltraScale+ MPSoCs (2021). Xilinx, XAPP1320, v4.0
14. Oberg, J., Hu, W., Irturk, A., et al.: Information flow isolation in I2C and USB. In: Proceedings of the 48th Design Automation Conference (2011)
15. Pellizzoni, R., Meredith, P., Nam, M.Y., et al.: Handling mixed-criticality in SoC-based real-time embedded systems. In: Proceedings of the Seventh ACM International Conference on Embedded Software (2009)
16. Runge, T., Kittelmann, A., Servetto, M., Potanin, A., Schaefer, I.: Information flow control-by-construction for an object-oriented language. In: Schlingloff, B.H., Chai, M. (eds.) SEFM 2022. LNCS, vol. 13550, pp. 209–226. Springer, Cham (2022). https://doi.org/10.1007/978-3-031-17108-6_13
17. Saeed, A., Dasari, D., Ziegenbein, D., et al.: Memory utilization-based dynamic bandwidth regulation for temporal isolation in multi-cores. In: 28th Real-Time and Embedded Technology and Applications Symposium (2022)
18. Schaefer, I., Runge, T., Knüppel, A., Cleophas, L., Kourie, D., Watson, B.W.: Towards confidentiality-by-construction. In: Margaria, T., Steffen, B. (eds.) ISoLA 2018. LNCS, vol. 11244, pp. 502–515. Springer, Cham (2018). https://doi.org/10.1007/978-3-030-03418-4_30
19. Suh, G.E., Lee, J.W., Zhang, D., Devadas, S.: Secure program execution via dynamic information flow tracking. SIGOPS Oper. Syst. Rev. **38**(5) (2004)

From Standard to Practice: Towards ISA/IEC 62443-Conform Public Key Infrastructures

Michael P. Heinl[1,2(✉)], Maximilian Pursche[1,2], Nikolai Puch[2], Sebastian N. Peters[2], and Alexander Giehl[1,2]

[1] TUM School of Computation, Information and Technology, Department of Computer Engineering, Technical University of Munich, Munich, Germany
[2] Department Product Protection and Industrial Security, Fraunhofer AISEC, Garching near Munich, Germany
michael.heinl@aisec.fraunhofer.de

Abstract. Public key infrastructures (PKIs) are a cornerstone for the security of modern information systems. They also offer a wide range of security mechanisms to industrial automation and control systems (IACS) and can represent an important building block for concepts like zero trust architectures and defense in depth. Hence, the ISA/IEC 62443 series of standards addresses the PKI paradigm, but there is little practical guidance on how to actually apply it to an IACS. This paper analyzes ISA/IEC 62443 for explicit and implicit requirements regarding PKI deployment to provide a guideline for developing and operating a standard-conform PKI. For this purpose, the analyzed requirements and IACS-specific constraints are combined with current research and best practices. To assess its viability, a tangible PKI use case is implemented in a test environment. The evaluation of this use case shows that common IACS components are capable of supporting PKI, but that important features are missing. For instance, the handling of PKI turns out to be time-consuming and involves many manual operations, a potential factor to render large-scale operations impractical at this point in time.

Keywords: PKI · ISA/IEC 62443 · IACS · Security Engineering · Zero Trust

1 Introduction

The Industrial Internet of Things (IIoT) increasingly connects operational technology (OT) involved in production processes to the business IT network and even the internet. This enables new types of value creation, ranging from more efficient processes to individual made-to-order production. However, it also leads to an increased risk of cyber attacks. The ISA/IEC 62443 series of standards provides guidance on the question *what* has to be done to reduce such risks to an acceptable level considering the special characteristics of Industrial Automation and Control Systems (IACS), e.g., the long life cycles and rigorous availability requirements. One measure is to apply the paradigm of public key infrastructures (PKIs) to IACS. This way, security services, such as encryption and

J. Guiochet et al. (Eds.): SAFECOMP 2023, LNCS 14181, pp. 196–210, 2023.
https://doi.org/10.1007/978-3-031-40923-3_15

authentication, can be utilized by OT devices as well, establishing a basis for security models like defense-in-depth and zero trust architectures. The goal of this paper is to specify *how* the different aspects related to PKI should be applied by providing a guideline with recommendations for a PKI concept that can be practically used to secure IACS in accordance with ISA/IEC 62443. For this, the paper synthesizes current research, best practices learned from the WebPKI, and requirements from ISA/IEC 62443. The viability is then evaluated by deriving a PKI concept for a tangible use case within a physical IACS testbed. The corresponding research questions answered in this paper are:

RQ 1: Which ISA/IEC 62443 requirements are relevant for PKIs?
RQ 2: Which aspects have to be considered when designing a PKI for IACS?
RQ 3: To what extent is PKI currently supported in industrial environments?

2 Related Work

Scientific papers dealing with PKI in industrial environments are scarce. In 2007, Hanke [11] analyzed PKI use cases within IACS without considering ISA/IEC 62443 or its predecessor ISA99. A more recent paper from 2021 [44] provides recommendations for PKI in IACS with regards to post-quantum security. While some general recommendations may overlap with the suggestions given during the course of this paper, their focus is solely on post-quantum attacks that PKI systems might face and which cryptographic algorithms are needed to correspondingly protect IACS environments. Apart from these publications, other relevant papers focus on ISA/IEC 62443 threat analysis [9] and its application in engineering projects [23,24] without specifically addressing PKIs. There are other sectors like automotive which address domain-specific challenges by applying PKI [22], e.g., the U.S. Department of Transportation's Security Credential Management System (SCMS) [42]. Although some challenges seem to be similar, a thorough comparison is out of scope of this paper.

3 Requirements Analysis

In order to develop a guideline for an ISA/IEC 62443-conform PKI, it is necessary to extract the requirements the standard places on a PKI. The focus of this paper is on technical requirements [18,19]. Depending on the fact whether the asset owner itself or a service provider operates the PKI, additional policy requirements may apply [15,16]. It is important to note that ISA/IEC 62443 does neither consider a PKI as a System under Consideration (SuC) nor a component of an IACS. Rather, a PKI is considered a security measure. Therefore, requirements are rarely imposed on the PKI itself but on how the IACS components and subsystems shall interact with it. Nevertheless, directions for the design of a PKI can be directly or indirectly derived from the selected requirements covered in Table 1. In addition, it is paramount that the architecture of the PKI does not hinder any security- or safety-related operation and that it integrates

Table 1. PKI Architecture Requirements.

Requirement		Name	PKI-related Content
PKI Architecture Requirements	SR/CR 1.8	Public key infrastructure certificates	A PKI has to follow best practices and the company's Certificate Policy (CP). A Certificate Policy (CP), for example, defines network locations of PKI entities or trust store configurations. RFC 3647 [32] is explicitly mentioned for guidance. Secure processes for the operation of the PKI need to be in place and should not negatively affect the system's performance
	SR/CR 1.9	Strength of public key authentication	The IACS and its components must, e.g., be able to validate signatures and the chain of trust up until a trusted (CA) certificate and check the certificate's revocation status. Components must also ensure that the used key and signature algorithm follows cryptographic guidelines. The number of roots of trust (RoT) has to be minimized and the secrecy of private keys ensured. There is an RE requiring *hardware security for public key authentication*, e.g., TPMs
	SR/CR 2.8	Auditable events	IACS and components must be able to produce audit logs for security-related events, incl. PKI operations. An RE requires the IACS to have the capability to maintain a *centrally managed, system-wide audit trail* and export in standardized formats
	SR/CR 4.2	Information persistence	It must be ensured that decommissioned systems and components do not leak confidential information. This implies the presence of certificate/key life cycle processes including proper sanitization
	SR/CR 4.3	Use of cryptography	Cryptography and key management shall follow international standards and best practices, e.g., by U.S. National Institute of Standards and Technology [26,28]. The strength of keys and algorithms should be chosen appropriate to the information it protects
	SR/CR 5.1	Network segmentation	Network segmentation must be possible to support the zone model and to break connections between segments during an incident without essential functions failing. This means that zones, components, and PKI entities like CA or RA should withstand being cut off from each other. Two REs require *independence from non-control system networks* and *logical and physical isolation of critical networks* from non-critical networks
	SR 5.2	Zone boundary protection	Traffic between zones must be monitored and only allowed if necessary. REs require communication between zones being preventable in case of an incident (*island mode*) or operational failure (*fail close*)
	SR/CR 7.3	Control system backup	IACS, components, and PKI entities must be able to perform backups without endangering confidential information like private keys. This implies either the exclusion of keys or encryption of backups
	SR/CR 7.4	IACS recovery and reconstitution	Recovery to a secure state after disruption must be possible, incl. configuration loaded from backup. A PKI must ensure the valid state of time-sensitive data after reconstitution, e.g., renew expired certificates before resuming operation
PKI End Entity/Relying Party Requirements	SR/CR 1.5	Authenticator management	IACS and components shall be able to initialize, change/refresh, and protect authenticators (including certificates and corresponding private keys) from disclosure in transit and storage. Components that are unable to meet this CR cannot use PKI security functionalities
	SR/CR 2.11	Timestamps	Timestamps are required for audit logs. Two REs require IACS-*internal time synchronization* with a central time source and the *protection of time source integrity*, which is also necessary to check expiration of time sensitive PKI information like certificates or CRLs
	SR/CR 3.3	Security functionality verification	Testing of security functions must be supported, e.g., authentication and proper handling of revoked certificates as well as test cases for further PKI-related functions
	SR/CR 3.7	Error handling	Error handling shall support remediation without revealing sensitive information to adversaries
	SR/CR 4.1	Information confidentiality	The confidentiality of sensitive information at rest and in transit must be protected. This extends SR/CR 1.5 by including PKI process information and configuration to impede reconnaissance
	SR/CR 7.1	Denial of service protection	IACS and components shall be able to maintain essential functions in case of Denial of Service events and *manage communication load from component* flooding according to an RE. Another RE requires IACS to *limit DoS effects to other systems or networks*. Hence, PKI-related communication must not cause DoS events
	SR/CR 7.2	Resource Mgmt.	Security functions have to be managed in a resource-efficient way to prevent overload and delay
	HDR/EDR/ NDR 3.12	Provisioning product supplier RoT	Host, embedded, and network components must be capable of being provisioned with supplier's RoT and protecting their integrity, authenticity, and confidentiality
	HDR/EDR/ NDR 3.13	Provisioning asset owner roots of trust	These types of components must also be capable of being provisioned with and protecting the owner's RoT without reliance external to their own security zone

well into the environment of the IACS. As a basis for the actual requirements, ISA/IEC 62443 defines three common control system security constraints generalizing the high availability and integrity requirements of IACS [18]. The first constraint, *support of essential functions*, is crucial when designing a PKI for IACS as it dictates that security measures shall not negatively impact health, safety, environment (HSE), or the availability of the system. As an example,

ISA/IEC 62443 stipulates that failure of a PKI service shall not interrupt or significantly delay essential IACS communication. *Compensating countermeasures* means that security requirements the system or component should fulfill can also be fulfilled by an external component if an appropriate interface is given. In case of an PKI, this could be, e.g., an Online Certificate Status Protocol (OCSP) OCSP responder providing Validation Authority (VA) services. Eventually, *least privilege* specifies that permissions concerning resources and information of the IACS must only be mapped to a specific role if they are necessary to fulfill the role's intended purpose. Based on this, ISA/IEC 62443 defines seven Foundational Requirements (FRs) [18,21]. Each FR is detailed by a set of System Requirements (SRs) [18] and Component Requirements (CRs) [19]. SRs/CRs consist of a baseline requirement and possible Requirement Enhancements (REs). SRs and CRs are often similar but there can be differences. Moreover, some CRs have specific variations depending on the type of the respective component, i.e., software application (SAR), embedded device (EDR), host device (HDR), or network device (NDR). Each SR/CR is associated with Security Levels (SLs) [18]. These SLs range from SL 1 to SL 4. The higher the SL, the better the corresponding protection. To accomplish a higher SL, a system/component often needs to meet REs in addition to the baseline requirement.

4 PKI Guideline

This section discusses the major structural and procedural aspects of a PKI. A PKI primarily targeting Machine-to-Machine (M2M) authentication faces different challenges than a PKI targeting Human-to-Machine (H2M) (H2M) authentication. While a lot of architectural requirements are similar, an H2M PKI requires additional processes during operation, e.g., addressing identification and fluctuation of employees. Although employment of a PKI targeting H2M is a valid use case for IACS, it is beyond the scope of this paper.

ISA/IEC 62443's two central requirements directed towards PKI are SR/CR 1.8 describing how to handle PKI operation and 1.9 detailing how systems and components should interact with certificates. The most important point in SR/CR 1.8 is the requirement to follow best practices and a Certificate Policy (CP). It points towards RFC 3647 [32] that assists in writing a CP and a Certificate Practice Statement (CPS). A CP is a document that defines roles, duties, and requirements for entities within a PKI, for example, Certificate Authority (CA), Registration Authority (RA), and End Entity (EE). It also provides legal and liability statements. A CPS, on the other hand, is more practical and provides details on how a PKI meets the set requirements. To employ a PKI in an IACS to the best possible extent, one would have to first define a CP based on security standards, best practices, and laws applicable to the industry. Subsequently, a CPS details how to fulfill each requirement set by the CP depending on the technical and organizational environment of the PKI. This paper discusses key points and gives recommendations to meet the requirements set by ISA/IEC 62443, enriched by best practices [4,5,7,8].

4.1 PKI Structure

A fundamental consideration that needs to be evaluated is whether to integrate the own PKI into a public PKI or to utilize a private PKI. SR/CR 1.8 mentions both possibilities. A public PKI has the advantage that a lot of the security recommendations are ideally already met, revocation procedures are in place, and certificates issued by most commercial CAs are publicly trusted, meaning their root CA certificate is present in most trust stores. Component and system configuration with such certificates is usually easier. This is especially advantageous, if EEs are to provide a public service or communicate with third-party entities. It means, however, to invest a certain amount of trust in the used CA. A way to qualitatively evaluate and compare the trustworthiness of CAs is to analyze their CPs and CPSs. Complementarily, Heinl et al. [12] demonstrated a method of assessing trustworthiness of CAs quantitatively. Utilizing a private PKI provides more sovereignty in terms of architecture, procedures, and operation. It is more transparent and thus trustworthy to the operator at the cost of securing such PKI is the operator's own liability and will take more effort.

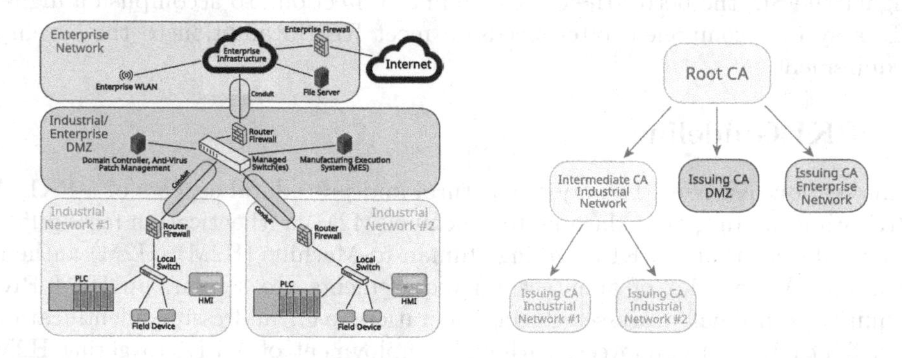

Fig. 1. Simplified CA hierarchy mapped to zone model based on ISA/IEC 62443 [18].

PKI Hierarchy. A PKI is inherently hierarchical with trust delegated from the root CA down to EE certificates. The recommended hierarchy structure includes at least three levels: a root CA, issuing CA(s), and EE(s). In a more complex PKI, there are often additional intermediate CAs. The root CA is the central trust source and should not be used to sign EE certificates [5]. Instead, it issues SubCA certificates that can be used as intermediate or issuing CAs. Intermediate CAs can, e.g., be used for different company branches. Their main advantage is the division of responsibility in a complex technical or organizational structure. Issuing CAs sign and issue EE certificates. In the context of ISA/IEC 62443, this hierarchy is advantageous, since the IACS environment is divided into zones according to use case and security requirements. If applicable, each zone should be provided with its own issuing CA, exemplified in Fig. 1. This allows revocation of an entire zone without affecting other zones in case of a security incident [44] as implicitly required by SR/CR 5.1 [18,19].

Roles and Responsibilities. When designing a PKI, trusted roles as well as their responsibilities and necessary permissions need to be identified on every level of the PKI hierarchy. While it may not be feasible that every role is covered by a different person, separation of duty and the principle of least privilege are crucial to prevent misuse of power and to provide non-repudiation among privileged roles [4,8]. For example, a role only entrusted with certificate issuance should only be able to use but not to read/export the issuing CA's private key. If relevant employees leave or change roles within the company, administration keys and passwords have to be changed [4]. Since PKI centralizes trust, CAs (especially the root CA) represent a potential single point of failure (SPoF) and special protection must be in place to secure them.

Certificate Profiles. Certificates should only be issued for a certain purpose defined in a certificate profile encompassing the X.509v3 certificate fields and extensions [33]. After an ISA/IEC 62443 risk assessment [17] comes to the conclusion which PKI services have to be employed in which zone, corresponding certificate profiles must be defined, depending on PKI hierarchy level as well as zone. There are three important extensions restricting what certificates can be used for. The first one is the *Basic Constraints* extension. It includes the `CA` field, indicating if the key pair can be used to sign other certificates. This must never be true for EE certificates, but needs to be set for CA certificates [5,33] along with the `KeyCertSign` and `CRLSign` bits of the second extension, *Key Usage*. If OCSP is utilized by this CA, `DigitalSignature` must be set, too. Other purposes should not be set for CA certificates to ensure that they are used for signing certificates and revocation procedures only [5]. For EE certificates, this extension should be adjusted to the intended application. For example, a typical TLS certificate will have `DigitalSignature` and `KeyEncipherment` set. Another use case leveraging `DigitalSignature` in combination with `NonRepudiation` would be signing audit logs sent to a central log server (cf. SR 2.8 RE 1). The final extension, *Extended Key Usage*, contains use case restrictions like server or client authentication within the TLS protocol and is not limited to the options specified in RFC 5280 [33]. While certificates that are used within a public context must adhere to this specifications, some extensions can be repurposed in a private IACS environment, e.g., to authenticate license keys. However, if an RFC 5280-compliant certificate parser cannot process an extension with the `critical` flag set to `true`, it will reject such certificate [33].

Network Segmentation. When a network utilizes PKI for a critical service, e.g., secure communication between critical components, it should not cause loss of availability when communication between network segments is broken or the zone goes into island mode. E.g., a network that should fulfill SR 5.1 RE 2 will need to have its issuing CA and revocation method within its own segment. To reduce the exposure of the PKI when CAs are deployed to every zone, it is useful to restrict the certificates a SubCA can issue. The field `PathLenConstraint` in the *Basic Constraints* extension enforces a maximum number of CA certificates that may follow in the certificate chain [33]. This should be 0 for all issuing CAs, since they shall only issue EE but not CA certificates. The extension *NameConstraints* restricts the name space for which a CA can issue certificates. To utilize it in an IACS, its name spaces should reflect the zone partitions.

Computer Security. Computers hosting PKI services must be hardened, e.g., running only tested and trustworthy software, changing or disabling default accounts/credentials [4,7], and require personalization and multi-factor authentication for all privileged roles [8]. Patch management shall be established [20] and security patches be implemented no later than six months after they became available, unless they conflict with other functionality or dependencies [4].

Physical Security. Zones or networks with high security requirements, e.g., an offline root CA, should be physically protected. E.g., only authorized personnel should have access, every entry and exit be logged, and portable media containing sensitive information not be brought out without authorization [4,8].

Monitoring and Logging. Monitoring must cover all PKI entities, incl. network reachability, CPU utilization, disk capacity, and logging processes [4,7]. In case of failure, responsible personnel should be alerted. Audit logs are a common way to detect security incidents and to provide accountability [7]. PKI-related events, e.g., certificate issuance, revocation, as well as general security events should be logged [5] and centrally aggregated (cf. SR 2.10 RE 1). This allows the detection and timely response to events. The retention period should be sufficiently long, e.g., at least two years [5], to enable post-incident forensics.

Cryptographic Recommendations. A PKI is built upon cryptographic primitives with a lot of research and development going on. In general, commonly accepted recommendations, e.g., by NIST [26–28] or the BSI [1–3], shall be followed to stay on top of these developments. However, there are environment-specific aspects, e.g., regarding key pair generation or signature algorithm, which have to be considered in the design phase of the PKI due to potential trade-offs between security and factors like performance, latency, and cost.

RSA has the advantage that most networked IACS components support it out of the box. Drawbacks are relatively long public keys and computationally expensive key operations like signature generation and decryption [25,43]. This stands in contrast to real-time constraints of IACS. The most prevalent alternative to RSA is Elliptic Curve Cryptography (ECC) needing a significantly smaller key size to achieve a similar effective key length (*security strength*), e.g., 256 bit (ECC) compared to 3072 bit (RSA) for a security strength of 128 bit [1,3,28]. ECC is often faster (with exceptions, e.g., signature validation), more energy-efficient, and the shorter key lengths make key handling easier for components [25,43]. Hence, depending on the application, RSA can be recommended for heterogeneous environments and a focus on time-critical signature validation whereas ECC can be recommended if there are constraints regarding storage, bandwidth, power consumption, and little to no legacy devices. It is recommended to use key lengths with at least 128 bit of security strength for both RSA and ECDSA as well as a SHA-2 or SHA-3 hash function with the same security strength for digital signatures [3,28]. For environments requiring post-quantum (PQ) security, entirely different algorithms have to be used [1,44]. Methods which can provide a smooth transition to PQ cryptography include hybrid certificates [30] and mixed certificate chains [31].

Long-term keys should be stored in a trusted platform module (TPM) or a hardware security module (HSM). They provide hardware-based protection and often functionality like binding a key pair to a device (TPM) or a multi-user authorization scheme (HSM) for very sensitive keys, e.g., of the root CA.

4.2 PKI Processes

Besides the structure, procedural aspects must also be evaluated.

Deploying EE Certificates. After defining a certificate profile, the key pair and a corresponding certificate signing request has to be generated. *Key generation* can either be done by the EE or by the CA in case the EE itself is not able to due to a missing cryptographically secure random number generator. The latter allows for key recovery, however, it also impairs non-repudiation. Once the certificate is signed, the EE certificate, the certificate chain, and the keys must then be transported to the EE via a secure channel [8]. While it seems desirable to deploy certificates to as many devices and utilize them as often as possible, their employment must be carefully considered. ISA/IEC 62443 classifies keys as authenticators and sensitive data which results in additional operational requirements. If a device does not handle sensitive data in terms of confidentiality or integrity, the operational effort by handling certificates may outweigh optional security functionality.

Long *validity periods* may be acceptable in IACS. If an ISA/IEC 62443 risk assessment [17] comes to a different conclusion for specific zones, then regular renewal of certificates may be necessary. In this case, automation protocols, like SCEP [37] or EST [36], which allow EEs to automatically obtain a certificate from a CA, should be taken into consideration. If components support such automatism, it can reduce operational effort and minimize human error while enabling short certificate life times. However, it must be ensured that the employed mechanisms meet the set requirements, especially regarding availability. Otherwise, manual renewal of certificates can mean serious operational overhead and even downtime. SR/CR 1.9 does not explicitly cover checks of a certificates' validity period because for some applications with very high availability requirements, communication with an expired certificate can be more acceptable than unsecured or no communication at all. In these cases, expired certificates may be accepted as a fallback under exceptional circumstances [29] as long as they are not revoked and their keys provide adequate protection.

Revocation. Revocation mechanisms like Certificate Revocation Lists (CRLs) [33] and OCSP [35] fulfill the purpose of indicating lost or otherwise compromised key material, that should not be accepted or used by any entity within the PKI [5,8]. While validity period checks may not be necessary within IACS, revocation status must be evaluated in every step of the certificate path (cf. SR/CR 1.9). CRLs are the basic form of revocation and are relatively independent from CA uptime or other PKI services by deploying the lists to every EE. Their main disadvantage is the update and maintenance process. A CRL can grow quite large and must be redeployed to every component once another certificate is added. Delta CRLs only containing certificates revoked since the last base CRL [33] can be an alternative to reduce overhead. In static environments, where communication only happens within a security zone, the disadvantages of CRLs might not matter as much, since they will be short and updates are unlikely. The most prevalent alternative to CRLs is OCSP, centralizing revocation checking in an OCSP reponder. EEs query the responder for the status of certificates they process, decreasing storage and computation effort of EEs but increasing network traffic and representing a potential SPoF. OCSP stapling [34] can solve some of the potential problems of OCSP by offering a signed and timestamped revocation status to certificate holders who can then present it to relying parties during authentication. This way, the revocation status can be verified without direct communication with the OCSP responder decreasing the communication load and availability constraints. Currently, OCSP stapling is only standardized as a TLS extension, but the principle could be used in other protocols as well.

Backup and Recovery. PKI entities need to be included in backup procedures following commonly accepted standards [4,15]. Confidential information has to be excluded from a component backup or encrypted [8,18,19]. Recovery procedures should be regularly tested, to ensure that IACS can resume operation [5] even if certificates are invalid at the time of restore. In such a case as well as for component backups excluding private keys, a certificate issuing process should be included into the restore procedure. There might be PKI entities depending on each other's configuration, for example an OCSP responder and the corresponding CA, that have to be backed up and restored together. Such dependencies should be analyzed and documented [7].

End of Life Procedures. When devices reach their end of service, processes must be in place to ensure that sensitive information, e.g., private keys, is purged (cf. SR/CR 4.2), including from backups and potential redundant systems [3,5,8]. If it cannot be ensured that all copies of a private key are destroyed, the corresponding certificate has to be revoked in addition.

Compliance and Auditing. Risk assessments, zone definitions, and usage evaluations should not only be done before initial deployment, but regularly [17].

Guidelines and requirements for PKIs will change and should be reviewed on a regular basis in order to incorporate them into the security program.

5 Implementation

This section describes a practical PKI implementation with the goal to test the feasibility of PKI usage in a representative IACS testbed. It builds on previous research [10], which identified possible attacks on the present IACS. The implementation focuses on preventing one of the identified attacks by implementing TLS on top of an existing communication protocol and outlines possible challenges that need to be considered when employing PKI in an IACS. Due to the feasibility study character and the limited scope of this format, a risk assessment as prescribed by ISA/IEC 62443 3-2 [17] is intentionally omitted.

The testbed represents a small production facility consisting of three production isles. Each isle is made up of a base and an application module. Modules are composed of components, with the base usually consisting of a main Programmable Logic Controller (PLC) *SIMATIC ET 200SP*, a Human-Machine Interface (HMI) *SIMATIC HMI TP700 Comfort*, and a conveyor belt. The application modules have components specific to their task and may contain additional PLCs. The main PLCs are connected to a router via Ethernet, which connects them to the Manufacturing Execution System (MES) which is the central control unit of the IACS. The MES runs the *MES4* software by Festo for managing manufacturing orders as well as the *TIA* portal (version 15.1) and *CODESYS* (version 3.5.14) to program the PLCs of the associated manufacturer. The implementation focuses on the station shown in Fig. 2, housing a storage application and marking the starting point of the manufacturing process. A smaller version of the main PLC from Siemens (*SIMATIC S7-1200*) controls the storage unit. The MES and the main PLC communicate via TCP/IP. The demonstrated attack [10] targets this communication channel by initially obtaining a MitM position using ARP spoofing and then altering data sent on the TCP layer. This results in full control of the application, including picking wrong starting materials from the storage without the MES noticing. To prevent this kind of attack, TLS is implemented.

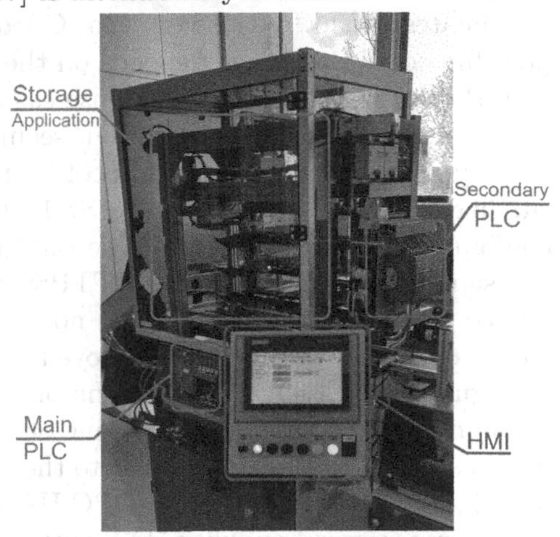

Fig. 2. Testbed setup.

5.1 Existing PKI Interfaces

Before implementing TLS, existing interfaces for certificate deployment are identified to provide an overview of the extent to which PKI can be deployed and which use cases are already implemented. The MES is Windows 10-based allowing to import any X.509v3 certificate to authenticate users, e.g., via smart card [13], using the Windows certificate utility. The MES software (MES SW) does neither have any documented certificate interfaces nor does it utilize the built-in Windows certificate store. The SIMATIC ET 200SP has a certificate store that can be configured via the TIA portal's central certificate manager [39]. It has two modes: the centrally managed, project-wide certificate store and an independent mode. When using the independent mode, certificates cannot be imported and only self-signed certificates can be created. The centrally managed certificate store comes with an own root CA and allows the issuance of device certificates signed by this very root CA or import of other certificates. A PLC's possible certificate usage depends on the built-in CPU. The used SIMATIC ET 200SP allows for four certificate use cases: TLS communication either as server or client, OPC UA authentication, securing PLC to HMI communication, and employment of the HTTPS protocol for the PLC-hosted web server [40,41]. The smaller SIMATIC S7-1200 has a S7-1200 CPU built in and does not have any configurable certificate store. The only documented certificate utilization is a self-signed certificate used for HTTPS access to the web server running on the device [38]. This certificate could not be configured, regardless of whether the global certificate manager is employed or not. The Festo PLC CECC-LK can be configured via CODESYS with the option to secure the production PLC code with an X.509 certificate, encrypting or signing the project by utilizing the Windows certificate manager. Similar to the TIA portal, the used CODESYS version 3.5.14 supports certificates for OPC UA but no other use cases. In 2020, version 3.5.16 was released enabling TLS with configurable certificates [6].

5.2 Selecting a PKI Tool

In order to make an informed tool selection, the landscape of open-source PKI software solutions is analyzed regarding *security*, *usability*, as well as *scalability and integration*. Table 2 shows the results of the four most promising candidates, namely Dogtag, EJBCA, OpenXPKI, and Step-CA, indicating that none of them can be seen as a clear favorite. Eventually, EJBCA is chosen for the implementation, especially due to its modular architecture.

Table 2. Open-source PKI tool decision matrix.

Groupings	Selection Criteria	Weights	Dogtag		EJBCA		Step-CA		OpenXPKI	
			Rating	Weighted	Rating	Weighted	Rating	Weighted	Rating	Weighted
Security	Confidentially (in transit and DB)	25,00 %	5	1,25	5	1,25	5	1,25	5	1,25
	Integrity in Database	25,00 %	5	1,25	5	1,25	5	1,25	5	1,25
	Access and Rights Management	25,00 %	5	1	5	1,25	2	0,5	5	1,25
	Release and Patch Cycle	25,00 %	4	1	5	1,25	5	1,25	4	1
Usability	GUI Usability	20,00 %	4	0,8	4	0,8	0	0	4	0,8
	CLI Functionality	20,00 %	4	0,8	4	0,8	5	1	5	1
	Documentation	35,00 %	4	1,4	4	1,4	4	1,4	3	1,05
	Vendor Support	25,00 %	1	0,25	5	1,25	5	1,25	5	1,25
Scalability and Integration	Variety of Supported Components	15,00 %	4	0,6	3	0,45	4	0,6	5	0,75
	Automation Possibilities	25,00 %	4	1	4	1	5	1,25	5	1,25
	Availability Concepts	30,00 %	3	0,9	4	1,2	2	0,6	3	0,9
	Multi Instance Operation	30,00 %	5	1,5	4	1,2	3	0,9	4	1,2
Total Score				11,75		13,1		11,25		12,95

5.3 PKI Installation and Configuration

EJBCA is installed including CA, RA, and VA functionality. Subsequently, the PKI is configured including the generation of certificate profiles as well as root CA and SubCA certificates. Eventually, two PKCS #12-formatted EE certificates are issued in the RA web GUI and the CA certificate chain is exported. Due to a lack of support for management protocols, e.g., SCEP or EST, the certificates have to be transferred to the TIA portal to assign them to the SIMATIC PLCs. For this purpose, the previously exported CA certificate chain as well as the PKCS#12 file containing the certificate and private key for the PLC are imported, requiring a restart of the device. The TIA portal limits the usage of certificates to RSA with SHA-1 or SHA-2. Certificates issued by the TIA portal's own CA use SHA-1 and 2048 bit RSA keys by default. The max. length for keys generated by the TIA portal is 3072 bit but it can handle larger keys if certificates are imported. Attempts to import ECDSA keys/certificates were rejected. Neither CRLs nor any other revocation method are supported [41].

5.4 Communication Configuration

The communication between PLC (client) and MES SW (server) utilizes two TCP/IP sessions. One session is used for order inquiries (query session) while the other transmits status information to the MES (state session). Once the query session is established, it is only used when a sled reaches the storage application. The PLC then queries the MES SW for instructions regarding the next operation. The state session communicates runtime information to the MES including production mode and error codes. Mutually authenticated TLS is implemented on both sessions to secure communication between MES SW and PLC. Since

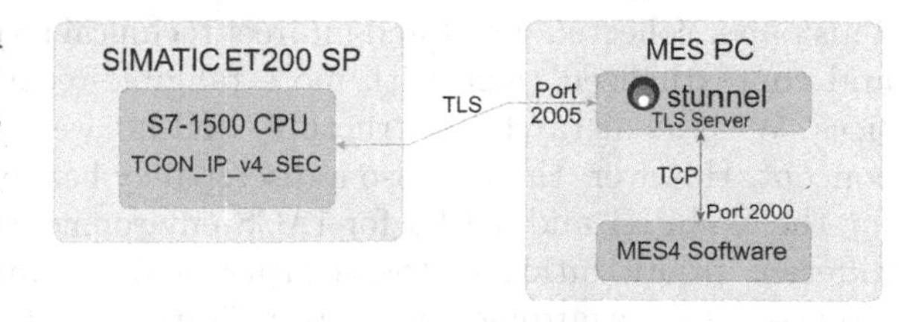

Fig. 3. TLS communication with the help of *stunnel.*

the existing proprietary MES SW does not provide TLS capabilities, the TLS wrapper *stunnel* is used on the MES to tunnel the unencrypted communication through a secure channel as illustrated by Fig. 3. The SIMATIC PLC is programmed via the TIA portal which utilizes STEP 7, an IEC 61131-compliant [14] software for programming PLCs. In STEP 7 V14, the datatype `TCON_IP_V4_SEC` was added to support TLS 1.2. The unsecured MES communication took place on TCP ports 2000 (query session) and 2001 (state session). These ports are now used internally via loopback interface, while TCP ports 2005/2006 are used for the external TLS connection via *stunnel*. Since revocation checking is not supported by the SIMATIC ET 200SP, it is done server-sided. *stunnel* is therefore configured with a CRL that is checked when a certificate is verified.

6 Evaluation

The probably most serious limitation for a PKI implementation within the present IACS, is the inability of the PLC to check certificate revocation status. With no standardized revocation method supported, the only available mechanism to limit the damage compromised keys can cause, are short certificate lifetimes. Maintaining short lifetimes, in turn, is not a trivial task because certificates can only be imported manually into the PLCs. Moreover, the PLCs need to reboot on reconfiguration, halting production. This makes short certificate lifetimes operationally unmanageable in a complex IACS comprising many similarly constrained devices. Another missing feature is the support of ECC. The shorter keys and faster computations could lower the strain on components' resources. Interestingly, the PLC uses ECDHE for the session key exchange during the TLS handshake which suggests that ECC operations are implemented in the PLC's firmware. Other security-relevant aspects are the missing crypto agility and weak default parameters of the used TIA portal version.

7 Conclusion

This paper collected direct and indirect technical requirements of ISA/IEC 62443 and contextualized them with more tangible recommendations and best practices. With its hierarchical structure, PKI fits well into the ISA/IEC 62443 zone concept. However, there is also a discrepancy between the security requirements for the WebPKI and a PKI for IACS environments. This is mainly due to the different prioritization of the security goals resulting in some interesting differences, e.g., regarding certificate validity periods. The implementation shows that it is possible to implement a PKI use case with common IACS components. However, it also reveals that PKI support is rudimentary and lacks important features, e.g., certificate revocation. Overall, it confirmed the impression that IACS are only slowly evolving due to their proprietary devices and long life cycles. However, the rapidly increasing importance of ISA/IEC 62443 suggests that stakeholders are aware of these circumstances, which in turn might also lead to more capable components enabling fully compliant PKI deployments. It must be considered that recommendations and requirements are not static. Once ISA/IEC 62443 or best practices change, this guideline has to be revised.

Acknowledgment. This work was supported by the German Federal Ministry for Economic Affairs and Climate Action (BMWK) under grant no. 13I40V010A.

References

1. BSI: Kryptographische Verfahren: Empfehlungen und Schlüssellangen (2022)
2. BSI: Kryptographische Verfahren: Empfehlungen und Schlüssellangen Teil 2 - Verwendung von Transport Layer Security (TLS) (2022)
3. BSI: Kryptographische Vorgaben für Projekte der Bundesregierung Teil 4: Kommunikationsverfahren in Anwendungen (2022)
4. CA/Browser Forum: Network & Certificate System Security Requirements (2021)
5. CA/Browser Forum: Baseline Requirements for the Issuance and Management of Publicly-Trusted Certificates (2022)
6. CODESYS GmbH: Features and Improvements CODESYS V3.5 SP16 (2020)
7. ETSI EN 319 401 V2.3.1: Electronic Signatures and Infrastructures; General Policy Requirements for Trust Service Providers (2021)
8. ETSI EN 319 411-1 V1.3.1: Electronic Signatures and Infrastructures; Policy and security requirements for Trust Service Providers issuing certificates; Part 1: General requirements (2021)
9. Fockel, M., et al.: Designing and integrating IEC 62443 compliant threat analysis. In: EuroSPI 2019 (2019)
10. Hagen, B.: Security analysis of an interconnected industrial automation testbed (production line). Master's thesis, Hochschule Augsburg (2022)
11. Hanke, M.: Embedded PKI in industrial facilities. In: ISSE/SECURE 2007 (2007)
12. Heinl, M.P., et al.: MERCAT: a metric for the evaluation and reconsideration of certificate authority trustworthiness. In: CCSW 2019 (2019)
13. Hughes, L.E.: Issue and manage windows logon certificates. In: Pro AD Certificate Services: Creating & Managing Digital Certificates for Use in MS Networks. Apress (2022)
14. IEC 61131-3:2013: Programming languages (2013)
15. IEC 62443-2-1:2010: Establishing an IACS security program (2010)
16. IEC 62443-2-4:2015: Sec. program requirements for IACS service providers (2015)
17. IEC 62443-3-2:2020: Security risk assessment for system design (2020)
18. IEC 62443-3-3:2013: System security requirements and security levels (2013)
19. IEC 62443-4-2:2019: Technical security requirements for IACS components (2019)
20. IEC TR 62443-2-3:2015: Patch management in the IACS environment (2015)
21. IEC TS 62443-1-1:2009: Terminology, concepts and models (2009)
22. Khan, S., et al.: Survey on issues and recent advances in vehicular public-key infrastructure (VPKI). IEEE COMST **24**(3) (2022)
23. Leander, B., et al.: Applicability of the IEC 62443 standard in Industry 4.0/IIoT. In: ARES 2019 (2019)
24. Maidl, M., et al.: A comprehensive framework for security in engineering projects - based on IEC 62443. In: IEEE ISSREW 2018 (2018)
25. Maletsky, K.: RSA vs. ECC Comparison for Embedded Systems (Microchip) (2020)
26. NIST: FIPS 140-3: Security Requirements for Cryptographic Modules (2019)

27. NIST: SP 800-57 Part 2 Rev. 1 - Recom. for Key Management: Part 2 - Best Practices for Key Management Organizations (2019)
28. NIST: SP 800-57 Part 1 Rev. 5 - Recom. for Key Management: Part 1 - General (2020)
29. OPC UA Foundation: Practical Security Recommendations for building OPC UA Applications. Whitepaper Security Working Group (2018)
30. Paul, S., et al.: Towards post-quantum security for cyber-physical systems: integrating PQC into industrial M2M communication. In: ESORICS 2020 (2020)
31. Paul, S., et al.: Mixed certificate chains for the transition to post-quantum authentication in TLS 1.3. In: ASIA CCS 2022 (2022)
32. RFC 3647: Internet X.509 PKI Certificate Policy & Certification Pract. Framew. (2003)
33. RFC 5280: Internet X.509 PKI Certificate and CRL Profile (2008)
34. RFC 6066: Transport Layer Security (TLS) Extensions: Extension Definitions (2011)
35. RFC 6960: X.509 Internet PKI Online Certificate Status Protocol (2013)
36. RFC 7030: Enrollment over Secure Transport (2013)
37. RFC 8894: Simple Certificate Enrolment Protocol (2020)
38. Siemens AG: SIMATIC S7-1200 Programmable controller (2015). https://cache.industry.siemens.com/dl/files/121/109478121/att_851433/v1/s71200_system_manual_en-US_en-US.pdf
39. Siemens AG: Using Certificates with TIA Portal (2019). https://support.industry.siemens.com/cs/attachments/109769068/109769068_CertificateHandlingTIAPortal_V1_0_en.pdf
40. Siemens AG: Config. of TLS-based PG/HMI Com. and the Protection of Confidential PLC Config. Data (2021). https://support.industry.siemens.com/cs/attachments/109772940/s71200_system_manual_en-US_en-US.pdf
41. Siemens AG: SIMATIC S7–1500, ET 200MP, ET 200SP, ET 200AL, ET 200pro Communication (2021). https://cache.industry.siemens.com/dl/files/942/84133942/att_1098064/v1/et200sp_manual_collection_en-US.pdf
42. U.S. Department of Transportation: Security Credential Management System (SCMS). https://www.its.dot.gov/factsheets/pdf/CV_SCMS.pdf
43. Vahdati, Z., et al.: Comparison of ECC and RSA algorithms in IoT devices. JATIT (2019)
44. Yunakovsky, S.E., et al.: Towards sec. recommendations for PKIs for production environments in the post-quantum era. EPJ Quantum Technol. **8**(1) (2021)

AI Safety

The Impact of Training Data Shortfalls on Safety of AI-Based Clinical Decision Support Systems

Philippa Ryan Conmy[1(✉)], Berk Ozturk[1], Tom Lawton[2], and Ibrahim Habli[1]

[1] Department of Computer Science, University of York, York, UK
{philippa.ryan,berk.ozturk,ibrahim.habli}@york.ac.uk
[2] Bradford Royal Infirmary, Bradford Institute for Health Research, Bradford BD9 6RJ, UK

Abstract. Decision support systems with Artificial intelligence (AI) and specifically Machine Learning (ML) components present many challenges when assuring trust in operational performance, particularly in a safety-critical domain such as healthcare. During operation the Human in/on The Loop (HTL) may need assistance in determining when to trust the ML output and when to override it, particularly to prevent hazardous situations. In this paper, we consider how issues with training data shortfalls can cause varying safety performance in ML. We present a case study using an ML-based clinical decision support system for Type-2 diabetes related co-morbidity prediction (DCP). The DCP ML component is trained using real patient data, but the data was taken from a very large live database gathered over many years, and the records vary in distribution and completeness. Research developing similar clinical predictor systems describe different methods to compensate for training data shortfalls, but concentrate only on fixing the data to maximise the ML performance without considering a system safety perspective. This means the impact of the ML's varying performance is not fully understood at the system level. Further, methods such as data imputation can introduce a further risk of bias which is not addressed. This paper combines the use of ML data shortfall compensation measures with exploratory safety analysis to ensure all means of reducing risk are considered. We demonstrate that together these provide a richer picture allowing more effective identification and mitigation of risks from training data shortfalls.

Keywords: Machine Learning · Training Data · Medical device safety

1 Introduction

Safety-related decision support systems incorporating Artificial intelligence (AI) and specifically Machine Learning (ML) components are increasingly being developed and deployed [18]. These can have many potential benefits, such as providing faster and richer computational support to complex tasks. However, developing a robust and fit-for-purpose ML algorithm is reliant on good training

J. Guiochet et al. (Eds.): SAFECOMP 2023, LNCS 14181, pp. 213–226, 2023.
https://doi.org/10.1007/978-3-031-40923-3_16

data, which reflects the required task. Even with a robust training regime, poor data will influence the performance, and safety of the output from the ML. Given that comprehensive verification of ML across all operating scenarios is typically impossible, these errors may be undetected until it is too late.

During operation the Human In/On The Loop (HTL) working with the system may need assistance in determining when to trust the ML output and when to override it, particularly in cases where there is a safety related outcome. For example, clinical advisory systems typically have a workflow allowing the clinician to override the output, but it may not be clear what the limitations or strengths are of the ML components, making it difficult to trust or ignore certain predictions [22]. This is particularly problematic where there is a difference of opinion between the ML predictor and clinician. Whilst there is research into the impact of different methods to manage training data shortfalls, these concentrate on maximising the ML performance with respect to certain metrics, and do not considering the different risks of varying performance at the system level. Thus the safety impact of data shortfalls is not well understood, nor are all means of reducing risk explored. We argue that taking a systems perspective is necessary for safety critical environments.

In this paper we examine how issues with training datasets, and means to compensate for them, can impact on safety performance. We combine the use of training data shortfall compensation methods and exploratory safety analysis to ensure all means of reducing risk are considered. We apply this combination to a diabetes comorbidity predictor (DCP), implemented using ML, used to support clinical decision making. The DCP is trained using a dataset which contains the real clinical records of the patients taken from the Connected Bradford database [21]. The dataset consists of over 42,000 rows of data for Type-2 diabetic patients from different backgrounds and over 14,000 different types of clinical records (features). Since the dataset records are obtained from different care centres, this causes differences in recorded data. When patients do not attend their visits regularly there can be changes/deficiencies in the recorded laboratory results, which causes the dataset to have a great number of missing values. This makes it critical to conduct systematic safety analysis to prevent and mitigate for misleading outcomes in the manner of patient safety.

This paper is laid out as follows. In Sect. 2 we describe this real world problem in more detail and describe some related work used to develop our approach. In Sect. 3 we describe the case study, training data issues and safety analysis. We discuss our results and findings in Sect. 4, and finally in Sect. 5 we present our conclusions.

2 ML Training Data and Safety

Increasingly, safety-critical systems with machine learning components are being developed and deployed [8]. Examples include autonomous cars (with or without a safety driver), drones, medical diagnosis systems and agricultural robots. There are many different approaches to machine learning, including supervised, semi-supervised and unsupervised training methods and models such as neural networks or decision trees. However, a core requirement for each is a set of valid training data which is pre-processed to tailor it for the task and model. For a safety-critical system, poor quality training data can lead to latent faults which can lead to hazardous behaviour. This is illustrated in Fig. 1. The top chain of elements indicates the ML training and system integration process. The lower row indicates how the error can propagate throughout the training lifecycle. A training data shortfall can mean the ML doesn't have complete or correct performance with respect to the system requirements. This may not be picked up in verification, as performing complete verification of ML is impossible in all but the most trivial cases. The same issue will affect testing during system integration but it may also be difficult to control the test space (e.g., real world testing of a drone cannot be done in controlled weather conditions), which means a latent failure could lead to a hazard during operation.

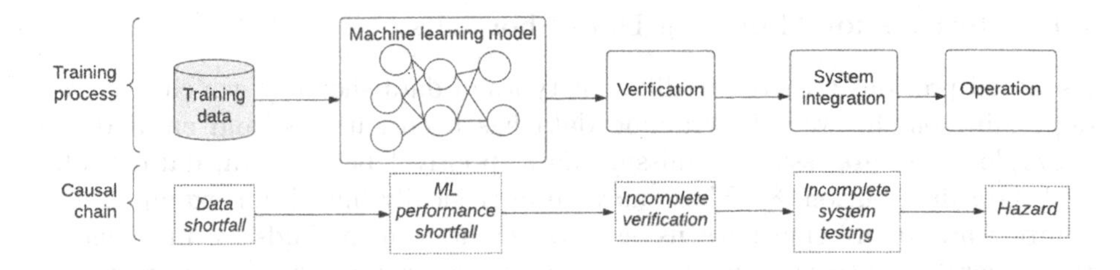

Fig. 1. Causal chain of failure events from training data shortfalls.

Consider the following examples. A classifier for an autonomous vehicle object detection system is trained using supervised learning. This uses a labelled set of training data, including images marked with labelled boxes. This data set includes a number of examples of dogs, but even though the training examples are properly labeled and framed, they only show dogs from a side view. The object detector may then fail to detect a dog facing forwards, contributing to a collision. A similar issue contributed to the fatal autonomous vehicle crash in Tempe, Arizona, 2018 [16], where a bicycle was not consistently recognised from the side, and the autonomous driving controller was unable to predict the path of the pedestrian pushing it quickly enough. Another example, would be a decision component determining whether an unmanned drone should return to base if the conditions are unfavourable may be trained using semi-supervised learning. For this, each training sample is marked as safe or unsafe, and contains a series of atmospheric readings on waypoints for the planned path. The ML training process is designed to allow it to look for patterns in the readings/predictions which can be matched to either safe or unsafe. However, the training set has very few samples where the temperature dipped below zero celsius, where icing could be a problem, and each of these samples contained very different sets of other readings making it hard to generalise. Therefore the ML might incorrectly decide it was safe to continue when in fact there was a severe risk the physical systems of the drone would fail and it would crash e.g., due to ice and low temperature impacting battery power.

When looking for shortfalls we need to consider the source of the data as this may impact on the types encountered. In some situations, the training data can be entirely user generated, such as when simulation software is used. The advantage of this approach is that the data scientist or engineer will have a very high degree of control over the data generated, but it may not be realistic without careful modelling and analysis. However, the opposite approach may be taken, where an off-the-shelf dataset is acquired and curated for the ML training. Real-world sampling will help assure the validity of the data, but the disadvantage is that there are likely to be missing cases or bias to certain situations, or even deliberate data poisoning. Our analysis considers both the normal case for data, where the sample may be valid but overall distribution introduces bias, and the failure case, where the sample may be corrupted in some way.

2.1 Methods for Managing Data Shortfalls

As noted previously there are different types of data shortfalls which may vary depending on the way the training data has been gathered and curated. For example, there are issues of missing data, poorly labelled data, data validity and data distribution [8]. These may be intrinsically linked, for example, if we compensate for missing data using data imputation methods we must ensure the generated data is valid. In this section we examine related literature on data imputation, concentrating on papers where it has been applied to similar prediction problems such as diabetes [7,11] and Covid-19 [4].

There are different types of data imputation methods to deal with the missing values, and these methods have been used for different domains. In [7,23] the authors take means of the full set of a particular feature to fill the missing values. However, only taking the average of the entire column and replacing the missing values with the average of the column may lead some bias or misleading outcomes. An issue with both these papers is that they focus purely on the ML performance indicators, and do not consider risk mitigation from a system safety perspective. Maximising the performance may not be required if other risk mitigation measures, such as explainability and transparency [13], are used.

In [5], the authors use kNN Imputation method to deal with the missing values in their dataset. In [1], four different imputation methods (case deletion, mean imputation, median imputation, and kNN Imputation) have been applied to compare these methods. Then, the authors concluded that the kNN Imputation performs better providing a better Mean Squared Error (MSE) value to deal with the missing values. In [24], multiple imputation methods have been compared, and it has been concluded that kNN Imputation has better than mean and median imputation methods. kNN imputation looks for similar cases and nearest neighbours, thus reduces bias from extreme outlying values or overall distribution. Again, the authors concentrate on ML fitness in isolation of the whole system.

An alternative method is described in [3,4,14] where the authors use the Bag Imputation method to fill the missing values in the dataset. This is a more sophisticated, and computer intensive, nearest neighbour method which uses additional ML to predict missing values, and to avoid overfitting and bias in the dataset [11]. Because we have a large amount of missing values and aim to prevent bias, we have decided to investigate bag imputation as a way to compensate for missing values in our dataset.

2.2 Safety Data Analysis Method

We argue that a system safety perspective is necessary to ensure that the risks associated with data shortfalls are methodically understood. By this we mean considering the impact on the effectiveness of the ML, and then considering how or if this might affect performance at the system level in combination with other information and actors. We also consider other activities during the training process which might reduce the risk. Further, we need to identify and assess the additional risks that using data imputation may introduce.

In Fig. 1 we show the ML training and operation lifecycle. There are opportunities to reduce risk at every stage, including controls on how training data is selected, adding specific verification/integration tests for known issues, and how information is presented to the operator, e.g., using explainability, so that they are given a richer picture for individual decisions.

To support the system safety perspective we need an exploratory safety analysis technique which could be effective in identifying types of data shortfall, such as for particular clinical features of importance and how they could propagate. Therefore, we considered bottom-up/inductive analysis safety analysis methods rather than top-down/deductive techniques e.g., Fault Tree Analysis. We argue that by concentrating on the data issues as a starting point we can understand their causal impact more holistically.

Typical inductive safety analysis methods include Failure Modes and Effects Analysis (FMEA) [12], and HAZard and OPerability Studies (Hazops) [15]. Hazops uses particular guidewords, e.g., more, less, early, to provide general categories of failures to engineers performing the analysis. It was originally used in the chemical processing industry but has been successfully used for computer based analysis, both on data flows (such as training data to ML training) and on control flows. On the other hand FMEA is more traditionally applied to physical system safety so we did not consider it further. In [15,19] the authors have successfully used Hazops to identify safety issues in systems with ML. An alternative version to Hazops is Software Hazard Analysis and Resolution in Design (SHARD) [12] is demonstrated in [9] for a medical decision support system. We note the findings in [15] that SHARD is better suited for scalar data, and given that we are interested in data quantities, using a Hazops type approach may be more meaningful. Therefore, we used Hazops guidewords for our approach.

To summarise, training data shortfalls can lead to latent faults in an ML system which can in turn lead to hazardous behaviour. Whilst there are many methods to manage training data shortfalls, they can themselves introduce further issues such as bias. It may be impossible to train the ML effectively without their use when we are dealing with real-world data. The approach for our case study uses a combination of ML data shortfall compensation methods and exploratory Hazop style system safety analysis to identify and consider means to reduce these risks.

3 Case Study: ML-Based Clinical Decision Support System for Type II Diabetes-Related Co-Morbidity Prediction

In this section we describe a clinical case study which uses our approach of combining ML data shortfall compensation methods and safety analysis. For this we used training data which contains real clinical patient data from the Connecting Bradford database [21]. The dataset consists of over 42,000 rows for patients with type 2 diabetes mellitus from different backgrounds and over 14,000 different types of clinical records (features). Type 2 diabetes is a life-long health condition and is the most common type of diabetes in the world. This health condition may cause the level of sugar (glucose) in the blood to become very high, and if not managed properly, it may progress by causing serious comorbidities [2]. When Type-2 diabetes progress, this causes numerous different comorbidities affecting the heart, brain, kidney, and other diseases.

The most frequently recorded disease/condition in our dataset is hypertension. It is known that hypertension is the precursor of the other potential diseases, and having both Type-2 diabetes and hypertension are synergistically dangerous. Hence, this is very critical to make a proper prediction for the risk level of having hypertension. High or low-risk thresholds are calculated using the National Institute for Health Care Excellence (NICE) guidelines used by clinicians [20].

The decision support system is designed to provide a clinician with an independent prediction of whether a patient is at high or low risk of hypertension (e.g., in the next six months), and hence support their decision of whether intervention is required. The clinical workflow is summarised in Fig. 2. The DCP will use the most recent patient data set provided as input. It will provide a prediction as to whether the patient is at high or low risk of hypertension, as well as explanation of the prediction. Additionally, the clinician will gather information through discussion with the patient. We discuss provision of contextual information later in the paper, as there are issues of patient confidentiality. In this paper we are specifically considering hypertension which increases the risk of other comorbidities, however the general safety analysis principles discussed can apply to any of the predictions training pathways. Our future work will consider other co-morbidity predictions.

The hazards related to the system are

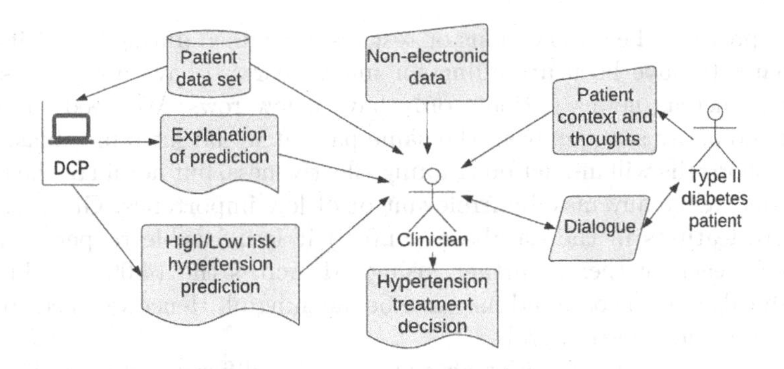

Fig. 2. Clinical workflow using DCP.

- **false positive** where a patient is categorised as high risk and given intervention that they do not require, possibly including medication with harmful side effects. For hypertension treatment may range from recommended lifestyle changes to specific medication. Common medications can have a range of minor side-effects (dizziness, headache, cough) to much more severe effects (e.g., angio-oedema). The clinician using the DCP would be making the decision of which medication to administer, and there is no requirement on the DCP to recommend treatment.
- **false negative** where a patient is categorised as low risk and no treatment is provided, leading to the condition not being managed. For hypertension this would mean medication specifically not being provided, potentially putting the patient at risk of severe outcomes such as heart attacks or stroke.

Classical risk analysis expects a combination of severity and likelihood to determine tolerability. Risk severity will depend on the particular comorbidity and potential outcomes of the false positive and false negative clinical decisions. In the case of hypertension, there is a potentially catastrophic outcome of heart attack and death if it is not treated. Calculating the likelihood of an incorrect prediction will require an understanding of the ML's performance for that particular comorbidity, but we note that there may be certain groups or individual patients where the predictions are less or more reliable. This may be due to weaknesses in the training data used or issues with specific information about an individual patient. An additional consideration is that the clinical decision may be influenced by other predictions from the DCP. As it is infeasible to assess likelihood of incorrect prediction for each individual patient, or to calculate overall likelihood with accuracy, we instead consider means to reduce the risk as far as possible at each stage of development and use.

3.1 The DCP Training Data

As noted, the training dataset consists of over 42,000 rows with 14,000 variables (features). A row represents a visit of the Type-2 diabetic patient and each feature represents the observations or test results gained during the visits. Some of the patients have been attending for many years so have many rows in the database, whereas newer patients only have a few rows. We need to consider whether too many samples from the same patient would introduce bias.

Data shortfalls will impact on training effectiveness, but not all of the features will impact safety any may be irrelevant or of low importance. Given the large number of features in the database (14,000) it is infeasible to perform safety analysis for each of them. Further, using ML across the 14,000 would have a very high calculation cost, and may not be meaningful. Hence we need to reduce this set to be more meaningful.

Since the record types differ according to the different sites, or if patients were not able to attend their appointments regularly, we have a large amount of missing values in our dataset (typically over more than half for each feature). We need to compensate for this during training, using data imputation, in order to train the ML. It is of note that missing data may itself be significant (see Sect. 4) however understanding the varying reasons for missing data, which could be clinically significant or simply due to different reporting practice across multiple clinics, would be difficult to infer without guidance, and lead to more uncertainty in the quality of outputs of the DCP.

An additional problem we cannot compensate for is that there may be groups of patients which are completely missing, e.g. from certain age groups or backgrounds. Also, we cannot compensate for validity issues, as whilst we can run some simple sanity checks e.g. for negative values for BMI, the issue of plausible but wrong data remains. Training data issues are illustrated in Fig. 3.

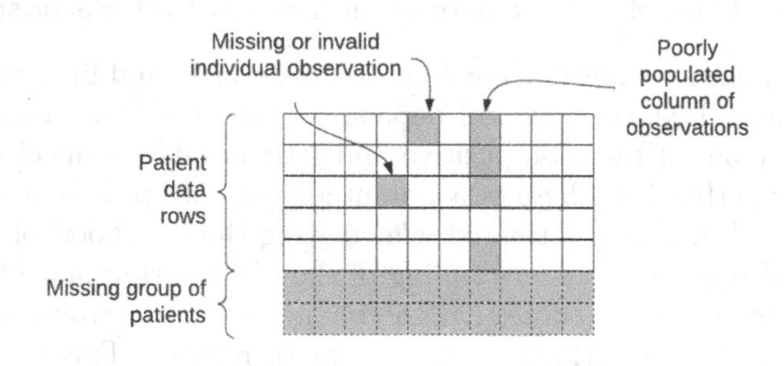

Fig. 3. Training data issues.

The systematic data pre-processing techniques applied for our study are shown in Fig. 4. First, a data frame has been prepared from the stored data. We have determined the most 20 frequent which are related to Type 2 diabetes FOIs and use these as a sub-set.

All the patients have been filtered by Type-2 diabetes, and duplicated or mistyped records have been deleted. When it has been ensured that we have unique records for each patient, the records are checked for the missing values. To fill all the missing values, we have used bag imputation method (see Sect. 2.1) incorporated in the R-Studio suite, as it reduces the risk of bias, and overfitting by predicting the missing values using ML. After dealing with all the missing values, we have normalized the dataset to fit all the values between 0 and 1 and to prevent from bias caused by the variation of the features. After finishing all the filtering and the necessary data pre-processing steps, the training dataset has been trained by the ML model.

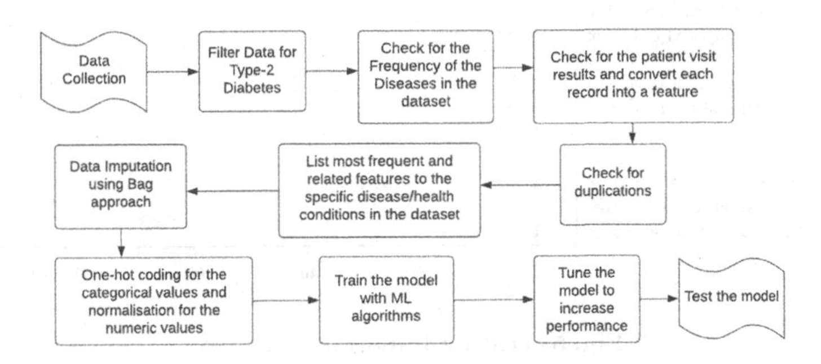

Fig. 4. Work-flow of the ML-based Type-2 Diabetes Progression Prediction.

After training the ML model, the feature importance of each variable has been calculated. Figure 5 shows us each variable's weighted importance levels to predict the output. This provides us some level of explainability of the model and also helps us to have a better understanding of the reasons behind of the model's predictions. Further, it allows us to focus the exploratory safety analysis on the FOIs. In order to ensure the validity at this stage, these FOIs were reviewed to confirm that they are plausible. Body Mass Index (BMI) is considered a good predictor, and was our highest FOI, so we have concentrated on that for this paper. Some of the other FOIs may be caused by hypertension, rather than being predictive. Note that the FOIs were gathered using an ensemble of different ML methods (including neural networks and random forests [17]), and future work is looking at comparing these individually.

3.2 Hazop Analysis

In this section we present an extract of the exploratory safety analysis of the FOIs as identified in the previous section and shown in Fig. 5. We have used a hazops style analysis (see Sect. 2.2) to consider how shortfalls in the training dataset could lead to hazards on output if not mitigated for each FOI, or groups of FOIs. The "flow" was interpreted as the flow of data into the ML training process. We used standard Hazop guidewords as inspiration for possible issues. We note that additional guidewords may be needed to capture unusual data shortfalls, although we did not identify any in this analysis. Some examples of how we interpret the guidewords are as follows.

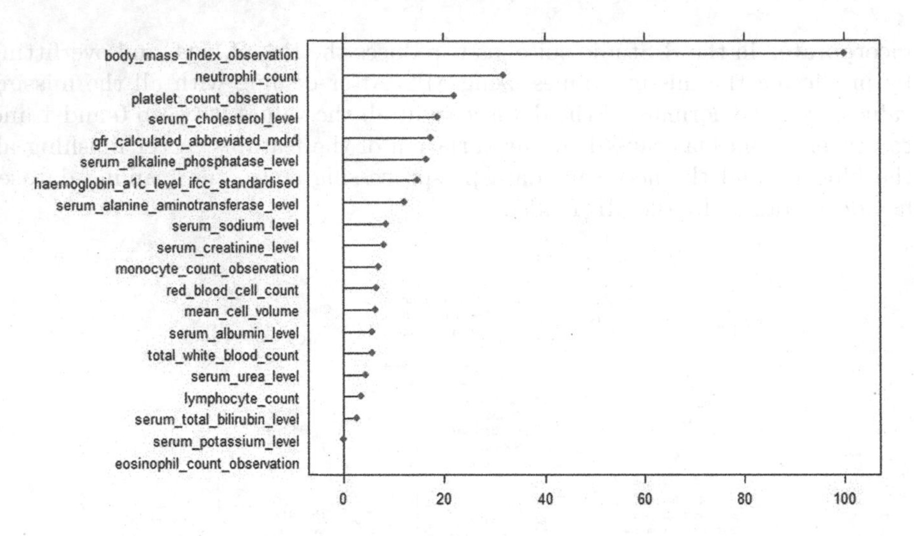

Fig. 5. Feature Importance Levels.

- More - indicates a bias in the data, e.g., over representation of particular patient group in the dataset
- No or Not - FOI or set of FOIs are missing
- Less - fewer examples of FOI than are desirable for good performance are present
- Early/Before - indicates that a FOI may be present but out of date with respect to the co-morbidity presenting itself
- Late/After - indicates that a FOI is present, but the overall patient data sample might be late in progression of the co-morbidity and hence not a useful predictor
- Part of - indicates missing data which needs to be compensated for
- Reverse - opposite diagnosis provided (i.e., False positive/negative)
- Instead - indicates the wrong FOI being used
- As well as - no interpretation

In Table 1 we show an extract of our analysis considering BMI as FOI, as this is the most critical. We list the guideword, identified deviation in the training data set, possible causes, effect on system safety and means to mitigate the deviation. Note that this analysis focuses on training data only, it may be useful to do a similar analysis for problems on the data used for an actual prediction as part of the overall safety assurance case, e.g. when there is no FOI in the patient record. The analysis has uncovered a number of risk mitigation measures which could be used, where practical, to reduce the risk of a latent failure caused by shortfalls in the training data leading to an incorrect prediction. These include technical approaches to data imputation and data sampling, but also manual data review, and explanations provided to the clinician. We have also included the discussion that the clinician would have with the patient as a mitigation

(Fig. 2) to emphasise that the ML decision is not used in isolation of other independent data sources. From the analysis we have a much richer understanding of risks and their mitigations.

Table 1. Extract of Hazop analysis of BMI FOI in Training Data.

Guideword	Deviation	Cause	Effect	Mitigation
No or not	Samples for ethnic group not included in training data (TD)	No/limited patients of ethnic group were patients	ML not trained or verified adequately for ethnic group with higher genetic risk of hypertension	Manual review of DB by expert, show clinician prototypical examples, patient discussion
Part of	Partially missing BMI in TD samples	BMI not consistently recorded	ML performance biased based on the data imputation method used, leads to poor performance for high or low BMI patients	Use bag imputation for TD records to reduce bias, recommend collection of BMI for future TD, show clinician prototype examples, patient discussion
More	Over representation in TD of high BMI patients	Most patients examined had high BMI	Prediction biased towards patients with high BMI, meaning patients with low BMI have less accurate predictions	Manual review of DB by expert, training samples picked across all ranges, show clinician prototype examples, patient discussion
More	Over representation in TD of certain ethnic group	Over diagnosis by trained ML for patients of other ethnic groups	TD dominated by ethnic group with genetic disposition to hyper tension	Manual review of DB by expert, show clinician prototype examples, patient discussion
Early/ Before and More	BMI data is out of date and training patients have changed BMI by time of diagnosis	DB not kept up to date, TD sampled from wrong part of patient history	ML underestimates likelihood of hypertension	TD selected from samples near to hypertension diagnosis, manual review of DB by expert, patient discussion
Instead	BMI value no longer highest FOI for some FOI distribution	Performance outlier from ML	Wrong prediction for hypertension	Show clinician FOI from training and for each prediction at point of use, patient discussion

4 Discussion

In the previous section we presented a case study combining system safety analysis with ML data shortfall compensation measures. In this section we discuss the findings in more depth.

It was infeasible to review all the potential features in the training data as there were over 14,000 of these. This meant that performing safety analysis prior on the data prior to pre-processing was not possible. Instead, it was necessary to reduce the set to 20 FOIs. The initial ML training (using data imputation to manage missing values) was performed to prioritise features and focus the safety analysis. However, it may be the case that training the ML using a much larger set of features would uncover a link or pattern of causes of hypertension which had not been considered previously. This is an avenue for further research.

When undertaking the safety analysis (Table 1) we suggested a number of risk mitigation methods which require further thought. One method for reducing the risks is a manual review of the patient database, for example to look for missing ethnic groups of patients or ensure up to date records have been kept. In practice this may be difficult to do effectively given the size of the database and some automation would be needed.

Another operational mitigation is to show the clinician similar patients from the TD to the one which the predictor has been applied to (i.e., prototypical examples as described in [10]). This would allow the clinician to review similar cases, their progression, and provides context to a particular prediction. However, the raw training data cannot be presented to the clinician for reasons of patient confidentiality and would need to be anonymised or obfuscated in some way. An avenue for further research is to consider whether using methods such as k-anonymise [6] would reduce the effectiveness such that this isn't a useful mitigation/explainability method for DCP.

Finally, it was noted by our clinical expert that missing data can be an important indicator of an underlying problem, for example if the patient was ill with another condition they may not have attended the clinic. In our clinical workflow (Fig. 2) we see this can potentially be considered via discussion with the patient. Another consideration is the progression of the comorbidity in the patient and whether this can improve predictive performance. Both cases may require a different and more complex ML model and training regime.

5 Conclusions

In this paper we have demonstrated that using a combination of ML data shortfall compensation measures, and exploratory safety analysis provides an effective method for the identification and mitigation of risks from training data shortfalls for a DCP. This takes a whole system perspective on risk identification and mitigation that is not found in similar literature in the area.

We have identified a number of avenues for further work including applying this methodology to an expanded predictor with multiple comorbidities (e.g., for brain diseases). Another is to review performance of different ML models with respect to bias from the different data imputation methods whilst balancing optimal performance against risk mitigations. Additional issues raised by the case study included balancing patient confidentiality with explainability, and wider contextual issues such as the clinical importance of missing data in the training data.

Acknowledgements. This work was supported by the Engineering and Physical Sciences Research Council (EP/W011239/1) and the Assuring Autonomy International Programme, a partnership between Lloyd's Register Foundation and the University of York.

References

1. Acuña, E., Rodriguez, C.: The treatment of missing values and its effect on classifier accuracy. In: Banks, D., McMorris, F.R., Arabie, P., Gaul, W. (eds.) Classification, Clustering, and Data Mining Applications. STUDIES CLASS, pp. 639–647. Springer, Heidelberg (2004). https://doi.org/10.1007/978-3-642-17103-1_60
2. Alonso-Morán, E., et al.: The prevalence of diabetes-related complications and multimorbidity in the population with type 2 diabetes mellitus in the Basque country. BMC Public Health **14**, 1059 (2014). https://doi.org/10.1186/1471-2458-14-1059
3. Bourdon, C., et al.: Metabolomics in plasma of Malawian children 7 years after surviving severe acute malnutrition: "ChroSAM" a cohort study. EBioMedicine **45**, 464–472 (2019)
4. Churpek, M.M., et al.: Hospital-level variation in death for critically ill patients with COVID-19. Am. J. Respir. Crit. Care Med. **204**(4), 403–411 (2021)
5. Driss, K., Boulila, W., Batool, A., Ahmad, J.: A novel approach for classifying diabetes' patients based on imputation and machine learning. In: 2020 International Conference on UK-China Emerging Technologies (UCET), pp. 1–4 (2020). https://doi.org/10.1109/UCET51115.2020.9205378
6. Emam, K.E., Dankar, F.K.: Protecting privacy using k-anonymity. J. Am. Med. Inform. Assoc. **15** (2008)
7. Hasan, M.K., Alam, M.A., Das, D., Hossain, E., Hasan, M.: Diabetes prediction using ensembling of different machine learning classifiers. IEEE Access **8**, 76516–76531 (2020). https://doi.org/10.1109/ACCESS.2020.2989857
8. Hawkins, R., Paterson, C., Picardi, C., Jia, Y., Calinescu, R., Habli, I.: Guidance on the assurance of machine learning in autonomous systems (AMLAS). arXiv preprint arXiv:2102.01564 (2021)
9. Jia, Y., Lawton, T., Burden, J., McDermid, J., Habli, I.: Safety-driven design of machine learning for sepsis treatment. J. Biomed. Inform. **117**, 103762 (2021). https://doi.org/10.1016/j.jbi.2021.103762. https://www.sciencedirect.com/science/article/pii/S1532046421000915
10. Jia, Y., McDermid, J., Lawton, T., Habli, I.: The role of explainability in assuring safety of machine learning in healthcare. IEEE Trans. Emerg. Top. Comput. **10**(4), 1746–1760 (2022). https://doi.org/10.1109/TETC.2022.3171314
11. Luo, F., et al.: Missing value imputation for diabetes prediction. In: 2022 International Joint Conference on Neural Networks (IJCNN), pp. 1–8. IEEE (2022)
12. McDermid, J.A., Nicholson, M., Pumfrey, D.J., Fenelon, P.: Experience with the application of HAZOP to computer-based systems. In: IEEE proceedings of the 10th Conference on Computer Assurance Systems Integrity, Software Safety and Process Security, pp. 37–48 (1997)
13. McDermid, J.A., Jia, Y., Porter, Z., Habli, I.: Artificial intelligence explainability: the technical and ethical dimensions. Phil. Trans. R. Soc. A **379**(2207), 20200363 (2021)
14. Modabbernia, A., Janiri, D., Doucet, G.E., Reichenberg, A., Frangou, S.: Multivariate patterns of brain-behavior-environment associations in the adolescent brain and cognitive development study. Biol. Psychiat. **89**(5), 510–520 (2021)
15. Molloy, J.J., McDermid, J.A.: Safety Assessment for Autonomous Systems' Perception Capabilities. arXiv abs/2208.08237 (2022)
16. National Transportation Safety Board: Collision Between Vehicle Controlled by Developmental Automated Driving System and Pedestrian, Tempe, Arizona,

March 18, 2018, NTSB/HAR-19/03 (2019). https://www.ntsb.gov/investigations/accidentreports/reports/har1903.pdf
17. Ozturk, B., Lawton, T., Smith, S., Habli, I.: Predicting progression of type 2 diabetes using primary care data with the help of machine learning. In: Medical Informatics Europe 2023 (2023)
18. Picardi, C., Hawkins, R., Paterson, C., Habli, I.: A pattern for arguing the assurance of machine learning in medical diagnosis systems. In: Romanovsky, A., Troubitsyna, E., Bitsch, F. (eds.) SAFECOMP 2019. LNCS, vol. 11698, pp. 165–179. Springer, Cham (2019). https://doi.org/10.1007/978-3-030-26601-1_12
19. Qi, Y., Conmy, P.R., Huang, W., Zhao, X., Huang, X.: A hierarchical HAZOP-like safety analysis for learning-enabled systems. In: AISafety 2022 (2022)
20. Ritchie, L.D., Campbell, N.C., Murchie, P.: New NICE guidelines for hypertension (2011)
21. Sohal, K., et al.: Connected bradford: a whole system data linkage accelerator. Wellcome Open Res. **7**, 26 (2022). https://doi.org/10.12688/wellcomeopenres.17526.2. https://europepmc.org/articles/PMC9682213
22. Sujan, M., et al.: Human factors challenges for the safe use of artificial intelligence in patient care. BMJ Health Care Inform. **26**(1), e100081 (2019)
23. Wei, S., Zhao, X., Miao, C.: A comprehensive exploration to the machine learning techniques for diabetes identification (2018). https://doi.org/10.1109/WF-IoT.2018.8355130
24. Zainuri, N.A., Jemain, A.A., Muda, N.: A comparison of various imputation methods for missing values in air quality data. Sains Malaysiana **44**(3), 449–456 (2015)

Data-Centric Operational Design Domain Characterization for Machine Learning-Based Aeronautical Products

Fateh Kaakai[1], Sridhar ("Shreeder") Adibhatla[2], Ganesh Pai[3](✉), and Emmanuelle Escorihuela[4]

[1] Thales, 3 Avenue Charles Lindbergh, 94628 Rungis, France
fateh.kaakai@thalesgroup.com

[2] Rockdale Systems LLC, Cincinnati, OH 45246, USA
shreeder@rockdalesystems.com

[3] KBR/NASA Ames Research Center, Moffett Field, CA 94401, USA
ganesh.pai@nasa.gov

[4] Airbus Operations (SAS), 316 Route de Bayonne, 31060 Toulouse, France
emmanuelle.escorihuela@airbus.com

Abstract. We give a first rigorous characterization of Operational Design Domains (ODDs) for Machine Learning (ML)-based aeronautical products. Unlike in other application sectors (such as self-driving road vehicles) where ODD development is scenario-based, our approach is *data-centric*: we propose the dimensions along which the parameters that define an ODD can be explicitly captured, together with a categorization of the data that ML-based applications can encounter in operation, whilst identifying their system-level relevance and impact. Specifically, we discuss how those data categories are useful to determine: the requirements necessary to drive the design of ML Models (MLMs); the potsince operating outside the flightential effects on MLMs and higher levels of the system hierarchy; the learning assurance processes that may be needed, and system architectural considerations. We illustrate the underlying concepts with an example of an aircraft flight envelope.

Keywords: Aeronautical products · Development assurance · Machine learning · Operational design domain · System safety

1 Introduction

Artificial Intelligence (AI)-enabling technologies like Machine Learning (ML) have the potential to transform the aviation industry by creating new products and services, and by enhancing the existing ones. However, ML introduces a new paradigm for design activities since the intended behavior of a function is

G. Pai—Contribution to the paper with support from the System-wide Safety project under the Airspace Operations and Safety Program of the NASA Aeronautics Research Mission Directorate.

J. Guiochet et al. (Eds.): SAFECOMP 2023, LNCS 14181, pp. 227–242, 2023.
https://doi.org/10.1007/978-3-031-40923-3_17

inferred from a body of data using statistical learning algorithms, rather than being specified and programmed. Data is thus central to the implementation of a final product design.

In traditional aviation domain systems engineering, operational requirements capture the conditions under which an end-product is expected to fulfill its missions. Those requirements, which are an expression of stakeholder needs, contain parameters and values that define an operational environment, or *operational domain* (ODs), in which an aviation system must properly operate. When requirements are elicited from and allocated to different layers of the system design—namely: *function* or *system*, *subsystem*, and eventually *item*[1]—the OD is also correspondingly allocated, resulting in *operational design domains* (ODDs) corresponding to those layers.

Motivation and Contributions. Specifying, refining, and allocating ODs to the system layers that will eventually integrate ML are activities not as well-understood as they are when going from a system/function layer to lower layers in conventional aviation systems development processes. Thus, a key challenge for the aviation systems domain is how to define, analyze, and manage the ODDs resulting from the allocation of the OD to the system layers integrating ML.[2] Addressing this challenge is especially important because it not only drives the data collection activities needed to ensure that a dataset representative of the intended operations is gathered, but also it influences the design of those layers and the underlying ML Models (MLMs).

To that end, this paper makes the following main contributions: (i) in Sect. 2, an identification of the dimensions along which the parameters that define an ODD for ML-based aeronautical products can be explicitly captured; (ii) in Sect. 3, a rigorous data-centric characterization of ODDs based on categorizing the data that ML-based functionality can encounter in operation. An aircraft flight envelope example also concretizes the underlying concepts; and (iii) in Sect. 4, an illustration of how the identified data categories can be used to determine the potential effects on the system layer integrating ML, along with the learning assurance activities and the system architectural considerations needed to mitigate those effects.

The approach in this paper is one of the cornerstones of a future process guidance document [15] for the development and certification/approval of safety-related aeronautical products implementing AI. That guidance is currently being developed through an aviation industry-based consensus process, jointly by the SAE Committee for AI in Aviation (G-34), and the EUROCAE working group for AI (WG-114).

[1] We use the standard aviation domain terminology for the layers of a system hierarchy/design.

[2] Additional related challenges (not in scope for this paper), such as the need to adapt requirements definition and validation processes to account for dataset requirements, have been comprehensively elaborated in [6].

Related Work. The concept of ODD was initially introduced and developed by the automotive systems industry [16]. As such, the current literature on specifying, developing, and using ODDs is largely in an automotive systems application context. For example, ODD specification for automated driving systems (ADSs) can be aided by a domain-specific language (DSL) using structured natural language founded on a formal, machine-processable domain model, to support both human comprehension and programmatic manipulation [7]. A *divide-and-conquer* approach to automotive function ODD development can be employed using a concept of so-called μODD [11] that partitions an ODD to place useful bounds on various safety-relevant parameters. Such partitions can then be tied to validation tests, whilst also encoding situation-specific parameter information. This approach is closest to our work, although the partitioning we achieve is data-centric, and orthogonal to μODD-based partitions. In [17], a hierarchical ODD definition is used to develop a scenario-based test framework for ADSs.

The ODD concept is being progressively matured in the automotive industry via ODD-related standards [1,9], as well as automotive system-centric safety standards concerning ML and AI [8,18]. Each of those guidance documents gives a mutually consistent definition for the ODD concept, emphasizing its relationship to safety. Nevertheless, automotive domain guidance cannot be directly applied to safety-critical aeronautical products owing to a variety of constraints, including: (a) differences in the regulatory approach between the automotive and aviation sectors; (b) the need for standards to be compatible with aviation regulations and regulatory acceptance of the associated compatibility arguments; (c) the stringency of assurance requirements in the aviation sector; and (d) consistency with the existing ecosystem of recommended engineering practices, e.g., for safety assessment [13], and aviation system development [14].

All of those factors, besides the key challenge discussed earlier, have additionally motivated the work in this paper. Next we give our notion of ODD.

2 System-Level Considerations

Operational Domains (ODs). When designing a product system, it is an established and well-understood aviation systems engineering practice to capture and analyze stakeholder needs at an early stage, along numerous dimensions such as the mission to be fulfilled, the expected performance in different system operating phases, and specific environmental conditions encountered. An OD is one of the results of such early-stage analysis, and it is embodied by the operational requirements for that system. In other words, the OD is captured in the form of requirements via a *specification* activity of a well-defined requirements development process. Thus, we consider an OD to be a specification of all foreseeable operating conditions under which an end-product is expected (and should be designed) to fulfill its missions. For instance, a *flight envelope* specifies, at a minimum, a combination of altitude and *Mach number*[3] values that define an operational environment in which an aircraft type must properly operate.

Operational Design Domains (ODDs). We define the *allocation of an OD* to be the *operational design domain* (ODD). This is largely aligned with other definitions of ODD [12,15,16,18]. Just as requirements are allocated across the different layers of the system design, and then refined with various criteria in mind, e.g., safety, architectural options, implementation choices, and physical considerations, an OD is also allocated to the lower design layers, and further refined so that each layer has its own ODD, i.e., the portion of the associated OD in which it should properly function. Such refinement can potentially (but not always) lead to rich and complex ODDs.[4] The principles and procedures governing OD allocation rely upon established aerospace practices [14]. As such, we can allocate the entire OD or a portion thereof to the subsystems that will be implemented using ML technologies (which we refer to, henceforth, as *ML-based subsystems*). Moreover, refining requirements as indicated earlier will bring forth corresponding enhancements of the OD reflecting the same considerations.

Describing ODs and ODDs. To describe an OD or ODD we elicit a variety of *parameters*, their range of admissible values, and, when relevant, distributions of occurrences over particular time intervals. In general, these define a multi-dimensional region. In practice, an OD or ODD is often likely to be a subset of that region. Although there are many ways to group parameters, the following is typical in practice:

- *Environmental Parameters*: These are variables outside the product (e.g., aircraft) system boundary, including weather conditions (ambient air temperature and pressure, wind conditions, humidity/rain/snow/ice, dust or sand levels, etc.) as well as application-specific parameters, e.g., brightness, contrast levels, and blur levels for optical sensor systems.
- *Operational Parameters*: These are parameters within the system boundary, examples of which include altitude and Mach number limits specified by a flight envelope, as well as ranges for angle of attack, pitch, roll, yaw angles, or their rates of change.
- *System Health Parameters*: These specify whether the system is expected to work only under nominal (non-failure) conditions, or whether it should be able to handle deterioration over time, sensor failures, or failures in specified system components (e.g., a failed actuator or a damaged flight control surface).

[3] Mach number is the ratio of true airspeed to the local speed of sound.
[4] Characterizing the complexity of an ODD is not in scope for this paper.

ML Constituent (MLC). Traditional systems engineering activities need to transition to ML activities at a certain stage of system development when integrating ML. In light of this, current regulatory guidance for introducing ML technologies into safety-related aeronautical applications [5], as well as ongoing standardization activities [15] have introduced a concept of *ML Constituent* (MLC) for systems integration purposes.

Effectively, an MLC represents the lowest-level of a functional decomposition from a system perspective that supports a subsystem function. It is a grouping of hardware and/or software items implementing one or more ML Models (MLMs) and their associated data pre- and post-processing items. Pre-processing may include (but is not restricted to) data cleanup, normalization, and feature computation. Similarly, post-processing may involve, among other actions, denormalization and blending of outputs from sub-models.

We qualify the ODD based on its allocation. Thus, allocating an OD to an MLC gives an MLCODD (i.e., the design space for an MLC), and likewise, the allocation to an MLM results in an MLMODD. An MLMODD may be identical to the MLCODD, though in practice it may be smaller. Additionally, an MLC can contain multiple MLMs each of which have their respective MLMODDs. Also, an MLMODD (or MLCODD) may be the same as the OD for the system, its superset (to provide robustness), or a subset thereof (to limit the design to a feasible region).

Thus, when a product will eventually integrate ML (e.g., as software whose design was learned through an ML training process) understanding the MLCODD is crucial to ensure that: (1) the data used for training is representative of that OD; and (2) the ML designer comprehends the complexity of the portion of the OD that has been allocated to machine learned functionality.

3 New ODD Concepts for Aviation

From the preceding narrative, it should be evident that developing an OD/ODD is itself not a new phenomenon in aviation systems engineering practice. However, it is the transition from an OD/ODD description to data collected for MLM training that is the major change relative to the way ODs are typically specified during conventional (i.e., non-ML based) product development. This change requires alternative approaches that are the focus of learning assurance processes [5,10].

We now give a data-centric conceptual characterization for ODDs, that partitions them based on *categories* and *kinds* of data. Henceforth, when we refer to "ODD" and "ML", we mean the MLMODD (or MLCODD), and the MLM (or MLC), respectively, and we will qualify our usage of those terms when it is not clear from context.

Categories and Kinds of Data. We define the following data categories:

(i) *Nominal*: Set of data points that lie in the interior of an ODD statistical distribution, that is *correct* with respect to the corresponding ML requirements.
(ii) *Outlier*: Set of data points outside an ODD. Some data can be mistaken to be *Outlier* data when they should have been *Nominal* data, had that ODD been correctly characterized by including at least one additional parameter.
(iii) *Edge Case*: Set of data points on an ODD boundary where exactly one ODD parameter has a valid extreme (maximum and minimum) value.
(iv) *Corner Case*: Set of data points where at least one ODD parameter is at their respective extremum (minimum and maximum value) of the range of values for those parameters that are admissible (or valid) for a given ODD (see Fig. 1 for examples). There are two types of *Corner Case* data:
   - *Feasible*: those that are part of the functional intent and, thus, within a given ODD (specifically at the vertices[5] of that ODD);
   - *Infeasible*: those that are not part of the functional intent and, thus, outside the ODD. Note that all *Infeasible Corner Case* data are a special case of *Outlier* data.
(v) *Inlier* (`InL`): Set of data that lie in the interior of the ODD following an error during data management, e.g., due to incorrect usage of units and dimensions. *Inlier* data are difficult to distinguish from *Nominal* data, and hence difficult to detect/correct.
(vi) *Novelty*: Set of data within an ODD according to the parameters used to describe that ODD, but which should have been considered to be *Outlier* data, had that ODD been correctly described by introducing at least one additional ODD parameter. In this sense, *Novelty* data points for an ODD could be seen as *duals* of those data points that are mistakenly considered to be *Outlier* data, when they should, in fact, have been *Nominal* data for that ODD. *Novelty* data usually arise from insufficient ODD characterization.

We can group the *Inlier*, *Outlier* (including *Infeasible Corner Case*), and *Novelty* categories into a single *Anomaly* data category. Data drawn from all the aforementioned categories may also be characterized as among the following kinds of sets:

(a) *In-Sample* (`InS`): Data used during MLM learning which the implementation of the MLM will have to process during inference in operation.
(b) *Out-of-Sample* (`OutS`): Data not used during MLM learning that the implementation of the MLM will have to process during inference in operation. It is on out-of-sample data that acceptable generalization behavior (and a corresponding guarantee) of the implemented MLM can be reasonably expected.

[5] ODDs without vertices e.g., an oval region, will therefore not have feasible corner cases.

(c) *In-MLMODD* (InMOD): Data that the implemented MLM will have to process during inference in operation. In-MLMODD data contribute to the intended function(s) of the MLM. We have: $\texttt{InMOD} = \texttt{InS} \cup \texttt{OutS}$ and $\texttt{InS} \cap \texttt{OutS} = \emptyset$

(d) *Out-of-MLMODD* (OutMOD): Data not seen during MLM learning that the implemented MLM *should not process* during inference in operation. Out-of-MLMODD data contributes to the intended function(s) of the MLC, e.g., specific processing to detect anomalies (see Sect. 4 for more details). We have: $\texttt{InMOD} \cap \texttt{OutMOD} = \emptyset$.

(e) *In-MLCODD* (InCOD): Data contributing to the intended function(s) of the MLC. We have: $\texttt{InCOD} = \texttt{InMOD} \cup \texttt{OutMOD}$.

(f) *Out-of-MLCODD* (OutCOD): Data not seen during MLM learning that the implemented MLC *should not process* during inference in operation. Out-of-MLCODD data contributes to the intended function(s) of the ML-based subsystem. We have: $\texttt{InCOD} \cap \texttt{OutCOD} = \emptyset$.

Real Data in Operation (and their associated statistical distributions), RDO, can now be defined as the set of all data seen in operation: $\texttt{RDO} \supseteq (\texttt{InCOD} \setminus \texttt{InL}) \cup \texttt{OutCOD}$.

The preceding concepts will serve as reference terms in forthcoming aviation industry specific guidance [15]. Nevertheless, we believe they are generic enough to be applicable in other domains, although there are some differences, e.g., our concept of *Edge Case* data differs from what is considered in [18].

Illustrative Example (Aircraft Flight Envelope). We now give an illustrative example of an aircraft flight envelope to concretize the preceding concepts. Informally, a flight envelope specifies the allowable combinations of two parameters—altitude (Alt) and airspeed, given here as a Mach number (Mach)—at which an aircraft design should function. Intuitively, this characterization of a flight envelope represents an Operational Domain (OD) of the aircraft system, and we refer to it, henceforth, as the *system* OD (SOD). This is closely related to a *functional* OD for the system which may include a specification of, for example, aircraft takeoff gross weights, the city pairs between which flight operations are intended, the routes (flight paths) that aircraft of a particular type design are expected to fly, the airports involved, the climb segments, and the landing approaches to be followed.

Figure 1 presents a notional flight envelope covering all phases of flight (shown as the irregular hexagonal region A). Mach and Alt values within this SOD are allowed, and therefore they are expected to be encountered in operation. Values of those parameters outside the SOD are disallowed since operating outside the flight envelope is usually dangerous in most circumstances.

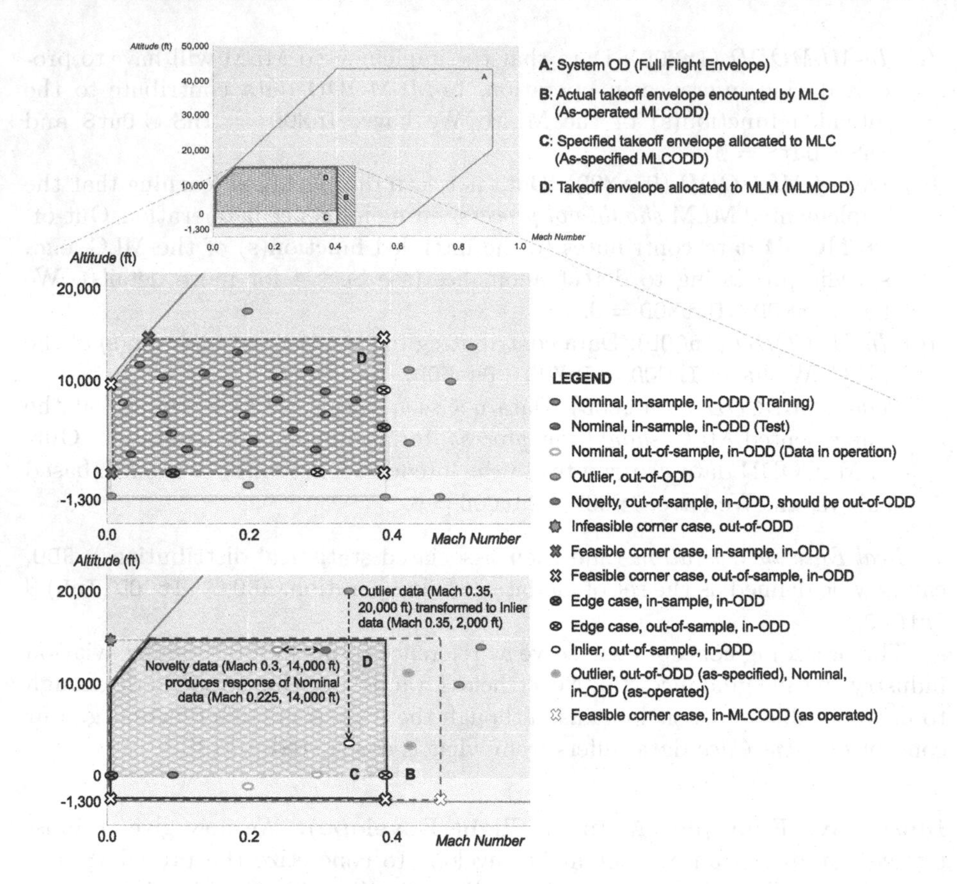

Fig. 1. An example flight envelope (region A) representing an aircraft system OD, whose refinement and allocation to an MLC and MLM give, respectively, an *As-operated* MLCODD (region B), containing an *As-specified* MLCODD (region C), itself containing the MLMODD (region D), for the takeoff regime. The shapes representing the different ODDs are practically congruent, but have been shown slightly offset here to differentiate each from the other. Zoomed-in views of the respective ODDs highlight the different categories and kinds of data used to characterize them.

Consider that a portion of this SOD is allocated to an ML-based subsystem to be used during the *takeoff* flight phase. Its ODD (not shown in Fig. 1) is the takeoff regime at the bottom of the SOD which, in turn, we refine and allocate to an MLC (contained by that ML-based subsystem). The resulting MLCODD parameters are: $0 \leq$ `Mach` ≤ 0.4 and $-1300\,\text{ft} \leq$ `Alt` $\leq 15000\,\text{ft}$. In Fig. 1, this *As-specified* MLCODD is the irregular pentagon with the solid dark border (region C).

For this ODD, observe that the upper bound for the airspeed parameter is `Mach` 0.4. However, aircraft with greater maximum takeoff weights, e.g., cargo aircraft, can often exceed this bound during takeoff. Thus, there are two possibilities: either the design was to be restricted to non-cargo aircraft, or there is a missing requirement that would be discovered in operation with cargo aircraft. For the latter case, the *as-operated* MLCODD would then have an increased upper bound on airspeed, e.g., $0 \leq \texttt{Mach} \leq 0.5$. In Fig. 1, this is the irregular pentagon (region B) that includes the earlier *As-specified* MLCODD (region C). Now, further consider that there is insufficient takeoff data for altitudes below sea-level to apply ML. Hence, we restrict the MLM to takeoff operations for $\texttt{Alt} \geq 0\,\text{ft}$. Thus, the MLMODD is a sub-region of the MLCODD allocated to the MLM contained in the MLC. In Fig. 1, this is the irregular pentagon with the dashed border (region D), with the same range for `Mach` as its containing MLCODD, but with sea-level as the lower bound on `Alt`. Figure 1 also zooms into these regions to give examples of the various categories and kinds of data described earlier.

Data inside the MLMODD (and/or MLCODD) can be drawn from the *Nominal*, *Edge Case*, *Feasible Corner Case*, *Inlier*, and *Novelty* data categories. The following observations are noteworthy: first, the MLM must demonstrate generalization from *Nominal*, *In-Sample* training data to *Nominal*, *In-Sample* test data, as well as to *Nominal*, *Out-of-Sample* data, all of which are *In-MLMODD*. Moreover, the MLM must exhibit *correct* behavior (i.e., the behavior meets the allocated requirements) on *Edge Case* as well as *Feasible Corner Case* data.

Next, the preceding data categories are disjoint relative to a specific allocation. For example, *Outlier* data for an MLM cannot also be a *Feasible Corner Case* for that MLM, though it can be one for the containing MLC. In Fig. 1, the data point (`Mach` 0.4, `Alt` : $-$ 1300 ft) is one such example of an *Outlier* for the MLMODD that is also an *Out-of-Sample*, *Feasible Corner Case* for the containing *As-specified* MLCODD.

We associate data points with specific categories relative to an allocation. In Fig. 1 for instance, the data point at (`Mach` 0.1, `Alt` : 0 ft) is an *Edge Case* for the MLMODD, but is *Nominal* data for the containing *As-specified* MLCODD. Similarly, each of the data points at (`Mach` 0, `Alt` : 0 ft), and (`Mach` 0.4, `Alt` : 0 ft) is a *Corner Case* from an MLMODD perspective but an *Edge Case* for the MLCODD.

Likewise, we can have *Outlier* data to the MLMODD that are within the MLCODD. In Fig. 1, examples of this case comprise any data point in the region of the *As-specified* MLCODD not included in the MLMODD, i.e., in the region defined by $0 \leq \texttt{Mach} \leq 0.4$, and -1300 ft $\geq \texttt{Alt} > 0$ ft. As shown, such points are *Outlier* data for the MLMODD, but can be *Nominal*, *Edge Case* or *Feasible Corner Case* data for the MLCODD. In the same way, points in the rectangular region of the takeoff envelope between $0.4 < \texttt{Mach} \leq 0.5$ and 0 ft $\leq \texttt{Alt} \leq 15000$ ft are *Outlier* data to both the MLMODD, and the *As-specified* MLCODD, but are within the *As-operated* MLCODD. For example, the data point at (`Mach` 0.5, `Alt` : $-$ 1300 ft) is a *Feasible Corner Case* for the *As-operated* MLCODD.

Recall that *Infeasible Corner Case* data are a special case of *Outlier* data that may not be reasonably encountered in operation, where two or more ODD parameters simultaneously take the extreme values admissible for that ODD.

Figure 1 (bottom right) shows one such example: the corner case at (`Mach` 0.0, `Alt` : 15000 ft) is infeasible for both the MLMODD and MLCODD because no airport runways exist at 15000 ft altitude.

Inlier data are within the MLCODD and/or MLMODD due to errors in data processing, scaling, normalization, and usage of incorrect units. In Fig. 1, the *Inlier* data point at (`Mach` 0.35, `Alt` : 2000 ft) is the result of a data preparation and scaling error of the *Outlier* data point at (`Mach` 0.35, `Alt` : 20000 ft). The result of processing such *Inlier* data is an incorrect response from the MLM, for example a flight control parameter value appropriate for the outlier data point is incorrectly produced at a lower altitude within the takeoff envelope.

Novelty data are within the MLMODD (and thus, also within the MLCODD), but are, in fact, data that should have been Out-of-MLMODD (or MLCODD). *Novelty* data are not excluded from the MLMODD due to an insufficiency in the number and variety of parameters used to specify the MLMODD. In Fig. 1, the data point (`Mach` 0.3, `Alt` : 14000 ft) is *Novelty* data producing a response appropriate for the *Nominal* data point at (`Mach` 0.225, `Alt` : 14000 ft). This occurs because the SOD and, in turn, the MLCODD and MLMODD have been specified using only two parameters (altitude and airspeed), either ignoring the effect of additional parameters such as air temperature, or implicitly assuming that the operations occur in the same environment as that in which the in-sample data were collected. In this example, operating in warmer air temperatures results in a lower Mach number, due to which the MLM receives an input that is invalid for the operating context, but is nonetheless *Nominal*.

In general, discovering data from the *Inlier*, *Novelty*, and *Outlier* categories that should be part of the required (or intended) MLCODD or MLMODD, occurs either during testing, during validation of the relevant ODDs, or from analysis of the data gathered from in-service experience. That usually results in re-defining the respective ODDs, e.g., by expanding its dimensions by including additional parameters, or modifying the admissible range of existing parameter values.

4 Support for System-Level Analysis

The combination of the category and kind of real data in operation, `RDO`, facilitates *partitioning* an MLMODD (and equivalently, an MLCODD) at a higher level than, say, partitioning by equivalence classes of inputs.[6] Then, from a safety standpoint for example, for each such partition we can analyze the contribution of the MLM (or the corresponding MLC) to system hazards in terms of the effects produced in response to inputs drawn from that partition. Examples of such effects include: an underperformance of the MLM; a hazardous failure condition; MLM or MLC malfunction; or, more generally, MLM and MLC failure modes and *hazard contribution modes* [3].

[6] In fact, we can combine those two ways of partitioning an ODD, e.g., by selecting an equivalence class of inputs within a *nominal*, *out-of-sample*, and *In-MLMODD* partition.

Subsequently, we can establish the (high-level) requirements that an MLM and its containing MLC should fulfill. These can include, for instance, restrictions on MLM behavior, constraints on data processing, limitations of use, as well as requirements necessary to manage the safety impact of the identified effects. The latter, in turn, also informs the selection of the mitigation measures appropriate for sufficient safety assurance. Such mitigations include the application of learning assurance processes (at the MLM layer), architectural mechanisms (at the MLC, ML-based subsystem, and system layers), as well as traditional development assurance processes as appropriate.

The tables given in Fig. 2 and Fig. 3 illustrate how we can use the partitions of an ODD to analyze the impact on an MLC and MLM: the row and column labels for a cell in the table correspond to the kinds and categories of data, respectively, and their combination is the partition of RDO we analyze. The content of a cell describes the results of a particular analysis for that partition, i.e., the effects

<table>
<tr><th colspan="3" rowspan="2">KIND OF DATA
(Real Data In Operation)</th><th colspan="3">DATA CATEGORIES</th></tr>
<tr><th>Nominal</th><th>Edge Case</th><th>Feasible Corner Case (CC)</th></tr>
<tr><td rowspan="6">In-MLCODD</td><td rowspan="2">In-MLMODD</td><td>In-Sample</td><td>E: MLM underperformance on particular known inputs

A
• Input detection and failover
• Input masking/filtering
• Input value replacement</td><td colspan="2">E
• MLM performance degradation
• Incorrect MLM response
• MLM Malfunction

A
• Extreme value monitoring
• Envelope protection and failover

L: Data augmentation</td></tr>
<tr><td>Out-of-Sample</td><td>E: MLM underperformance in localized regions

A
• Detection of regions of MLM underperformance
• Distribution drift monitoring
• Input routing/switching to alternative function
• MLM output range monitoring and failover
• MLM output masking
• MLM output value replacement</td><td colspan="2">E
• MLM performance degradation
• MLM malfunction

A
• Extreme value monitoring
• Envelope protection and failover
• MLM output range monitoring and failover
• MLM output masking
• MLM output value replacement</td></tr>
<tr><td colspan="2" rowspan="2">Out-of-MLMODD</td><td colspan="3">R: MLM shall not receive inputs from these data categories
R: MLC shall receive and process input from these data categories</td></tr>
<tr><td>A
• Input masking/filtering using pre-processing items of MLC
• OOD detection (of Out-of-MLMODD inputs) at ML-based subsystem level
• Input routing/switching to alternative function</td><td colspan="2">A
• Input masking/filtering using pre-processing items of MLC
• Extreme value monitoring
• OOD detection (of Out-of-MLMODD inputs) at ML-based subsystem level
• Input routing/switching to alternative function</td></tr>
<tr><td colspan="3" rowspan="2">Out-of-MLCODD</td><td colspan="3">E: MLC malfunction
R: MLC shall not receive inputs from these data categories</td></tr>
<tr><td>A
• Input masking/filtering at ML-based subsystem level
• Input routing/switching to alternative function</td><td colspan="2">A
• Extreme value monitoring
• OOD detection (of Out-of-MLCODD inputs) at ML-based subsystem level
• Input routing/switching to alternative function</td></tr>
</table>

Fig. 2. Assessing the impact of an ODD on an MLM and the corresponding MLC in relation to the partitions induced by the categories and kinds of real data in operation (specifically the *Nominal*, *Edge Case*, and *Feasible Corner Case* data categories) described in terms of the potential *effects* (**E**) of the data, the *requirements* (**R**) induced, the *learning assurance* (**L**) processes that may be needed, and candidate *architectural* (**A**) options for mitigation.

of encountering data from that partition, and the considerations that emerge on the requirements, architectural mitigations, and on learning assurance. When the analysis is common to multiple partitions, we show this in a cell that spans multiple columns. Note that these kinds of analyses can be applied to any ML-based subsystem, MLC, or MLM, and is agnostic to their allocated function. Also note that Fig. 2 and Fig. 3 are mainly examples, hence they are not comprehensive or complete. Thus, some effects (and the corresponding architectural options) can be common to the different partitions.

For brevity, here we highlight some specific example options from a combination of analyses. In practice, however, each analysis would be separately undertaken since the identified learning assurance techniques only apply during design, whereas the identified architectural options are primarily relevant in use.

Figure 2 shows an analysis from a safety standpoint: the potential effects of the ODD partition characterized by *In-MLMODD*, *In-Sample*, *Nominal* data include MLM underperformance on specific inputs (as observed during training and testing). In some applications, the exact inputs from that partition may also be encountered in operation. Thus, architectural mitigations for such data can include monitoring to detect those specific inputs, together with input value replacement, masking, or filtering, and/or failover.

Similarly, other partitions of the ODD can be characterized by *In-MLMODD*, *Out-of-sample*, *Edge Case* (or *Feasible Corner Case*) data. Figure 2 shows these combined into a single partition since the high-level effects (such as MLM malfunction), as well as the corresponding architectural mitigations (e.g., extreme value monitoring, or envelope protection and failover) are similar for each. However, we note that for particular applications involving a specific MLM, the individual effects (and therefore the necessary architectural mitigations) from *Edge Case* inputs are likely to differ from those resulting from *Feasible Corner Case* inputs.

Likewise, a common requirement is induced by the ODD partition(s) formed by (each of) the *In-MLCODD*, *Out-of-MLMODD*, *Nominal* data (and *Edge Case*, or *Feasible Corner Case* data respectively). For example, an MLM shall not receive and process input data drawn from those partitions of the ODD. Consequently, the architectural options available are also largely similar, although extreme value monitoring mainly applies to *Edge Case* and *Feasible Corner Case* data, rather than to *Nominal* data. Additionally, note that for *Out-of-MLCODD* kind of data, there is no distinction between *Nominal*, *Edge Case* and *Corner Case* data from an MLC standpoint. However, those categories are distinct from the perspective of the OD allocated to the containing ML-based subsystem, which induces distinct architectural mitigations as shown.

Figure 3 shows a similar analysis from a system development standpoint for the *Novelty*, *Outlier* (including *Infeasible Corner Case*), and *Inlier* data categories. Such data are not part of the functional intent, and therefore a requirement on MLM development is to exclude such data for model training. As such, the learning assurance process must include data selection and management activities to assure that the training data indeed excludes inputs drawn from

<table>
<tr><th colspan="3" rowspan="2">KIND OF DATA
(Real Data in Operation)</th><th colspan="3">DATA CATEGORIES</th></tr>
<tr><th>Novelty</th><th>Outlier
(including Infeasible CC)</th><th>Inlier</th></tr>
<tr><td rowspan="3">In-MLCODD</td><td rowspan="2">In-MLMODD</td><td>In-Sample</td><td colspan="3">R: MLM training data shall not include inputs from these data categories (since functional intent excludes such data)
L: Data selection and management processes, including pre-processing</td></tr>
<tr><td>Out-of-Sample</td><td>E
• Incorrect MLM response (MLM does not meet its requirements)
• MLM malfunction
A
• Envelope protection and failover
• MLM output range monitoring and failover
• MLM output masking
• MLM output value replacement
L: ODD parameter identification</td><td>Excluded by definition: Outlier and Infeasible CC data are Out-of-MLMODD, therefore they are not In-MLMODD</td><td>E
• Incorrect MLM response (MLM does not meet its requirements)
• MLM malfunction
A: Dissimilar inputs with cross-checking</td></tr>
<tr><td colspan="2">Out-of-MLMODD</td><td>• Excluded by definition: Novelty data are In-MLMODD, therefore they are not Out-of-MLMODD</td><td>E: MLM malfunction
R: MLM shall not receive inputs from this data category
A
• MLC preprocessing based input masking/filtering
• OOD detection (of Out-of-MLMODD inputs) at ML-based subsystem level
• Input fault flags
• Input masking or replacement
• Input routing/switching to alternative function
L
• Learning assurance processes shall analyze outlier data for ODD modification.</td><td>Excluded by definition: Inlier data are In-MLMODD, therefore they are not Out-of-MLMODD</td></tr>
<tr><td colspan="3">Out-of-MLCODD</td><td colspan="3">R
• MLC shall not receive inputs from these data categories
• ML-based subsystem containing MLC shall receive and process inputs from these data categories
A
• OOD detection (of Out-of-MLCODD inputs) at ML-based subsystem level
• Input routing/switching to non-ML items / alternative function</td></tr>
</table>

Fig. 3. Assessing the impact of ODDs characterized by *Anomaly* data, i.e., *Novelty*, *Outlier* (including *Infeasible Corner Case*), and *Inlier* data categories, similar to the assessment in Fig. 2.

those data categories to preclude an MLM from producing responses that are inconsistent with the functional intent.

The partitions of the ODD characterized by *In-MLMODD*, *Out-of-sample*, *Novelty* (and likewise *In-MLMODD*, *Out-of-Sample*, *Inlier*) data need special attention: specifically, *Novelty* data may not be detectable through operational monitoring. Indeed, if such data could be detected at runtime, the relevant features would then have been included in the set of MLMODD parameters, rendering such data *Nominal* rather than *Novelty*. *Inlier* data are also similarly difficult to detect in operation. To mitigate MLM failure conditions resulting from the former, learning assurance activities are particularly important, especially those facilitating a rigorous and comprehensive identification of MLMODD parameters and features.

In some circumstances, it may be possible to detect and recover from the *effects* of *Novelty* data if the responses produced result in a range violation. For those situations a range of output monitoring, masking, replacement, and failover mechanisms offer an architectural solution to risk mitigation. To mitigate the effects of *Out-of-Sample*, *In-MLMODD*, *Inlier* data, dissimilar and/or independent inputs with cross-checking is a candidate architectural pattern.

An MLM cannot receive *In-MLMODD*, *Out-of-Sample*, *Outlier* data since those are, by definition, *Out-of-MLMODD*. However, in a similar vein to *Novelty* and *Inlier* data, the ODD partition characterized by *In-MLCODD*, *Out-of-MLMODD*, *Outlier* data also needs particular attention: as we saw earlier (Sect. 3, Fig. 1), it is possible to encounter *Outlier* data that *ought to have been included* in the MLCODD—and by allocation, also in the MLMODD—but were not. This situation can occur due to an error in the requirements, a deficiency in the data collection process, or a lack of knowledge (epistemic uncertainty). This induces a learning assurance feedback step (see Fig. 3) to analyze *Outlier* data to validate and potentially update both the MLMODD and the MLCODD from in-service experience.

5 Conclusions and Future Work

We have clarified the dimensions along which the parameters that define an ODD for an ML-based aeronautical product can be captured, whilst identifying the categories and kinds of data that can be encountered in operation. We have concretized the underlying concepts using an aircraft flight envelope example considering its allocation to an ML Model (MLM) for the takeoff regime. Our data-centric ODD characterization gives a useful framework to identify and organize system development, safety, and assurance activities, which we have illustrated through examples of some high-level effects of the data both on the MLM and its containing ML Constituent (MLC), along with the architectural options available for mitigation.

The work described here has emerged from an ongoing, aviation industry-led, consensus based effort. As such, validating the relevance, applicability, and utility of the underlying concepts and approach largely relies on a committee consensus agreement and, eventually, regulatory endorsement. To that end, aviation industry practitioners are applying the approach to a variety of real-world applications such as airborne collision avoidance [2], safe flight termination[7], and time-based separation of transport aircraft in terminal approaches [4]. These use cases corroborate our earlier assertion (see Sect. 4) that the work in this paper is sufficiently generic to be applicable to ML-based aeronautical products used both in airborne systems, and for air traffic management/navigation services. As a key avenue of future work, we are committed to take the lessons learned from those validation efforts—of the successes, insights, and possible gaps—to refine and further mature our approach. A related, crucial aspect of our future effort is to leverage the concepts and approach presented here to define a rigorous process

[7] See: https://safeterm.eu/.

for MLCODD development and validation, and MLCODD coverage verification (to be elaborated in a forthcoming paper). Such a process does not yet exist in the prevailing aviation standards and guidance on recommended practices. Thus, it will represent a concrete extension to the state-of-the-practice.

Our data-centric ODD characterization (Sect. 3), though rigorous, would benefit from a formalization of the identified categories and kinds of data, and their interrelations. This could facilitate assessing whether certain desirable properties hold, e.g., that the data categories *cover* an ODD in some formally defined sense, and that they are internally *complete*. This paper has mainly considered singleton MLMs and MLCs. We intend to extend our approach to the situations of multiple MLCs within a single ML-based subsystem, and multiple sub-MLMs within a single MLC. These cases have interesting safety and architectural implications from which we expect to gain a deeper insight into hazardous behavior emerging from the interactions of multiple MLMs and MLCs. In a similar vein, the support for system-level analysis (Sect. 4) can be further elaborated towards a more comprehensive and complete description of the potential effects of real data encountered in operation, together with the requirements induced, architectural options available for mitigation, and the learning assurance activities necessary.

This paper has given a new data-centric characterization for ODDs that is not an extension, enhancement, or tailoring of prior automotive domain ODD concepts. A related avenue of future work is to compare and contrast our ODD concept and principles with those of other safety-critical domains such as automotive, healthcare, rail, and space. We remain cautiously optimistic that our work is sufficiently general to be adopted, extended, and applied in those domains by the associated subject-matter experts.

Acknowledgments. We thank the members of the ODD working group of the joint EUROCAE WG-114 and SAE G-34 committees who contributed to the discussions that shaped the concepts and approach in this paper. We are additionally grateful to the anonymous reviewers whose comments aided us in improving the paper.

References

1. BSI Standards Ltd.: Operational Design Domain (ODD) Taxonomy for an Automated Driving System (ADS) - Specification. BSI PAS 1883:2020, August 2020
2. Damour, M., et al.: Towards certification of a reduced footprint ACAS-XU system: a hybrid ML-based solution. In: Habli, I., Sujan, M., Bitsch, F. (eds.) SAFECOMP 2021. LNCS, vol. 12852, pp. 34–48. Springer, Cham (2021). https://doi.org/10.1007/978-3-030-83903-1_3
3. Denney, E., Pai, G., Smith, C.: Hazard contribution modes of machine learning components. In: Espinoza, H., et al. (eds.) Proceedings of the AAAI Workshop on Artificial Intelligence Safety (SafeAI). AAAI, CEUR Workshop Proceedings (2020)
4. EUROCONTROL: COAST (Calibration of Optimised Approach Spacing Tool) with Use of Machine Learning Models. White Paper V1.1, April 2021
5. EASA: First Usable Guidance for Level 1 Machine Learning Applications. EASA Concept Paper Issue 01 (December 2021)

6. G-34, Artificial Intelligence in Aviation Committee: AIR 6988, Artificial Intelligence in Aeronautical Systems: Statement of Concerns. SAE International, April 2021
7. Irvine, P., Zhang, X., Khastgir, S., Schwalb, E., Jennings, P.: A two-level abstraction ODD definition language: part I. In: 2021 IEEE International Conference on Systems, Man, and Cybernetics (SMC), pp. 2614–2621 (2021)
8. ISO/TC 22/SC 32: Road Vehicles - Safety and Artificial Intelligence. ISO/AWI PAS 8800 (Under development) (2021)
9. ISO/TC 22/SC 33: Road vehicles - Test Scenarios for Automated Driving Systems - Taxonomy for Operational Design Domain. ISO/DIS 34503 - Draft International Standard (2023)
10. Kaakai, F., Adibhatla, S., et al.: Toward a machine learning development lifecycle for product certification and approval in aviation. SAE Int. J. Aerosp. **15** (2022)
11. Koopman, P., Osyk, B., Weast, J.: Autonomous vehicles meet the physical world: RSS, variability, uncertainty, and proving safety. In: Romanovsky, A., Troubitsyna, E., Bitsch, F. (eds.) SAFECOMP 2019. LNCS, vol. 11698, pp. 245–253. Springer, Cham (2019). https://doi.org/10.1007/978-3-030-26601-1_17
12. NHTSA, US Department of Transportation: Automated Driving: A Vision for Safety. Report No. DOT HS 812 442, September 2017
13. S-18, Aircraft and System Development and Safety Assessment Committee: ARP 4761, Guidelines and Methods for Conducting the Safety Assessment Process on Civil Airborne Systems and Equipment. SAE International, December 1996
14. S-18, Aircraft and System Development and Safety Assessment Committee: ARP 4754A, Guidelines for Development of Civil Aircraft and Systems. SAE International, December 2010
15. SAE G-34 Committee for AI in Aviation and EUROCAE WG-114 for AI: Process Standard for Development and Certification/Approval of Aeronautical Safety-Related Products Implementing AI. AS 6983 Draft Standard Work In Progress, February 2023
16. SAE International: Taxonomy and Definitions for Terms Related to Driving Automation Systems for On-Road Motor Vehicles. Surface Vehicle Recommended Practice J3016 (2018)
17. Thorn, E., Kimmel, S., Chaka, M.: A Framework for Automated Driving System Testable Cases and Scenarios. Report No. DOT HS 812 623, National Highway Traffic Safety Administration, September 2018
18. Underwriter Laboratories Inc.: ANSI/UL 4600 Standard for Safety for the Evaluation of Autonomous Products, April 2020

Online Quantization Adaptation for Fault-Tolerant Neural Network Inference

Michael Beyer[1,2](✉), Jan Micha Borrmann[1], Andre Guntoro[1], and Holger Blume[2]

[1] Bosch Corporate Research, Robert Bosch GmbH, Renningen, Germany
{michael.beyer2,janmicha.borrmann,andre.guntoro}@de.bosch.com
[2] Institute of Microelectronic Systems, Leibniz University Hannover, Hannover, Germany
blume@ims.uni-hannover.de

Abstract. Neural networks (NNs) are commonly used for environmental perception in autonomous driving applications. Safety aspects in such systems play a crucial role along with performance and efficiency. Since NNs exhibit enormous computational demands, safety measures that rely on traditional spatial or temporal redundancy for mitigating hardware (HW) faults are far from ideal. In this paper, we combine algorithmic properties with dedicated HW features to achieve lightweight fault tolerance. We leverage that many NNs maintain their accuracy when quantized to lower bit widths and adapt their quantization configuration during runtime to counteract HW faults. Instead of masking computations that are performed on faulty HW, we introduce a fail-degraded operating mode. In this mode, reduced precision computations are exploited for NN operations, as opposed to fully losing compute capability. This allows us to maintain important synapses of the network and thus preserve its accuracy. The required HW overhead for our method is minimal because we reuse existing HW features that were originally implemented for functional reasons. To demonstrate the effectiveness of our method, we simulate permanent HW faults in a NN accelerator and evaluate the impact on a NN's classification performance. We can preserve a NN's accuracy even at higher error rates, whereas without our method it completely loses its prediction capabilities. Accuracy drops in our experiments range from a few percent to a maximum of 10%, confirming the improved fault tolerance of the system.

Keywords: Fault Tolerance · Neural Network Hardware · Neural Networks · Quantization · Approximate Computing · Automotive

1 Introduction

The exceptional performance of neural networks (NNs) in visual perception tasks comes at a high computational cost. For environmental perception systems in autonomous driving, dedicated hardware (HW) accelerators are a suit-

J. Guiochet et al. (Eds.): SAFECOMP 2023, LNCS 14181, pp. 243–256, 2023.
https://doi.org/10.1007/978-3-031-40923-3_18

able solution to fulfill energy efficiency and throughput requirements. Consequently, they are often selected over more general purpose HW such as CPUs or GPUs. Since such systems are vital for object detection and classification tasks, integrity of processed data as well as availability of compute resources have to be ensured. However, with shrinking transistor structure sizes, integrated circuits have become more susceptible to hardware faults [1]. Besides transient faults such as bit flips, process variation and aging can lead to permanent HW faults [4]. These faults are irreversible and can lead to a complete loss of functionality, which is not acceptable in the targeted safety-critical applications. However, ensuring availability by means of traditional temporal or spatial redundancy-based approaches is not desirable due to the increased complexity and costs (e.g., area, power).

In this paper we tackle the challenge of efficiently improving fault tolerance of NN accelerators by introducing a fail-degraded operating mode. Our approach is motivated by the fact that NNs tolerate computations with reduced precision and maintain their accuracy. Instead of masking computations that are performed on faulty HW, we adapt the precision of computations during runtime and reuse faulty HW. This dual-use via reusing existing HW features allows us to introduce lightweight fault tolerance and thus improve overall availability of the accelerator system. To summarize, our contributions are:

- We propose a lightweight HW extension for circumventing hardware faults by adapting a NN's quantization configuration during inference. By reusing existing HW features of NN accelerators, we keep the required HW overhead low. Our method increases the fault tolerance of the system without retraining the network or adapting how it is mapped on HW.
- We analyze different rounding techniques for reducing bit widths of values online during NN processing. We assess their effectiveness for maintaining the contributions of highly relevant synapses of the NN when computing with reduced precision.
- Using extensive simulations, we evaluate our proposed method on a scalable vector processor accelerator. The performance of our method is assessed based on the classification accuracy of ResNet18 and VGG16 NNs with varying hardware defect rates on two datasets. The results confirm the benefits of our method.

This document is structured as follows. We introduce our error mitigation method in Sect. 2. Then we outline our experimental setup and evaluate our method with two image classification NNs. Finally, we compare our method with related work in Sect. 4 and conclude our paper in Sect. 5.

2 Online Quantization Adaptation

To achieve efficient fault tolerance, we leverage that NNs can often tolerate computations with reduced precision without deteriorating their algorithmic performance (e.g., classification accuracy). Our method adapts the precision with which computations are performed during runtime. A typical NN accelerator consists of several processing elements (PE) that perform multiply-accumulate (MAC) operations. PEs of recent accelerators can support computations with varying precision to increase flexibility [3,8,21]. This feature enables HW support for mixed-precision NNs, where data of individual layers can be quantized differently. Compared to a uniform bit width for the whole NN, mixed-precision allows for a better tradeoff between a smaller memory footprint and NN accuracy [23]. A reduced bit width also results in fewer data transfers to the accelerator, which is key to improving the energy efficiency of the system [25]. Furthermore, the compute performance is increased by processing multiple low-precision data words simultaneously. Support for computations with different bit widths can be achieved by integrating several low-precision functional units. For a given input data word, each functional unit executes operations on subword level. Computations requiring higher precision (i.e., the full width of the input word) are performed by combining partial results of multiple functional units.

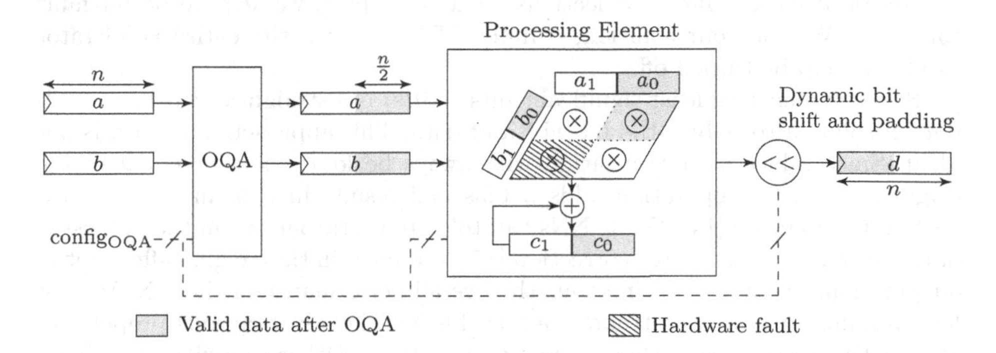

Fig. 1. Hardware schematic for online quantization adaptation (OQA). Processing element supports computations with the full or half the original bit width. After HW faults are detected, remaining still functioning low-precision functional units within the PE are used to uphold compute capability. The OQA block adapts bit widths of values and adjusts how inputs are fed into the PE. Flexibly remapping the inputs of the PE allows to circumvent faulty subunits. Output values are padded and shifted to match the original fixed-point format before OQA.

In Fig. 1, a PE is shown that supports computations with either the full or half the original bit width. For simplicity, we opt for two bit widths in this example. Our method, however, is not limited to this design or configuration. Considering the two inputs a and b shown in Fig. 1, the PE performs

$$c = c + a \cdot b \tag{1}$$

for computations using the full precision and

$$c_{1,0} = \begin{cases} c_1 := c_1 + a_1 \cdot b_1 \\ c_0 := c_0 + a_0 \cdot b_0 \end{cases} \tag{2}$$

for computations with reduced precision. Regardless of the desired precision, operations are always performed on subword level. Computations requiring full precision are obtained by combining partial results:

$$a \cdot b = a_1 \cdot b_1 \cdot 2^n + a_1 \cdot b_0 \cdot 2^{\frac{n}{2}} + a_0 \cdot b_1 \cdot 2^{\frac{n}{2}} + a_0 \cdot b_0, \tag{3}$$

where n is the bit width and multiplications with 2^i are bit shifts. PEs essentially have built-in redundancy for the reduced precision functional units. Consequently, after HW faults have been detected, remaining still functioning units can be used to uphold the compute capability, albeit with a lower quality. Our proposed method is to truncate bits of original full-width values and round toward zero before they are fed to partially faulty PEs. Computations are then performed with reduced precision on fault-free subunits within the PE. In contrast to simply masking computations that are performed on faulty HW [30], the key idea is to maintain highly relevant synapses of the NN, i.e., connections with a large weight value associated to it. By keeping contributions to feature extraction made by those connections, at least in part, we achieve better fault tolerance. Without our solution, complete PEs or even the entire accelerator would have to be turned off.

Simply truncating least significant bits (LSBs) is a straightforward and well-known solution to reduce the bit width of data. This approach, however, is not ideal. Since values after truncation will always be lower than or equal to the original value, this operation adds a bias and results in a quantization error with non-zero mean [14]. While NNs can tolerate perturbations in the processed data without requiring error correction [17], variances in the weights affect entire output channels of a layer and thus the overall performance of the NN. We opt for rounding to zero after truncation, as this has the advantageous property of attenuating values rather than overestimating them. Other rounding techniques such as convergent rounding [14], where ties are rounded to the nearest even value, can lead to higher-than-original contribution of individual weights. This skews the distribution of a layer's parameters and thus can affect the performance of the NN.

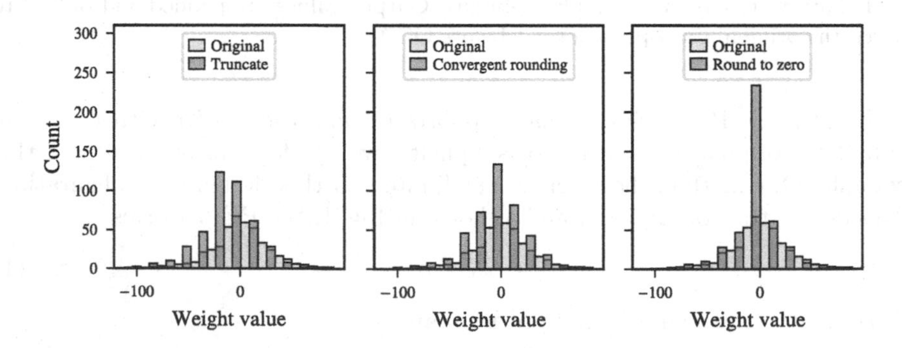

Fig. 2. Comparison of different rounding modes for reducing precision of values.

In Fig. 2, we show a histogram of a given NN layer's weights that is quantized from floating-point to an 8-bit fixed-point format. We compare different modes where weight values are truncated to 4 bits and, depending on the rounding mode, rounded and rescaled to 8-bit for comparison. When rounding to zero, a lot more values become zero, but the overall distribution of weights is preserved. Truncating without rounding and convergent rounding lead to a distortion and over-/underestimation of values.

In Fig. 1 our proposed modification to a PE is shown. During normal operation, the PE's input values are not modified by the online quantization adaptation (OQA) block. If a HW fault is detected by an arbitrary error detection mechanism, e.g., periodic checks or checksums [16], the fail-degraded operating mode is configured using the $\text{config}_{\text{OQA}}$ configuration parameter. For clarity, we bundle several sub-configuration data in this configuration signal parameter in Fig. 1. The signal specifies whether the OQA HW block is active and configures the PE such that HW error(s) are circumvented. When OQA is active, values are truncated and rounded toward zero. The number of truncated bits is a design time parameter and depends on the capabilities of the PE. Rounding to zero is accomplished by adding one to negative values after truncation. Afterward, functional units detected as faulty are circumvented by adjusting how inputs are fed into the PE after OQA. In Fig. 1, the PE has four multipliers that operate on $n/2$ bits each and the multiplier on the bottom left is affected by a HW fault. Consequently, the input values a and b are truncated to $n/2$ bits and placed into the lower subwords of the PE's input registers. The top right multiplier is then selected by $\text{config}_{\text{OQA}}$ as the result of the multiplication.

Since other, potentially still fully functional PEs in the accelerator expect input values with the original fixed-point format and precision, the output c has to be rescaled to the initial format. This is done by means of shifting and padding. For our example in Fig. 1, the $n/2$-bit result is padded with $n/2$ zeros as LSBs to obtain the final n-bit result d. The fixed-point format can be different for each layer in a quantized NN. Consequently, reducing the bit width and thus modifying the fixed-point format has to be done with care as some operations require fixed points to be aligned. During NN inference, PEs have no information on the fixed-point representation of the processed data. Therefore, correct handling of the data has to be ensured during processing when adapting the quantization configuration online. The number of truncated bits $n_{truncated}$ when switching to a lower precision is a design-time parameter. Consequently, any required bit shifts during processing can be translated to the new fixed-point format with the following equation:

$$s_{new} = s_{desired} - n_{truncated}, \tag{4}$$

where $s_{desired}$ is the desired shift amount and s_{new} the new shift amount. A positive number means shifting to the left, a negative number shifting right.

In Fig. 1 we show only one possible fault configuration. However, by flexibly remapping the inputs of the PE after OQA, also other fault configurations can be addressed. This allows us to effectively reuse remaining functional units to improve the fault tolerance and thus availability of the accelerator. By enabling dual use of HW features that were originally added for performance and efficiency reasons, the added HW complexity of our method is kept low. OQA is realized using MUXes and an adder circuit for each input. Since the addition is targeting low bit widths, additional latency from this operation is minimal. Selecting the PE's output can also be realized with MUXes and shifting the PE output is achieved by extending existing shifting logic.

3 Experiments

3.1 Experimental Setup

We evaluate our method using the two popular classification benchmarks CIFAR-10 [11] and the German Traffic Sign Recognition Benchmark (GTSRB) [24]. CIFAR-10 consists of a training and test subset with 40 000 and 10 000 32 × 32 RGB images. They are categorized in 10 different classes ranging from animals to objects such as cars or airplanes. GTSRB contains images of 43 different German traffic signs and is split into training and testing subsets with 39 209 and 12 630 RGB images. Since the images exhibit an unbalanced variation in brightness, we normalize them across RGB channels and resize each to a uniform size of 32 × 32 pixels. We train the two state-of-the-art neural network architectures ResNet18 [9] and VGG16 [22] on both CIFAR-10 and GTSRB and use 20% of the training dataset for validation during training. The training data is normalized to have zero mean and unit variance and is augmented by horizontal flipping of images and adding Cutout patches of size 16 [6]. Cutout is a simple regularization technique that masks sections of the input to NNs which has shown to improve the model's robustness and classification performance. For GTSRB, horizontal flipping is not performed as this can change the meaning of some images. The networks are trained with stochastic gradient decent with a learning rate of 0.025 and a momentum of 0.9. During training, the learning rate is annealed down to 10^{-8} using cosine annealing [13]. Each network is trained for 100 epochs and a batch size of 96. Afterward, the models are quantized to 8-bit [29].

Our proposed OQA method is implemented as a custom TensorFlow layer in our simulation framework ProxSim [18] which enables fast and efficient evaluation of quantized NNs using GPUs. The custom layer models convolution operations in fixed-point format as they would be performed on HW. Additionally, we implement a fine-granular selection of different quantization modes for each element of its output tensor using a configurable mask. This approach allows us to simulate how convolutions are mapped on an accelerator. Consequently, different accelerator HW architectures and fault configurations can be modeled efficiently by defining the appropriate mask.

As a hardware target, we use a scalable vector processor accelerator. Figure 3 shows an overview of our accelerator system [15,28]. The accelerator consists of clusters of vector units. Vector units have two vector lanes, the actual PEs, and a configurable local memory that can be accessed by lanes in the unit. Additionally, each vector lane has a register file scratchpad. We consider a PE architecture where computations with different bit widths are realized by combining partial

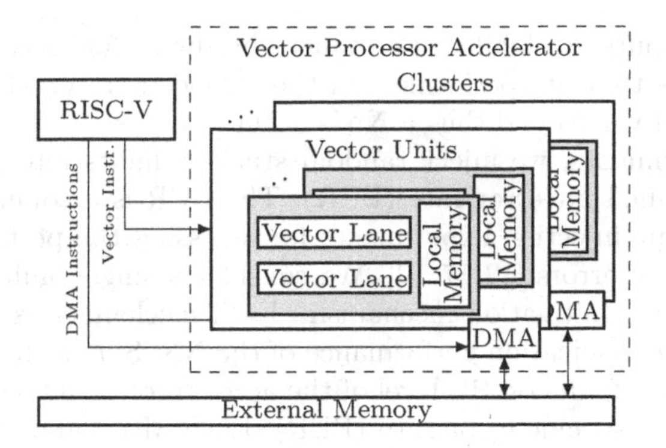

Fig. 3. Schematic of our targeted vector processor for NN acceleration [28].

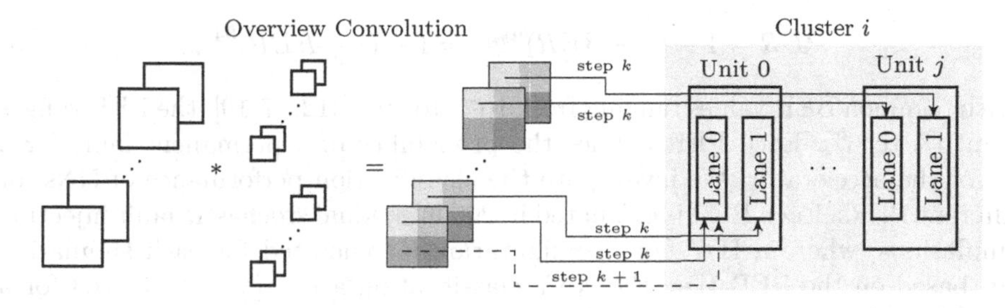

Fig. 4. Mapping of convolutions on our accelerator. Arrows indicate the assignment of feature map segments to vector lanes. As there are usually more segments to process than there are lanes, each lane is responsible for multiple segments. For clarity, segment for step $k + 1$ is only shown for one lane.

results of several low bit-width functional units (e.g., 8-bit precision is realized by combining multiple 4-bit computations. See Fig. 1, cf. [3]). Computations are orchestrated by a RISC-V processor, which generates DMA and vector instructions that are broadcast to all clusters, units, and lanes. We map convolutions on the accelerator using an output stationary data flow. Larger feature maps that do not fit in on-chip memory are split into multiple segments. Individual segments of a layer's output are then computed on each vector lane as shown in Fig. 4. Since there are usually more segments to process than there are vector lanes, each lane is responsible for multiple segments.

3.2 Error Model

To analyze the effectiveness of our method, we perform stochastic fault injection experiments. In this work, we focus on permanent hardware faults modeled as stuck-at faults, where individual bit values are stuck at either one or zero. Transient errors, such as bit-flips, are not the focus of our experiments. They are non-permanent and non-destructive errors that can be fully removed by resetting the system or rewriting affected memory cells [4]. Consequently, they have no effect on compute availability. Permanent hardware faults have an impact on all computations that are performed on the erroneous HW, which significantly affects computed values and thus a NN's accuracy.

In our experiments, we inject random stuck-at faults into registers of PEs based on a specified bit error rate (BER). The BER is a common metric used in stochastic fault injection experiments for assessing the performance of NNs under the effect of errors [12,17,19]. We consider a single fault to be sufficient to trigger the error mitigation mechanism. Each mechanism is evaluated based on the resulting classification performance of the NN. Similar to [12], we analyze the impact of hard faults on PE level of the accelerator architecture and convert the BER to a PE specific error rate (PER). Each MAC unit has a total of 64 bits, comprising two 8-bit input registers, one 16-bit intermediate register, and a 32-bit accumulator. Consequently, the PE specific error rate is defined as:

$$PER = 1 - (1 - BER)^{n_{bits}} = 1 - (1 - BER)^{64} \ . \tag{5}$$

With common BER values ranging from 10^{-7} to 10^{-3} [12,17,19], the PER ranges from 0% to 6%. Each PER defines the probability of a permanent fault for a PE of the accelerator. We investigate the classification performance of NNs for different PERs. Each PER is evaluated in 200 individual stochastic fault injection simulations, where a HW fault configuration is generated for each simulation run based on the PER (i.e., the NN's classification accuracy is evaluated for a given HW fault configuration and the configuration is reset after each run). In this work, the focus is not on reproducing accurate error rates or real errors in hardware. Instead, a worst-case behavior is investigated which can help making design decisions and evaluating the fault tolerance of the system.

In our experiments we investigate two error mitigation mechanisms for computations that are performed on faulty PEs:

1. *Online Quantization Adaptation*: Values are requantized online, as described in Sect. 2, and computations are performed with reduced precision.
2. *Discard*: Values are discarded and set to zero.

We use the second case as a baseline to compare our method. Similar to related work [30], this approach is essentially masking all computations that are performed on faulty PEs. For OQA, we assume that reducing precision from 8-bit to 4-bit is always possible for computations performed on faulty PEs. As outlined in Sect. 2, this is achieved using remaining functional units in the PE.

3.3 Results

The results of our fault injection experiments for ResNet18 and VGG16 are shown in Figs. 5 and 6. The results of all our experiments are presented using box plots for the NN's classification accuracy on the respective test subset over 200 stochastic fault injection simulations for each PER. We opt for box plots instead of solely focusing on the mean accuracy. This allows us to draw further conclusions about the robustness of the NN's classification performance over

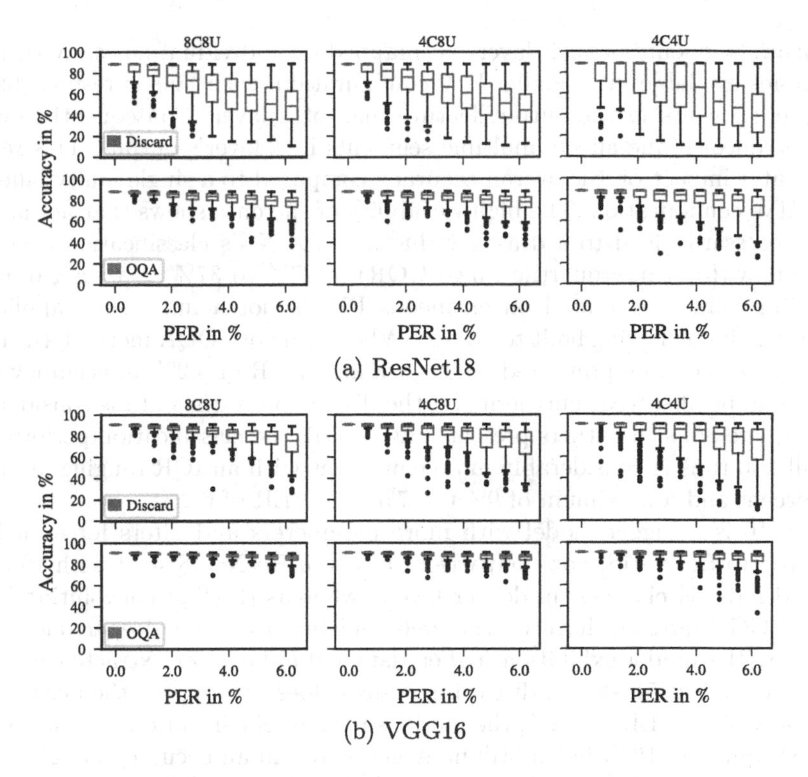

Fig. 5. Results of the fault injection simulations for CIFAR-10. Higher median and lower variability is better. Abbreviations at the top indicate the hardware configuration, where nCkU represents n clusters and k units.

all performed fault injection simulations. In each plot, the NN's classification accuracy is denoted on the y-axis and the PER on the x-axis. We investigate three different hardware configurations of our accelerator, ranging from a larger one with eight clusters and eight units (8C8U, 128 vector lanes in total) to a smaller one with four clusters and four units (4C4U, 32 vector lanes in total).

In Fig. 5a the results for ResNet18 trained on CIFAR-10 are shown. When discarding values computed on erroneous PEs, the accuracy decreases rapidly. At a PER of 2.1%, the median accuracy drops already by more than 10% for the configurations 8C8U and 4C8U. The predictions become random guesses at a PER of 5.5% where the accuracy drops to 50% for all HW configurations. For the small configuration 4C4U, the median accuracy is still equal to the baseline without faults at lower error rates. Since each PE's probability for a HW fault is based on the PER in our experiments, fewer total PEs mean a lower fault count for a smaller accelerator configuration compared to a larger one at a given PER. Overall, however, hardware faults have a greater impact on the classification accuracy of a NN for the 4C4U configuration. The whiskers and outliers in the plots show that a complete loss of classification performance is possible already at a small error rate where a single lane is faulty. As outlined in Fig. 4, computations for output segments of each layer are mapped on individual lanes. With fewer total lanes available in a smaller HW configuration, each lane is responsible for multiple segments in the output feature map of a layer. Consequently, even a single erroneous lane affects multiple segments in a layer's output. This results in a greater impact on the overall accuracy compared to a single faulty lane in a larger HW configuration. The high variability of the data shows that not all HW faults necessarily lead to a drastic reduction of a NN's classification accuracy. However, with an interquartile range (IQR) of 27% to 37% at a PER of 6.2%, discarding values computed on erroneous PEs is not a universally applicable mechanism for achieving fault tolerance. When using our OQA method, the rapid accuracy drop can be prevented. Only at a high PER of 6.2% an accuracy drop of 10% can be observed. Furthermore, the dispersion of the data is considerably lower. There are still outliers present, but overall, the classification performance over all 200 runs is considerably more consistent with an IQR ranging around a few percent and a maximum of 9% to 17% at a PER of 6.2%.

VGG16 is a bigger model with more parameters and filters for each layer compared to ResNet18. For comparison, layers of ResNet18 start with 16 channels and have 64 channels in deeper layers, whereas the first convolution layers of the VGG16 already have 64 channels and reach 512 for deeper ones. Consequently, the model exhibits a higher data redundancy of extracted features. Therefore, as Fig. 5b shows, discarding values does not impact the accuracy as drastically. At a PER of 6.2%, the model's median classification accuracy drops by approximately 16%, but it still manages to retain an accuracy of 75% for all HW configurations. Nevertheless, the variability of the model's classification performance and the number of outliers are also increasing rapidly with increasing PER. With OQA, the accuracy decreases by only 4% in the worst case for the tested configurations. Furthermore, the dispersion of the data is significantly lower with a maximum IQR of only 4% to 7% compared to the 17% to 28% without applying OQA.

The results for GTSRB shown in Fig. 6 are similar to CIFAR-10 for both models. Overall, the accuracy does not degrade as fast for this classification task, especially for VGG16. Both models have a higher baseline accuracy on this dataset. The regular shapes of traffic signs compared to the varying shapes and colors of objects and animals make GTSRB an easier classification task. This has a positive effect on the model's robustness to perturbations, which has also been observed by related work focusing on transient HW faults [2]. ResNet18 exhibits an accuracy loss of approximately 10% at a PER of 2.8% with an IQR of 21%–29% when values are discarded. Similar to CIFAR-10, the variability of ResNet18's prediction quality starts to increase quickly with an increasing error rate. At higher error rates, well-founded predictions are no longer possible. Discarding values is more beneficial for VGG16 on this dataset. The accuracy is decreased by approximately 4% with an IQR of 4%–7% at a PER of 6.2%. However, the high number of outliers illustrates that discarding is not a suitable error mitigation mechanism in this case either. With OQA, the accuracy of ResNet18 drops only by 5%–7% at a PER of 6.2% with an IQR of 9%–17%. For

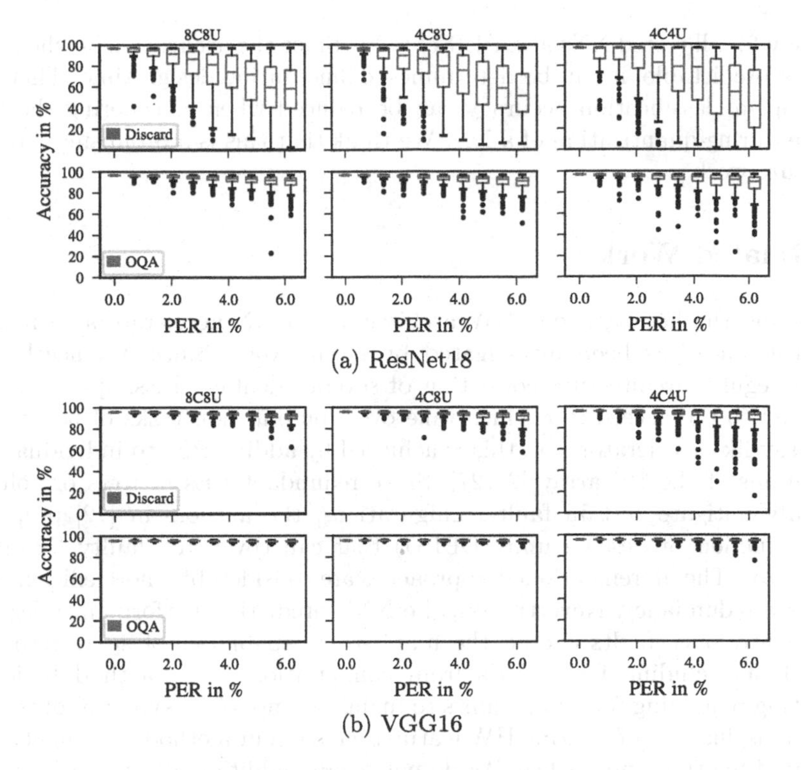

Fig. 6. Results of the fault injection simulations for GTSRB. Higher median and lower variability is better. Abbreviations at the top indicate the hardware configuration, where nCkU represents n clusters and k units.

VGG16, we are able to practically absorb all faults. The accuracy loss at a PER of 6.2% is less than a percent with a maximum IQR of 1%.

Overall, the results show that our OQA method can preserve a NN's classification performance and thus can significantly improve the fault tolerance of the system. Discarding values computed on erroneous HW is generally a valid approach where NNs can maintain their accuracy. However, the variability of the data shows that a complete loss of classification accuracy is possible even at low error rates. Ensuring that important connections of the NN contribute at least partially to feature extraction has shown to be highly beneficial. For all tested NN's, the accuracy can be preserved consistently. Furthermore, the variability of a NN's prediction performance in our experiments is decreased considerably. Compared to discarding values, the IQR for OQA is smaller for both NNs and both datasets. This allows for higher confidence in predictions made by a NN when executed on faulty HW. Although outliers also occur with our method, they can be easily handled in subsequent processing steps. In video-based perception systems, for example, objects are tracked over time and predictions can be made over multiple input frames. Considering the top two predictions is also beneficial. In this scenario, we are able to retain at least the original error-free accuracy for all tested NNs at a PER of 6.2%. Since the accuracy loss when using OQA is predictable, it can be determined offline during design time. Therefore, the drop in classification accuracy can be reduced when considering the OQA feature during quantization of a NN. We think that this is a promising direction for future work.

4 Related Work

Leveraging the homogeneous HW architecture of NN accelerators to improve fault tolerance has been investigated by recent work. Since NN accelerators have a regular architecture consisting of several identical PEs, spare PEs can be added to the accelerator which take over computation tasks of faulty ones. For array-like accelerators [5], this is achieved by adding PEs to individual rows or columns of the PE array [26,27]. Since redundant PEs in rows or columns can only mitigate certain fault configurations, the authors in [12] propose a dot-production processing unit (DPPU) that can cover any faulty PE in the accelerator. The aforementioned approaches are considerably more efficient than complete redundancy. Nevertheless, the NN's prediction performance degrades significantly once faults exceed the number or capabilities of spare resources. Instead of excluding faulty PEs from computations, our method builds on exploiting remaining functional units to maintain important synapses of the network. This dual-use of existing HW features enables our method to be lightweight compared to other approaches. We do not require additional HW resources that are not necessarily used during normal, fault-free operation. Consequently, additional costs (e.g., power or silicon area) can be avoided.

Returning zeros for computations performed on faulty PEs is a straightforward approach to mask HW faults [30]. As our experiments show, this does not guarantee that the NN's prediction accuracy is maintained. Fault-aware training has been proposed to alleviate this issue [10,30]. However, current accelerators are designed for fast and efficient inference of NNs. Consequently, retraining a NN for a given fault configuration is not possible and especially not during runtime. In contrast, our method does not require retraining of the NN to achieve fault tolerance. We can adapt to a given fault configuration during runtime. The possible performance degradation of the NN (e.g., accuracy) when using OQA can be determined offline. Therefore, optimizing a NN's quantization configuration while considering OQA could further minimize accuracy degradation.

Adapting how NNs are mapped on HW such that faulty PEs are not involved in computations has been proposed in [7]. This is a more complex solution which may cause that safety or system related timing constraints are no longer met. Our method is completely transparent to software and does not require rescheduling of the workload. However, we think that this approach can be used in addition to our method. Rescheduling can be limited to critical computations as suggested in [20]. Mapping computations that are highly relevant for preserving a NN's accuracy on still functioning PEs and more resilient ones to PEs with reduced precision can further limit the impact on classification accuracy.

5 Conclusion

In this paper, we have proposed a new method to effectively mitigate the impact towards compute availability of HW faults in NN accelerators. To the best of our knowledge, we are the first to exploit HW features that are generally added for functional reasons to improve fault tolerance. This dual-use of existing HW makes our method lightweight compared related approaches. We evaluate our method with stochastic fault injection simulations using two NNs trained on two datasets. The results confirm the effectiveness of our method in preserving a NN's classification performance.

Acknowledgements. This work is supported by the German federal ministry of education and research (BMBF), project ZuSE-KI-AVF (grant no. 16ME0062).

References

1. Baumann, R.: Radiation-induced soft errors in advanced semiconductor technologies. IEEE Trans. Device Mater. Rel. **5**(3), 305–316 (2005)
2. Beyer, M., Schorn, C., Fabarisov, T., Morozov, A., Janschek, K.: Automated hardening of deep neural network architectures. In: ASME International Mechanical Engineering Congress and Exposition (IMECE), vol. 13 (2021)
3. Camus, V., Mei, L., Enz, C., Verhelst, M.: Review and benchmarking of precision-scalable multiply-accumulate unit architectures for embedded neural-network processing. IEEE J. Emerg. Sel. Top. Circuits Syst. **9**(4), 697–711 (2019)
4. Castano, V., Schagaev, I.: Resilient Computer System Design. Springer, Cham (2015). https://doi.org/10.1007/978-3-319-15069-7
5. Chen, Y.H., Krishna, T., Emer, J.S., Sze, V.: Eyeriss: an energy-efficient reconfigurable accelerator for deep convolutional neural networks. IEEE J. Solid-State Circuits **52**(1), 127–138 (2017)
6. DeVries, T., Taylor, G.W.: Improved regularization of convolutional neural networks with cutout. arXiv preprint arXiv:1708.04552 (2017)
7. Gambardella, G., et al.: Efficient error-tolerant quantized neural network accelerators. In: 2019 IEEE International Symposium on Defect and Fault Tolerance in VLSI and Nanotechnology Systems (DFT), pp. 1–6. IEEE (2019)
8. Ghodrati, S., Sharma, H., Young, C., Kim, N.S., Esmaeilzadeh, H.: Bit-parallel vector composability for neural acceleration. In: 2020 57th ACM/IEEE Design Automation Conference (DAC), pp. 1–6. IEEE (2020)
9. He, K., Zhang, X., Ren, S., Sun, J.: Deep residual learning for image recognition. In: IEEE Conference on Computer Vision and Pattern Recognition (CVPR), pp. 770–778. IEEE (2016)
10. Kim, S., Howe, P., Moreau, T., Alaghi, A., Ceze, L., Sathe, V.S.: Energy-efficient neural network acceleration in the presence of bit-level memory errors. IEEE Trans. Circuits Syst. I **65**(12), 4285–4298 (2018)
11. Krizhevsky, A.: Learning Multiple Layers of Features from Tiny Images. Technical report (2009)
12. Liu, C., et al.: HyCA: a hybrid computing architecture for fault tolerant deep learning. IEEE Trans. Comput.-Aided Des. Integr. Circuits Syst. **41**(10), 3400–3413 (2022)

13. Loshchilov, I., Hutter, F.: SGDR: stochastic gradient descent with warm restarts. In: International Conference on Learning Representations (2017)
14. Menard, D., Novo, D., Rocher, R., Catthoor, F., Sentieys, O.: Quantization mode opportunities in fixed-point system design. In: 2010 18th European Signal Processing Conference, pp. 542–546. IEEE (2010)
15. Nolting, S., Giesemann, F., Hartig, J., Schmider, A., Paya-Vaya, G.: Application-specific soft-core vector processor for advanced driver assistance systems. In: 2017 27th International Conference on Field Programmable Logic and Applications (FPL), pp. 1–2. IEEE (2017)
16. Ozen, E., Orailoglu, A.: Sanity-check: boosting the reliability of safety-critical deep neural network applications. In: 2019 IEEE 28th Asian Test Symposium (ATS), pp. 7–75. IEEE (2019)
17. Ozen, E., Orailoglu, A.: Boosting bit-error resilience of DNN accelerators through median feature selection. IEEE Trans. Comput.-Aided Des. Integr. Circuits Syst. **39**(11), 3250–3262 (2020)
18. De la Parra, C., Guntoro, A., Kumar, A.: ProxSim: GPU-based simulation framework for cross-layer approximate DNN optimization. In: 2020 Design, Automation and Test in Europe (DATE), pp. 1193–1198. IEEE (2020)
19. Reagen, B., et al.: Ares: a framework for quantifying the resilience of deep neural networks. In: 2018 55th ACM/ESDA/IEEE Design Automation Conference (DAC), pp. 1–6. Association for Computing Machinery (2018)
20. Schorn, C., Guntoro, A., Ascheid, G.: Accurate neuron resilience prediction for a flexible reliability management in neural network accelerators. In: 2018 Design, Automation and Test in Europe (DATE), pp. 979–984. IEEE (2018)
21. Sharma, H., et al.: Bit fusion: bit-level dynamically composable architecture for accelerating deep neural network. In: 2018 ACM/IEEE 45th Annual International Symposium on Computer Architecture (ISCA), pp. 764–775. IEEE (2018)
22. Simonyan, K., Zisserman, A.: Very deep convolutional networks for large-scale image recognition. arXiv preprint arXiv:1409.1556 (2014)
23. Song, Z., et al.: DRQ: dynamic region-based quantization for deep neural network acceleration. In: 2020 ACM/IEEE 47th Annual International Symposium on Computer Architecture (ISCA), pp. 1010–1021 (2020)
24. Stallkamp, J., Schlipsing, M., Salmen, J., Igel, C.: Man vs. computer: benchmarking machine learning algorithms for traffic sign recognition. Neural Netw. **32**, 323–332 (2012)
25. Sze, V., Chen, Y.H., Yang, T.J., Emer, J.S.: Efficient processing of deep neural networks: a tutorial and survey. Proc. IEEE **105**(12), 2295–2329 (2017)
26. Takanami, I., Fukushi, M.: A built-in circuit for self-repairing mesh-connected processor arrays with spares on diagonal. In: 2017 IEEE 22nd Pacific Rim International Symposium on Dependable Computing (PRDC), pp. 110–117 (2017)
27. Takanami, I., Horita, T.: A built-in circuit for self-repairing mesh-connected processor arrays by direct spare replacement. In: 2012 IEEE 18th Pacific Rim International Symposium on Dependable Computing (PRDC), pp. 96–104 (2012)
28. Thieu, G.B., et al.: ZuSE-KI-AVF: application-specific AI processor for intelligent sensor signal processing in autonomous driving. In: 2023 Design, Automation and Test in Europe (DATE) (2023)
29. Vogel, S., Springer, J., Guntoro, A., Ascheid, G.: Self-supervised quantization of pre-trained neural networks for multiplierless acceleration. In: 2019 Design, Automation and Test in Europe (DATE), pp. 1094–1099. IEEE (2019)
30. Zhang, J.J., Basu, K., Garg, S.: Fault-tolerant systolic array based accelerators for deep neural network execution. IEEE Des. Test **36**(5), 44–53 (2019)

Neural Networks and Testing

Evaluation of Parameter-Based Attacks Against Embedded Neural Networks with Laser Injection

Mathieu Dumont[1,2], Kevin Hector[1,2], Pierre-Alain Moëllic[1,2(✉)], Jean-Max Dutertre[3], and Simon Pontié[1,2]

[1] CEA Tech, Centre CMP, Equipe Commune CEA Tech - Mines Saint-Etienne, 13541 Gardanne, France
{mathieu.dumont,kevin.hector,pierre-alain.moellic,simon.pontie}@cea.fr

[2] Univ. Grenoble Alpes, CEA, Leti, 38000 Grenoble, France

[3] Mines Saint-Etienne, CEA, Leti, Centre CMP, 13541 Gardanne, France
dutertre@emse.fr

Abstract. Upcoming certification actions related to the security of machine learning (ML) based systems raise major evaluation challenges that are amplified by the large-scale deployment of models in many hardware platforms. Until recently, most of research works focused on API-based attacks that consider a ML model as a pure algorithmic abstraction. However, new implementation-based threats have been revealed, emphasizing the urgency to propose both practical *and* simulation-based methods to properly evaluate the robustness of models. A major concern is parameter-based attacks (such as the Bit-Flip Attack - BFA) that highlight the lack of robustness of typical deep neural network models when confronted by accurate and optimal alterations of their internal parameters stored in memory. Setting in a security testing purpose, this work practically reports, for the first time, a successful variant of the BFA on a 32-bit Cortex-M microcontroller using laser fault injection. It is a standard fault injection means for security evaluation, that enables to inject spatially and temporally accurate faults. To avoid unrealistic brute-force strategies, we show how simulations help selecting the most sensitive set of bits from the parameters taking into account the laser fault model.

Keywords: Hardware Security · Fault Injection · Evaluation and certification · Machine Learning · Neural Network

1 Introduction

The massive deployment of Machine Learning (ML) models in a large spectrum of domains raises several security concerns related to their integrity, confidentiality and availability. For now, most of the research efforts are essentially focused on models seen as abstractions, i.e. the attack surface is focused on so-called

M. Dumont—At SGS Brightsight after paper submission: mathieu.dumont@sgs.com.

J. Guiochet et al. (Eds.): SAFECOMP 2023, LNCS 14181, pp. 259–272, 2023.
https://doi.org/10.1007/978-3-031-40923-3_19

API-based attacks, excluding threats related to their implementation in devices that may be physically accessible by an adversary, as it the case for embedded ML models. The flaws intrinsically related to the models and the *physical* ones may be used jointly to attack an embedded model or exploit data leakages.

A typical and well-known API-based attack are the *adversarial examples* [1] that aims at fooling the prediction of a model with input-based alterations at inference time. However, recent works demonstrated that physical attacks are realist threats by targeting critical elements of a model such as activation functions [2] or its parameters (parameter-based attacks) [3] as studied in this work.

In such a security context, with the demonstration of worrying attack vectors, there is an urgent need of certification for AI-based systems more especially for *critical* ones[1]. Therefore, alongside the demonstration of new attacks and the development of defenses, an important challenge relies on the availability of proper robustness evaluation and accurate characterization methods in addition to both simulation and experimental tools and protocols. Model robustness evaluation is one of the most important challenge of modern artificial intelligence and several remarkable works from the Adversarial Machine Learning community have already raised major issues for adversarial examples [4], with many defenses relying on weak evaluations [5].

Dealing with embedded neural network models and weight-based adversarial attacks, this challenge encompasses both *safety* and *security* concerns: parameters stored in memory may be altered by random faults because of hostile environments or strong energy consumption limitations [6] and also be the target of an adversary that aims at optimally threatening a model.

2 Related Works and Objectives

2.1 Parameter-Based Attacks

Implementation-Based Threats. An important part of the ML security literature concerns *algorithmic* or so-called *API-based attacks* that exploit input/output pairs and additional knowledge from the model in case of white-box attacks. A large body of work shows that these threats concern every stage of the ML pipeline [7] and threaten the confidentiality (model and data), integrity and availability of the models. However, these attacks do not consider that the adversary may have a direct interaction with the algorithm as it can be the case for an embedded AI system. *Implementation-based attacks* precisely exploit software or hardware flaws as well as the specific features of the device. For example, *side-channel analysis* [8] have been demonstrated for model extraction as an efficient way to extract information from the model architecture or the values of the parameters [9]. Alongside safety-related efforts that evaluate the robustness of ML models against random faults [6], some works demonstrate that models are highly sensitive to *fault injection analysis* [10] that alter the data, the parameters as well as the instructions flow [2,11].

[1] See the European AI Act: https://artificialintelligenceact.eu/.

Weight-Based Adversarial Attacks. New attack vectors have been highlighted and more essentially parameter-based attacks (also named *weight-based adversarial attacks*). Let's consider a supervised neural network model $M_W(x)$, with parameters W (also referred as *weights*), trained to optimally map an input space $\mathcal{X} = \mathbb{R}^d$ (e.g., images) to a set of labels $\mathcal{Y}$. M is trained by minimizing a loss function $\mathcal{L}\big(M_W(x), y\big)$ (typically the cross entropy for classification task) that quantifies the error between the prediction $\hat{y} = M_W(x)$ and the correct label y. As formalized in [3] or [6] with Eq. 1, a parameter-based attack aims at maximaizing the loss (i.e., increase mispredictions) on a small set of N test inputs. As for the imperceptibility criterion of adversarial examples, the attacker may add a constrain over the perturbation by bounding the bit-level Hamming distance (HD) between the initial (W) and faulted parameters (W'), corresponding to an *adversarial budget* S.

$$\underbrace{\max_{W'} \sum_{i=0}^{N-1} \mathcal{L}\Big(M(x_i; W'), y_i\Big)}_{\text{mispredictions}} \text{ s.t. } \overbrace{HD(W', W) \leq S}^{\text{advbudget}} \tag{1}$$

A state-of-the-art parameter-based attack is the Bit-Flip Attack (hereafter BFA) [3] that aims to decrease the performance of a model by selecting the most sensitive bits of the stored parameters and progressively flip these bits until reaching an adversarial goal. In [3] or [12], the objective is to ultimately degrade the model to a random-guess level. The selection of the bits is based on the ranking of the gradients of the loss w.r.t. to each bit $\nabla_b \mathcal{L}$, computed thanks to a small set of inputs. First, each selected bit is flipped (and restored) to measure the impact on the model accuracy. Then, the most sensitive bit is permanently flipped according to the gradient ascendant as defined in [3].

Adversarial Goals. Parameter-based attacks are not limited to the alteration of the target model integrity. BFA has been recently demonstrated for powerful model extraction in [13] with an Intel i5 CPU platform: RowHammer is used to perform a BFA on the parameters of a model stored in DRAM (DDR3). The threat model in [13] follows a typical model extraction setting: the adversary knows the model's architecture but not the internal parameters and has only access to a limited portion of the training dataset ($< 10\%$). His goal is to build a *substitute* model as close as possible as the target model. Interestingly, this joint use of RowHammer and BFA is performed in a side-channel analysis fashion: the observation of the induced faults enables to make assumptions on the value of some bits of the parameters. Then, the knowledge of these bits enables to efficiently train a substitute model, by constraining the value range of the parameters, with high *fidelity* compared to the target model. We discuss this goal in Sect. 4.4.

2.2 Scope and Objectives

Because parameter-based attacks are the basis of both powerful integrity and confidentiality threats, their practical evaluation on the different platforms where fault injection may occur is becoming a critical need for present and future standardization and certification actions of critical AI-systems. We position our work on a different set of platforms than the main works related to the BFA (CPU platforms with DRAM), that is MCU platforms (Cortex-M with Flash memory), yet a very important family of embedded AI systems regarding the massive deployment of ML models on MCU-based devices for a large variety of domains. Our main positioning is as follows:

- We set this work in a security evaluation and characterization context. Therefore, we do not position ourselves through an *adversary* but an *evaluator* point of view.
- Our scope is parameter-based threats for neural network embedded in 32-bit microcontroller (hereafter MCUs).
- For that purpose, we use Laser Fault Injection (hereafter LFI) as an advanced and very spatially and temporally accurate injection means, a reference technique that is used in many security evaluation centers.

State-of-the-art is focused on simulation-based evaluations or on RowHammer attacks (i.e., exclusively DRAM platforms that excludes MCU). To the best of our knowledge, this work is the first to demonstrate the practicability and suitability of the characterization of a weight-based adversarial perturbation against Cortex-M MCU thanks to LFI.

2.3 Related Works

Since the presentation of the BFA in [3], several works analyzed the intrinsic mechanisms of the attack as well as potential protections [12,14] and evaluated its properties according to the threat model, training parameters and model architecture [15]. The standard BFA is *untargeted* since the induced misclassifications are not chosen by the adversary. Therefore, some works also proposed *targeted* versions [16] with specific target inputs and/or labels. Other recent works propose alternative methods to efficiently select the most sensitive parameters to attack [6]. For our work, we use and adapt (for LFI) the standard BFA of [3] as it is the state-of-the-art baseline for weight-based adversarial attacks.

To the best of our knowledge, the only work related to laser injections for MCU-based platform against embedded neural networks has been proposed in [2]. Our work differs significantly by the target device and the elements we target in the model. In [2], the authors used an 8-bit microcontroller (`ATMega328P`) and a 32-bit neural network implemented in C (a multilayer perceptron trained on the MNIST data set). They only focused on the activation functions by inducing instruction skips (i.e., the faulted instructions are not executed, as if they were skipped). They used a laser and only targeted the last hidden layer. We used a Cortex-M 32-bit microcontroller and embedded 8-bit quantized neural network thanks to a state-of-the-art open source library (*Neural Network on Microcontrollers*[2], hereafter NNoM) for model deployment. Our attack vector

[2] https://github.com/majianjia/nnom/.

(BFA-like attack) and fault model enable to evaluate the robustness of a model against an advanced adversary that aims at significantly altering the performance of a model with a very limited number of faults or extract information about parameter values for model extraction.

Thus, our work is also closely linked to [11] that demonstrated at USENIX'20 a complete exploitation of the BFA with RowHammer on an Intel i7-3770 CPU platform. Although we target a different type of platforms with 32-bit MCU and another fault injection means (LFI), we share the same objective to go beyond simulations and propose a complete practical evaluation of a parameter-based attacks against embedded quantized DNN models.

3 Evaluator Assumptions and Experimental Setups

3.1 Goal and Evaluator Assumptions

Objectives. The main objective of the evaluator is to evaluate the robustness of a model against precise fault injections by decreasing the average accuracy on a labelled test set. More precisely, the scenario corresponds to a generic *untargeted* case (i.e., the incorrect labels are not controlled by the evaluator). Note that a targeted scenario (for specific test samples or target labels) is possible [16] but out of the scope of our experiments. A secondary objective is to minimize the evaluation cost with a strategy that reduces the number of faults to be injected (i.e., avoid an exhaustive search that may be unrealistic according to the complexity of the target model).

Evaluator Hypothesis. Classically for security testing, the evaluator simulates a worst-case adversary that has a perfect knowledge of the model (white-box attack) and is able to query the model without limitation. The evaluator has a full access to the device (or clones of the device) and can perform elementary characterizations to adapt and optimize the fault injection set-up.

3.2 Single Bit-Set Fault Model on Flash Memory

We consider an accurate fault model relevant for LFI previously explained and demonstrated for NOR-Flash memory of Cortex-M MCU by Colombier and Menu [17]: the bit-set fault model. As its name suggests, the fault sets a targeted bit to a logical `1`: when the bit was already at `1` the fault has no effect. When targeting a Flash memory at read time with a laser pulse, the induced bit-set is transient: it affects the data being read at that time while the stored value is left unmodified. Authors from [17] explained the underlying mechanism of the bit-set fault injection with the creation of a *photoelectric current* induced by the laser in a floating-gate (FG) transistor that flows from its drain to the ground. This current is added to the legitimate one so that the total current is above the reference that makes the bit read as a logical `1`.

LFI is a local fault injection means: its effect is restricted to the bit line connected to the FG transistor inside the laser spot area. More precisely several current components are induced in the affected transistors, depending on the laser spot diameter: up to two adjacent bits can be faulted simultaneously [17].

3.3 Target and Laser Bench Setup

Device Under Test (DUT). Our target board embeds an ARM Cortex-M3 running at 8 MHz. It includes 128 kB of Flash memory and is manufactured in the 90 nm CMOS technology node. The dimension of the chip is 3 x 2.5 mm. Since LFI requires the surface of the die being visible, the microcontroller packaging was milled away with engraving tools to provide an access to its laser-sensitive parts. The chip was then mounted into a test board compatible with the ChipWhisperer CW308 platform.

Laser Platform. Our laser fault analysis platform integrates two independent laser spots with a near infrared (IR) wavelength of 1, 064 nm, focused through the same lens. Each laser spot has a diameter ranging from 1.5 to 15 μm depending on the lens magnification. Both spots can move inside the whole field of view of the lens with minimum distortion. The laser source can reach a maximum given power of 1, 700 mW. The delay between the trigger and the laser shot can be adjusted with a step of a few nanoseconds. An infrared camera is used to observe the laser spot location on the target and a XY stage enables to move the objective above the entire DUT surface.

3.4 Datasets and Models

Although simulation-based works exclusively used complex deep neural networks trained for vision tasks (e.g., ResNet on ImageNet), it does not represent a large part of real-world applications that take benefit from classical fully-connected architecture (or multilayer perceptron, hereafter MLP) that fit and perform well on the widespread constrained platforms we studied in this paper.

We considered two classical datasets. The **IRIS** dataset consists of 150 samples, each containing 4 real-value inputs and labelled according to 3 different iris species. We trained two simple models that provide an easy insight on the bit-set fault model effect: `IRIS_A` and `IRIS_B` are both composed with one hidden layer with one neuron and four neurons respectively. `IRIS_B` has 96% accuracy on the test set. **MNIST** dataset is composed of gray-scale handwritten digits images (28×28 pixels) from `0` to `9`. Our model (noted `MNIST`) is a MLP with one hidden layer of 10 neurons. Inputs are compressed to $\mathbb{R}^{50}$ with a classical principal component analysis. The resulting model has 620 trainable parameters and reaches 92% of accuracy on the test set. All models use ReLU as activation function.

3.5 Model Implementation on MCU

Models were trained with TensorFlow. Few tools are available to embed previously trained models in microcontroller boards such as TensorFlow Lite, X-Cube-AI or NNoM. We chose NNoM as it is an efficient and convenient open source platform that fits our security testing objectives. NNoM offers 8-bit model quantization (with a standard uniform symmetric powers-of-two quantization scheme) and a complete white-box access to the inference code that enables to

```
1  while (rowCnt){
2      //pA : address, stored input
3      //pB : address, stored weight
4      for (int j = 0; j < dim_vec; j++)
       { //loop on all neuron
       parameters
5          q7_t inA = *pA++;      //load
       input to inA, address increment
6          q7_t inB = *pB++;     //load
       weight to inB, address increment
7          ip_out += inA * inB; //neuron
         weighted sum
8      }
9      *pO++ = (q7_t)__NNOM_SSAT((ip_out
         >> out_shift), 8);
10     rowCnt--;}
```

(a) C code of the weighted-sum computation in a fully-connected layer.

```
1 ;q7_t inB = *pB++              ;Weight n+1
          initialization
2 ldr       r3, [r7, #80]  ;Loading the
          address of the weight n
3 adds      r2, r3, #1         ;Next weight
          address
4 str       r2, [r7, #80]   ;Input value
          loading into r2 reg
5 ldrsb.w r3, [r3]             ;Weight value
          loading. LASER SHOT
6 strb      r3, [r7,#23]   ;Store of the
          weight in SRAM reg
```

(b) Assembler code of line 6 of listing 1a. Our target is the load instruction, line 5.

Fig. 1. C and Assembler codes from the NNoM inference. (Color figure online)

draw a timing profile of the sensitive operations we target (reading values from the Flash memory).

Listing 1a is an extract of the C code source of the core calculation of an inference from NNoM, that is the weighted sum between the inputs (i.e. the input data or the outputs of the previous layer) and the model parameters before the non-linear activation is applied. It consists in loading the neuron input and weight values (`inA` and `inB`, line 5 and 6), then process the multiplication and accumulation in an intermediate output value (`ip_out`, line 7). Line 9 corresponds to the quantization. The assembler code in Listing 1b corresponds to the weight initialization of Listing 1b, line 6. The weight value, stored into the Flash memory, is loaded into register `r3` (line 5 in red). Based on our single bit-set fault model, if a laser beam is applied during the execution of the load instruction, a bit-set could be induced directly on the loaded value. To characterize the impact of the laser pulse, we synchronize the laser thanks to a trigger signal in the C code before line 6 and monitor the parameter value before and after the trigger using UART communication.

4 A Parameter-Based Attack with LFI

To analyze the efficiency and practicality of LFI we first demonstrated the accuracy of the induced faults on a single neuron composed of four weights. Then, we scaled up to functional models trained on IRIS and MNIST to analyze the impact at a model-level.

4.1 Initial Characterization on a Neuron

An important preliminary experience is to set up our laser bench on our DUT. For conciseness purpose, we do not detail all the Flash memory mapping procedure neither the selection of the laser parameters. For that purpose, we thoroughly followed the experimental protocol from [17] and fixed the laser power to

170 mW and the pulse width to 200 ns. By selecting the lens magnification ×5, we chose a spot diameter of 15 µm to have a wide laser effective area.

Then, we implemented a 4-weights neuron. We set the Y-position to 100 µm to only focus on the X-axis motion. Figure 2a shows that, when moving along the X-axis of the Flash memory, bit-sets are induced one after another on the whole 32-bit word line and the four quantized weights are precisely faulted. We repeated this experience with different weight values and noticed a perfect reproducibility, as also reported in [17]. By noticing positions of weights and their most significant bit (MSB), we easily conclude the little endian configuration of the Flash memory. Figure 2b illustrates how the weights are stored according to the laser bench X-axis.

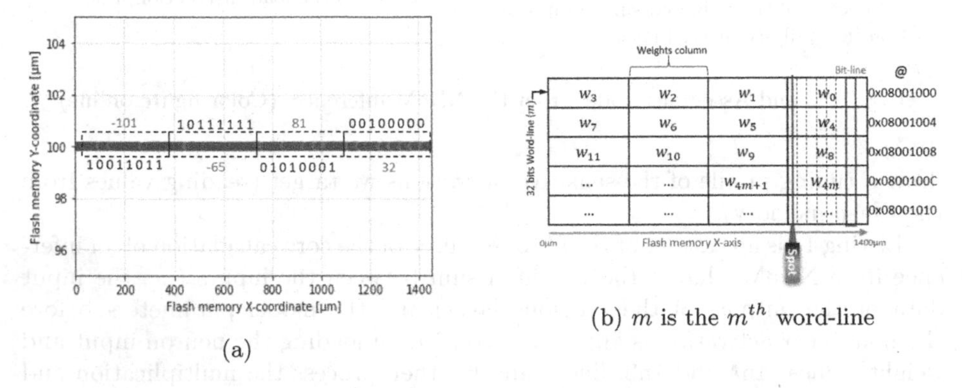

Fig. 2. (a) Bit-set faults (red dots) induced on the 4 weights of a neuron (IRIS_A model). (b) Flash memory schematic section filled with quantified weights. (Color figure online)

4.2 Target Multilayer Perceptron Models

Since we can change the value of parameters related to a neuron, the next step consisted in analyzing the impact of such faults on the integrity of a target model and in measuring the potential drop of accuracy. For that purpose, we used the IRIS_B model. For each X-step, we evaluated the robustness of the model by feeding it with 50 test samples (i.e., 50 inferences). Classically, we computed the adversarial accuracy by comparing the output predictions to the correct labels and measure the accuracy drop in comparison to the nominal performance without faults.

During one inference, all the weight loading instructions triggered a laser shot (i.e. in total 40 shots for the hidden layer). By targeting one bit line at a given X-position of the laser, only weights on the same address column are faulted with a bit-set. For illustration, as pictured in Fig. 2b, LFI actually induced bit-sets in the MSBs of weights w_0, w_4, w_8, w_{4m} (with m the mth word-line), which belong to the targeted bit line.

Figure 3a shows the impact of the laser shots on the model accuracy (blue curve) on the test set. The red curve provides the number of faulted bits. The

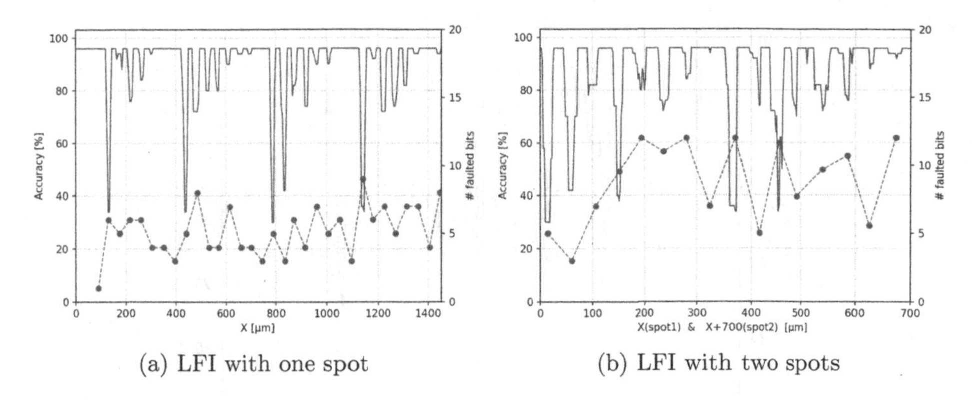

(a) LFI with one spot

(b) LFI with two spots

Fig. 3. Impact on the accuracy of LFI on `IRIS_B` model. Average accuracy over 50 inferences (blue) and number of faulted bits (red). (Color figure online)

X-axis corresponds to the position of the laser in the Flash memory. Four regular patterns of accuracy drop corresponding to the four stored weights are clearly visible. The observed accuracy drops (5 to 6 per pattern) match the coordinates of the MSBs with a decreasing correlation. For MSB bits, the model is close to a random-guess level (i.e. around 30%). On average, around 6 bit-sets are induced and the most harmful configuration corresponds to only 5 bit-sets (at 790 µm) leading to an adversarial accuracy of 30%.

One limitation is that only 1/4 of all weights could be faulted during one inference. However, since our laser platform integrates a second spot, we experimented a configuration where one spot targets the first bit line (i.e the MSB of the weight column w_3 in Fig. 2b) and the second spot targets the bit line #16 (i.e. the MSB of the weight column w_1). As a result, two weight columns can be targeted during one inference, and half of the weights are likely to be faulted. As reported in Fig. 3b more bit-set faults were injected (up to 12), and the accuracy drop for non-MSB bits is more important.

To extend the previous experiments, we embedded the `MNIST` model that gathers 620 weights (i.e. 4,960 bits) and strictly kept the same experimental setup. Results are presented in Fig. 4. On average, we observe more induced bit-sets with the most important drop of accuracy (65.5%) reached with only 25 bit-sets at $X = 1,100\,\mu m$. We also point out that, in some cases, a very slight improvement of the accuracy appeared.

In comparison with `IRIS_B` (Fig. 3a), we note that bit lines are less distinctive. Indeed, even if the position of the most significant bits can be identified, other bits are difficult to distinguish. Actually, in Fig. 2b the bit line (green box) represents all single bit lines connected to the same bit index for different 32-bit words addresses. Since the `MNIST` model has more parameters, almost all bit lines are linked to a weight parameter value stored in Flash memory. The laser spot, with a size of 15 µm, is larger than a single bit line, explaining why bit line indexes are hardly discernible. Moreover, depending on the laser position, the effective area of the spot can encompass two different bit line indexes.

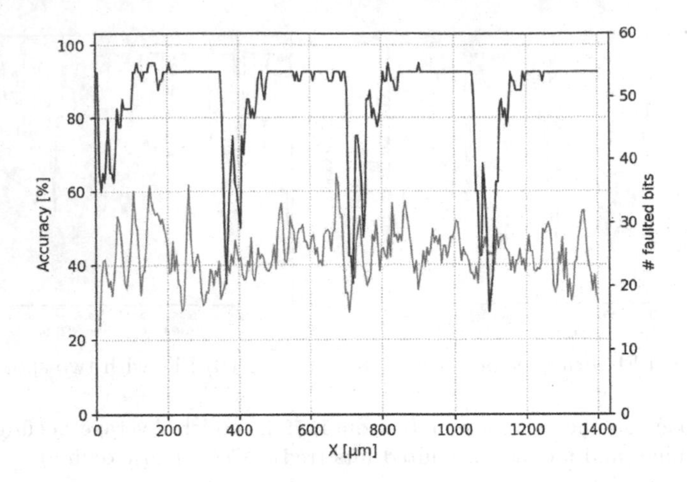

Fig. 4. Impact on the accuracy of LFI on a MLP model trained on MNIST. Average accuracy over 100 inferences (blue) and number of faulted bits (red). (Color figure online)

4.3 Advanced Guided-LFI

Adapt BFA to LFI. So far, we exhaustively attacked all the parameters stored in memory. This *brute-force* strategy may be impractical with deeper models. Therefore, an important objective for an evaluator is to optimally select the most sensitive bits to fault. For that purpose, as in [11], we adapted the bit selection principle of the standard BFA[3] to our fault model. We refer to this adapted attack as BSCA (*Bit-Set Constrained Attack*). We also adapted the adversarial objective by introducing an *adversarial budget* [6,15] representing a maximum of faults the evaluator is able to process. Indeed, the random-guess level objective of [3] overestimates the number of faults and raises variability issues because of the last necessary bit-flips needed to ultimately reach the objective [15]. Therefore, we set the adversarial budget to 20 bit-sets.

As inputs, the BSCA has the target model M_W, its parameters W, the weight column index m in the Flash memory, the bit line index b and the adversarial budget S. The output of the BSCA is a new model $M_{W'}^{m,b}$ with W' and W that differ only by at most S bit-sets. Then, we can compute an adversarial accuracy on $M_{W'}^{m,b}$ and measure the accuracy drop compared to the nominal performance of $M_W^{m,b}$. The BSCA proceeds through the following steps:

1. The BFA ranks the most sensitive bits of W according to $\nabla_b \mathcal{L}$.
2. Exclude the bits already set to 1 and not related to m and b.
3. Pick the best bit-set and perform the fault permanently in M.
4. Repeat the process until reaching S. The output is the faulted model $M_{W'}^{m,b}$.

[3] [3]: https://github.com/elliothe/BFA/.

Finally, we perform BSCA over all the weight columns and bit line and keep the faulted model, simply referred as $M_{W'}$, with the worst accuracy (Acc) evaluated on a test set (X^{test}, Y^{test}): $M_{W'} = \arg\min_{m,b} Acc(M_{W'}^{m,b}, X^{test}, Y^{test})$

We used the `MNIST` model and simulated the BSCA to find the 20 most significant weights to fault for each weight column. For illustration purpose, Fig. 5a shows the effect of bit-set induced on the 8 bits of a weight only for the second weight column (with the MSB refered as Bit #0). Among the most sensitive weights from the second column, only few bit-sets on the most significant bits efficiently alter the model performance, bit-sets on other bits have less or even no influence.

Experiments and Results. The basic idea is to use the BSCA to guide the LFI. For that purpose, we need to put to the test that LFI reach (near) identical performance than what expected by simulations. We ran a BSCA simulation (in Python) over all the weight columns and bit lines that pointed out the MSB of the second column weight as the most sensitive. Therefore, contrary to the previous experiments, the laser source was triggered only when the 20 most sensitive weights were read from the Flash. The laser location was set accordingly ($X = 760\,\mu\text{m}$) and the power increased to 360 mW to ensure a higher success rate on weights stored in distant addresses.

The blue curve in Fig. 5b represents our experimental results (mean accuracy over 100 inferences) while the red one is the BSCA simulations for the MSB. First, we can notice that experimental and simulation results are almost similar, meaning that we can guide our LFI with high reliability and confidence[4]. For an adversarial budget of only 5 bit-sets (0.1% faulted bits) the embedded model accuracy drops to 39% which represents a significant loss and a strong integrity impact compared to the nominal performance of 92%. Moreover, after 10 bit-sets (accuracy to 25%), the most effective faults have been injected and the accuracy did not decrease anymore. In a security evaluation context, this observation positions the level of robustness of the model according to the adversarial budget.

4.4 Exploitation of LFI for Model Extraction

To illustrate the diversity of the adversarial goal an evaluator aims to evaluate (Sect. 2.1), we propose a first insight of the use of LFI and BSCA for model extraction. At S&P 2022, Rakin *et al.* [13] demonstrated how to exploit BFA with RowHammer (i.e., exclusively DRAM platforms) in a model extraction scenario. The goal is to steal the performance of a model of which an adversary knows the architecture and a small part of the training data but not the internal parameters. The idea of [13] is to guess the value of the maximum of MSBs of the model thanks to rowhammer (bit-flip) and then to train a substitute model with these values as a training constraint.

[4] The fact that experimental results are slightly more powerful than simulations may be explained by the impact of the width of the laser spot on nearby memory cells.

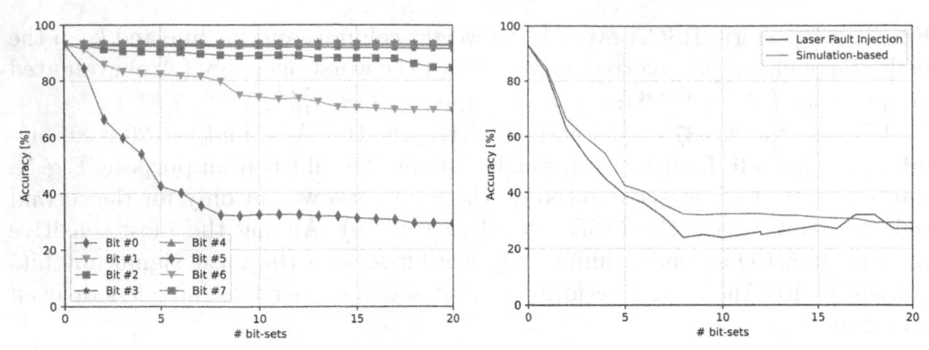

(a) Influence of the faulty bit number on the accuracy for the weight column 2 (simulation results).

(b) LFI BSCA attack targeting the 20 most sensitive MSB of the 2nd weight column.

Fig. 5. Guided-LFI on a MLP model trained on MNIST. (Color figure online)

Even though we rely on a different fault model (bit-set with LFI), it is relatively straightforward to follow the same extraction method as in [13]. Indeed, by comparing the outputs of the model on a set of inputs (that could be random ones) with and without LFI we can guess the value of the targeted bit: if the *faulted* outputs are not strictly identical than the nominal ones, that means the bit-set has an effect and therefore the bit was at 0. Otherwise, the bit value was already to 1 or the bit-set has no influence on the outputs. We simulated the BSCA to a deeper MNIST model with 3 layers with respectively 128, 64 and 10 neurons (109K parameters) and used 500 random input images to compare the outputs with and without bit-set faults on the MSB of each weight. To remove ambiguity, we made the simple assumption that working with enough inputs, a bit-set on a MSB at 0 has always an impact on the outputs. With that method, we managed to extract 91.9% of the MSB of the model parameters, which is enough to efficiently trained a substitute model, consistently to [13]. Further analysis and experiments (more particularly, impact of the model architecture and the input set) should be the subject of future dedicated publications.

5 Conclusion

This work aims to contribute to the development of reliable evaluation protocols and tools for the robustness of embedded neural network models, a growing concern for future standardization and certification actions of critical AI-based systems. We conclude by highlighting some research tracks that may pursue this work and fill some limitations with further analysis on other model architectures and platforms (e.g., system-on-chip).

First, our results are not limited by the model complexity (BFA has been demonstrated on state-of-the-art deeper models with millions of parameters) but by the complexity of the MCUs. Indeed, the challenge is to characterize how the

Flash memory is organized so that the evaluator can precisely target the weight columns and bit lines. Therefore, further experiments would be focused to other targets (e.g., Cortex M4 and M7).

Second, contrary to adversarial examples with recent attacks specifically designed for robustness evaluation purposes [18], parameter-based attack still lacks of maturity and recent works highlight limitations of the BFA [6,15] and propose improvements or alternatives. Thus, considering or combining more attack methods would improve the evaluation by simulating a more powerful adversary.

Finally, this work could be widen to the practical evaluation of protections against weight-based adversarial attacks. Additionally to traditional generic countermeasures against fault injection [10], specific defense schemes against BFA encompass weight clipping, clustering-based quantization [2], code-based detectors [19] or adversarial training [6]. To the best of our knowledge, none of these defenses have been practically evaluated against fault attack means such as RowHammer, glitching or LFI. As for adversarial examples (with so many defenses regularly *broken* afterwards) the definition of proper and sound evaluations of defenses against parameter-based attacks is a research action of the highest importance.

Acknowledgment. This work is supported by (CEA-Leti) the European project InSecTT (www.insectt.eu, ECSEL JU 876038) and by the French ANR in the *Investissements d'avenir* program (ANR-10-AIRT-05, irtnanoelec); and (MSE) by the ANR PICTURE program (https://picture-anr.cea.fr). This work benefited from the French Jean Zay supercomputer with the AI dynamic access program.

References

1. Szegedy, C., et al.: Intriguing properties of neural networks. In: International Conference on Learning Representations (2014)
2. Hou, X., Breier, J., Jap, D., Ma, L., Bhasin, S., Liu, Y.: Security evaluation of deep neural network resistance against laser fault injection. In: Proceedings of the International Symposium on the Physical and Failure Analysis of Integrated Circuits, IPFA, pp. 1–6 (2020)
3. Rakin, A.S., He, Z., Fan, D.: Bit-flip attack: crushing neural network with progressive bit search. In: Proceedings of the IEEE/CVF International Conference on Computer Vision (ICCV), October 2019
4. Carlini, N., et al.: On evaluating adversarial robustness. arXiv preprint arXiv:1902.06705 (2019)
5. Tramer, F.: Detecting adversarial examples is (nearly) as hard as classifying them. In: International Conference on Machine Learning, pp. 21692–21702. PMLR (2022)
6. Stutz, D., Chandramoorthy, N., Hein, M., Schiele, B.: Random and adversarial bit error robustness: energy-efficient and secure DNN accelerators. IEEE Trans. Pattern Anal. Mach. Intell. (2022)
7. Papernot, N., McDaniel, P., Sinha, A., Wellman, M.P.: SoK: security and privacy in machine learning. In: IEEE European Symposium on Security and Privacy (EuroS&P), pp. 399–414 (2018)

8. Mangard, S., Oswald, E., Popp, T.: Power Analysis Attacks: Revealing the Secrets of Smart Cards. Springer, New York (2007). https://doi.org/10.1007/978-0-387-38162-6
9. Joud, R., Moëllic, P.-A., Pontié, S., Rigaud, J.-B.: A practical introduction to side-channel extraction of deep neural network parameters. In: Buhan, I., Schneider, T. (eds.) CARDIS 2022. LNCS, vol. 13820, pp. 45–65. Springer, Cham (2023). https://doi.org/10.1007/978-3-031-25319-5_3
10. Barenghi, A., Breveglieri, L., Koren, I., Naccache, D.: Fault injection attacks on cryptographic devices: theory, practice, and countermeasures. Proc. IEEE **100**(11), 3056–3076 (2012)
11. Yao, F., Rakin, A.S., Fan, D.: DeepHammer: depleting the intelligence of deep neural networks through targeted chain of bit flips. In: 29th USENIX Security Symposium, pp. 1463–1480 (2020)
12. He, Z., Rakin, A.S., Li, J., Chakrabarti, C., Fan, D.: Defending and harnessing the bit-flip based adversarial weight attack. In: Proceedings of the IEEE/CVF Conference on Computer Vision and Pattern Recognition, pp. 14095–14103 (2020)
13. Rakin, A.S., Chowdhuryy, M.H.I., Yao, F., Fan, D.: Deepsteal: advanced model extractions leveraging efficient weight stealing in memories. In: 2022 IEEE Symposium on Security and Privacy (SP), pp. 1157–1174. IEEE (2022)
14. Liu, L., Guo, Y., Cheng, Y., Zhang, Y., Yang, J.: Generating robust DNN with resistance to bit-flip based adversarial weight attack. IEEE Trans. Comput. (2022)
15. Hector, K., Moëllic, P.-A., Dumont, M., Dutertre, J.-M.: A closer look at evaluating the bit-flip attack against deep neural networks. In: IEEE 28th International Symposium on On-Line Testing and Robust System Design (IOLTS), pp. 1–5 (2022)
16. Rakin, A.S., He, Z., Li, J., Yao, F., Chakrabarti, C., Fan, D.: T-BFA: targeted bit-flip adversarial weight attack. IEEE Trans. Pattern Anal. Mach. Intell. 1 (2021)
17. Colombier, B., Menu, A., Dutertre, J.M., Moellic, P.A., Rigaud, J.B., Danger, J.L.: Laser-induced single-bit faults in flash memory: instructions corruption on a 32-bit microcontroller. In: IEEE International Symposium on Hardware Oriented Security and Trust, HOST (2019)
18. Liu, Y., Cheng, Y., Gao, L., Liu, X., Zhang, Q., Song, J.: Practical evaluation of adversarial robustness via adaptive auto attack. In: Proceedings of the IEEE/CVF Conference on Computer Vision and Pattern Recognition, pp. 15105–15114 (2022)
19. Javaheripi, M., Koushanfar, F.: Hashtag: hash signatures for online detection of fault-injection attacks on deep neural networks. In: IEEE/ACM International Conference On Computer Aided Design (ICCAD), pp. 1–9 (2021)

Towards Scenario-Based Safety Validation for Autonomous Trains with Deep Generative Models

Thomas Decker[1,2(✉)], Ananta R. Bhattarai[1,3], and Michael Lebacher[1]

[1] Siemens AG, Munich, Germany
[2] Ludwig Maximilians Universität, Munich, Germany
[3] Technical University of Munich, Munich, Germany
{thomas.decker,michael.lebacher}@siemens.com, ananta.bhattarai@tum.de

Abstract. Modern AI techniques open up ever-increasing possibilities for autonomous vehicles, but how to appropriately verify the reliability of such systems remains unclear. A common approach is to conduct safety validation based on a predefined Operational Design Domain (ODD) describing specific conditions under which a system under test is required to operate properly. However, collecting sufficient realistic test cases to ensure comprehensive ODD coverage is challenging. In this paper, we report our practical experiences regarding the utility of data simulation with deep generative models for scenario-based ODD validation. We consider the specific use case of a camera-based rail-scene segmentation system designed to support autonomous train operation. We demonstrate the capabilities of semantically editing railway scenes with deep generative models to make a limited amount of test data more representative. We also show how our approach helps to analyze the degree to which a system complies with typical ODD requirements. Specifically, we focus on evaluating proper operation under different lighting and weather conditions as well as while transitioning between them.

Keywords: Operational Design Domain (ODD) · Safety Validation · Deep Generative Models · Autonomous Train · Rail-Scene Segmentation

1 Introduction

Artificial Intelligence (AI) enables technologies that can process vast amounts of data from various sources in real time and its potential for autonomous vehicles is progressively transforming the transportation industry. This is especially true for the railway domain, where driverless trains are associated with various economic and societal benefits [14]. Moreover, fully automated trains are already in service for many years in constrained and well-controlled environments such as metro lines with platform screen doors [2]. However, enabling operation in general open settings is significantly more demanding as trains are constantly required to perceive and interact with the current environment. While AI has

J. Guiochet et al. (Eds.): SAFECOMP 2023, LNCS 14181, pp. 273–281, 2023.
https://doi.org/10.1007/978-3-031-40923-3_20

shown promising capabilities in this regard [12], it is still unclear how to rigorously assure the safety of such systems from a regulatory and legal perspective [2]. A popular approach to conduct safety validation of automated vehicles is scenario-based testing [10]. Ideally, fully automated trains are expected to handle any environmental conditions and even unexpected events in a safe and robust manner, but the resulting space of possible scenarios is infeasible to test globally. As a consequence, scenario-based testing is typically performed considering a predefined Operational Design Domain (ODD) [5] which refers to all specific conditions under which a system is strictly required to behave properly including physical, geographical and regulatory constraints [4]. While there already exist proposals regarding ODD specifications for railway applications [13], collecting sufficient test cases covering all relevant aspects and systematically conducting appropriate evaluations still remains challenging. However, AI-powered data generation in the form of deep generative models has demonstrated remarkable capacities to realistically simulate complex data structures [1]. In this work, we propose a framework to systematically leverage deep generative models for scenario-based testing and summarize our practical experiences. Specifically, we create high-resolution image data with conditional Generative Adversarial Networks (cGANs) [15] allowing us to fix high-level image contents, such as the position of rails or other objects while altering different ODD-related characteristics during simulation. In this way, we can make a limited number of test cases more representative for the purpose of safety validation. We further apply our approach to test a camera-based rail-scene segmentation model that is implemented via a deep neural network [17]. Such systems enable accurate perception of the frontal environment which is crucial for safe train operation and obstacle detection [11]. We demonstrate how to perform a systematic model evaluation under natural perturbations like different lighting and weather conditions as well as while transitioning between them. Such an analysis complements classical robustness certification [6,9] and provides an additional tool to validate system safety in a comprehensive way.

2 Background

GANs Generative adversarial networks (GANs) are a popular category of deep generative models that have been extensively studied in computer vision and demonstrate remarkable capabilities to simulate realistic images and videos [7]. GANs consist of two neural networks, a generator and a discriminator, that are trained in concert to create new samples resembling the training data. Conditional GANs (cGANs) are extended versions that allow controlling the properties of generated data via additional input arguments. For images, cGANs enable semantic editing, style translation or creating images with specific details [8].

Semantic Segmentation and RailSem19 Semantic segmentation describes the task of dividing an image into semantically distinct sub-regions and assigning them a corresponding label. Deep neural networks attain state-of-the-art performances for this purpose and have also been applied in corresponding railway

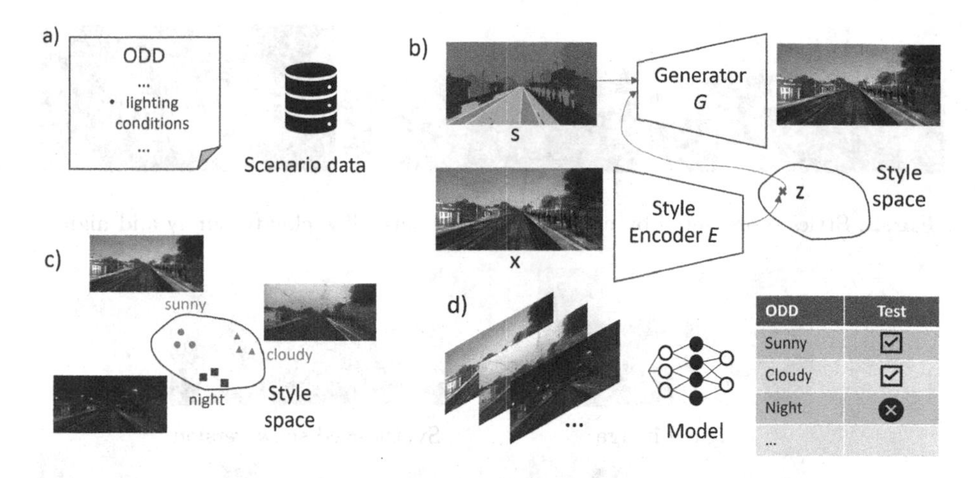

Fig. 1. Proposed approach for scenario-based ODD validation with cGANs.

applications [11]. Such models are typically trained via labeled training data comprising images and matching ground truth semantic label masks. A popular metric to evaluate segmentation performance is the Intersection over Union (IoU) score ranging from 0 to 1, where a score of 1 indicates a perfect match between ground truth and the predicted regions and 0 means no overlap. RailSem19 [16] is a publicly available dataset for semantic segmentation of railway scenes. It contains 8500 high-resolution images of real train and tram front views together with pixel-wise semantic labels corresponding to 19 different classes. The provided labels allow to distinguish a variety of different safety-critical objects such as rails, cars, humans or other on-rail vehicles. The dataset also covers various different operation environments, illuminations and weather conditions which all resemble typical components of ODD descriptions for railway applications [13].

3 Proposed Methodology

The goal of our approach is to leverage deep generative models in a systematic way to validate if an AI-powered model fully complies with specific ODD requirements given only a limited amount of test cases. Our proposed methodology is illustrated in Fig 1. As usual for safety validation, we suppose access to a predefined ODD description as well as a set of representative scenario data (a). In our use case, an ODD might among other things also require models to work well under changing lighting conditions and the extensive RailSem19 dataset provides corresponding scenarios. As a second step, we utilize the scenario data for training a cGAN to enable conditional generation of new relevant scenarios (b). In particular, we choose the pix2pixHD architecture [15] that enables the creation of high-resolution images via a generator G receiving two distinct inputs. First, the semantic structure of the desired image can be controlled by providing a semantic label mask s informing G where in the image specific objects or

Fig. 2. Styles represented by cluster centers of class Sky: cloudy, sunny and night.

Fig. 3. Synthesizing snowfall by altering features of different semantic categories.

structures should appear. Second, a separate encoder network E was designed to grasp the stylistic characteristics of different semantic categories. More precisely, E encodes low-level details of regions in x into low-dimensional feature vectors z forming a numerical style space. This setup allows us to semantically manipulate a given scenario to increase test capacities and improve ODD coverage. To do so, we first run the trained encoder on all instances in the training set and save the resulting feature vectors. Following [15], we perform clustering on these features for each semantic category to localize ODD-related concepts in the style space (c). For instance, the cluster centers for the category Sky can encode styles such as sunshine, cloudiness or night. This enables us to synthesize new realistic images with identical high-level structures determined by s but exhibiting different stylistic properties, like the same railway scene under varying lighting conditions. Moreover, we can also simulate continuous transitioning between two styles by interpolating the corresponding style encodings during image generation. To systematically test how well a model complies with an ODD requirement we can semantically manipulate available scenarios to exhibit specific properties and evaluate its effect on the model's performance (d). In the case of rail-scene segmentation this methodology allows us for instance to explicitly validate if a model works sufficiently well under sunny, cloudy or nighttime illumination.

4 Results and Experiences

Scenario Simulation To simulate test scenarios we trained the proposed cGAN on RailSem19 based on the default implementation provided by the authors [15].

Fig. 4. Top: Original image, and synthesized versions with minor artifacts. **Bottom**: Original image, and synthesized versions with significant artifacts.

Applying k-means clustering to all style encoding indeed enables us to locate k=10 distinct regions in the style space that correspond to different lighting and weather conditions. Figure 2 shows some styles represented by cluster centers for the sky class, which we refer to as prototypical cloudy, sunny, and night during our experiments. To manipulate illumination, we replace the feature vector of the sky instance in a given image with desired cluster center and synthesize a new image as outlined in Sect. 3. Changing the weather to snowfall involves manipulating several instances in the railway scene individually. Therefore, we replace the original style features of all semantic classes with their respective cluster centers that best depict the snowfall weather condition. Figure 3 shows how the features of each semantic category are altered to translate an original weather condition into snowfall. Overall, images with a significant amount of sky, vegetation, terrain or rails are of high quality, e.g. Fig 4. However, we also observed significant artifacts while encoding images with buildings, people and cars. Figure 4 shows such an example where also the simulation of snow fails. Hence, model evaluations on synthesized examples should ideally be complemented by manual human inspection on a case-by-case basis to ensure sound conclusions.

Model Evaluation For our experiments we use the PSPNet [17], which we train on RailSem19 similarly to the procedure in [16]. Out of the 8500 available images we randomly selected 7140 for training and fine-tuning leaving us only 1360 for rigorous testing. On this test set, we achieve a mean IoU of 0.65 over all classes which is comparable with the reference performance reported in [16]. To validate if the model also complies with the ODD-related requirements of proper operation under different lighting conditions and snow we applied our proposed methodology to create 4 new versions of the original test set where we modified the style of all images accordingly. The corresponding IoU scores per segmentation class are reported in Fig. 5. Our evaluation reveals that in all scenarios the model performs well with respect to the detection of tram/rail tracks or trackbeds but seems to struggle with traffic lights/signs or trucks. Also, simulating nighttime conditions seems to be particularly detrimental to the model performance, as for instance indicated by the significant IoU drops for segmenting cars, humans, construction sites or other on-rail vehicles. Since accurate detection of corresponding objects is potentially safety-critical our evaluation possibly reveals a crucial deficiency. To verify if this is indeed the case or just due to simulation artifacts we can also evaluate the model behavior on individual examples transitioning from their original style to night mode. Figure 6 displays an image with an on-rail vehicle in front of the train, that is accurately detected under the original illumination. Progressively moving to night causes the model to miss the object, but its visual appearance also becomes unnatural requiring closer inspection by a human auditor. Moreover, by evaluating other images under style transition we can demonstrate other deficiencies. In Fig. 7 the model performs well on the original image. Since it is already sunny, the synthesized sunny version is quite similar but the rail tracks are perceived as tram tracks by the model. Surprisingly, when transitioning towards night conditions the prediction suddenly turns correct at some point, although the visual appearance of the rails changes only marginally. Similarly in Fig. 8, moving to snow causes the model to suddenly confuse rail and tram tracks despite the high visual similarity of the tracks in all pictures.

	Road	Side-walk	Con-struction	Tram-track	Fence	Pole	Traffic-light	Traffic-sign	Vege-tation	Terrain	Sky	Human	Rail-track	Car	Truck	Trackbed	On-rails	Rail-raised	Rail-embed.
Original	0.59	0.58	0.73	0.74	0.52	0.56	0.40	0.43	0.86	0.67	0.95	0.62	0.89	0.75	0.34	0.74	0.73	0.68	0.55
Cloudy	0.54	0.53	0.68	0.75	0.47	0.53	0.33	0.25	0.82	0.63	0.96	0.50	0.90	0.68	0.13	0.72	0.63	0.70	0.56
Sunny	0.53	0.53	0.68	0.73	0.48	0.55	0.36	0.27	0.82	0.61	0.96	0.48	0.89	0.66	0.11	0.72	0.58	0.69	0.53
Night	0.40	0.42	0.29	0.59	0.25	0.32	0.04	0.07	0.52	0.46	0.14	0.32	0.84	0.47	0.04	0.65	0.17	0.61	0.45
Snow	0.45	0.43	0.61	0.64	0.40	0.49	0.35	0.22	0.78	0.49	0.95	0.47	0.86	0.63	0.09	0.55	0.56	0.64	0.51

Fig. 5. Class-wise IoU results of the trained segmentation model on test data.

(a) IoU: 0.89 (b) IoU: 0.90 (c) IoU: 0.0 (d) IoU: 0.0

Fig. 6. Change in IoU for an on-rail object when changing the original lighting to night-time. Huge performance decline when going from 6b to 6c.

(a) IoU: 0.95 (b) IoU: 0.01 (c) IoU: 0.06 (d) IoU: 0.91

Fig. 7. Change in IoU of rail tracks when moving from original to night illumination. Unstable performance when going from 7a to 7b and 7c to 7d.

(a) IoU: 0.93 (b) IoU: 0.93 (c) IoU: 0.26 (d) IoU: 0.10

Fig. 8. Change in IoU of the rail tracks when changing the original weather condition to snow. Despite similarity, performance drops from 8b to 8c.

5 Conclusion

In this work we report our experiences with cGANs to validate if an AI-powered model complies with typical ODD requirements, especially varying weather and lighting conditions. We intend to expand the approach to also enable the rendering of new objects such as obstacles, persons or vehicles on the rails. Comparing the simulation quality of similar generative model types, such as variants of recently popularized Diffusion Models [3] is also relevant for future work.

Acknowledgement. We acknowledge the support from the Federal Ministry for Economic Affairs and Climate Action (BMWK) via grant agreement 19I21039A.

References

1. Bond-Taylor, S., Leach, A., Long, Y., Willcocks, C.G.: Deep generative modelling: A comparative review of VAEs, GANs, normalizing flows, energy-based and autoregressive models. IEEE Trans. Pattern Anal. Mach. Intell. **44**, 7327–7347 (2021)
2. Flammini, F., De Donato, L., Fantechi, A., Vittorini, V.: A vision of intelligent train control. In: Reliability, Safety, and Security of Railway Systems. Modelling, Analysis, Verification, and Certification: 4th International Conference, RSSRail 2022, Paris, France, June 1–2, 2022, Proceedings. pp. 192–208. Springer (2022) https://doi.org/10.1007/978-3-031-05814-1_14
3. Kawar, B., Zada, S., Lang, O., Tov, O., Chang, H., Dekel, T., Mosseri, I., Irani, M.: Imagic: Text-based real image editing with diffusion models. arXiv preprint arXiv:2210.09276 (2022)
4. Koopman, P., Fratrik, F.: How many operational design domains, objects, and events? Safeai@ aaai 4 (2019)
5. Koopman, P., Wagner, M.: Toward a framework for highly automated vehicle safety validation. SAE Technical Paper, Tech. Rep (2018)
6. Li, L., Xie, T., Li, B.: Sok: Certified robustness for deep neural networks. In: 44th IEEE Symposium on Security and Privacy, SP 2023, San Francisco, CA, USA, pp. 22–26 (2023). IEEE (2023)
7. Liu, M.Y., Huang, X., Yu, J., Wang, T.C., Mallya, A.: Generative adversarial networks for image and video synthesis: algorithms and applications. Proc. IEEE **109**(5), 839–862 (2021)
8. Pang, Y., Lin, J., Qin, T., Chen, Z.: Image-to-image translation: methods and applications. IEEE Trans. Multimedia **24**, 3859–3881 (2021)
9. Paterson, C., Wu, H., Grese, J., Calinescu, R., Păsăreanu, C.S., Barrett, C.: DeepCert: Verification of Contextually Relevant Robustness for Neural Network Image Classifiers. In: Habli, I., Sujan, M., Bitsch, F. (eds.) SAFECOMP 2021. LNCS, vol. 12852, pp. 3–17. Springer, Cham (2021). https://doi.org/10.1007/978-3-030-83903-1_5
10. Riedmaier, S., Ponn, T., Ludwig, D., Schick, B., Diermeyer, F.: Survey on scenario-based safety assessment of automated vehicles. IEEE Access **8**, 87456–87477 (2020)
11. Ristić-Durrant, D., Franke, M., Michels, K.: A review of vision-based on-board obstacle detection and distance estimation in railways. Sensors **21**(10), 3452 (2021)
12. Tang, R., De Donato, L., Besinović, N., Flammini, F., Goverde, R.M., Lin, Z., Liu, R., Tang, T., Vittorini, V., Wang, Z.: A literature review of artificial intelligence applications in railway systems. Transp. Res. Part C Emerging Technol. **140**, 103679 (2022)
13. Tonk, A., Boussif, A., Beugin, J., Collart-Dutilleul, S.: Towards a specified operational design domain for a safe remote driving of trains. In: Proceedings of the 31st European Safety and Reliability Conference, Angers, France. pp. 19–23 (2021)
14. Trentesaux, D., et al.: The autonomous train. In: 2018 13th Annual Conference on System of Systems Engineering (SoSE). pp. 514–520. IEEE (2018)
15. Wang, T.C., Liu, M.Y., Zhu, J.Y., Tao, A., Kautz, J., Catanzaro, B.: High-resolution image synthesis and semantic manipulation with conditional GANs. In: Proceedings of the IEEE conference on computer vision and pattern recognition. pp. 8798–8807 (2018)

16. Zendel, O., Murschitz, M., Zeilinger, M., Steininger, D., Abbasi, S., Beleznai, C.: Railsem19: A dataset for semantic rail scene understanding. In: Proceedings of the IEEE/CVF Conference on Computer Vision and Pattern Recognition Workshops. pp. 0–0 (2019)
17. Zhao, H., Shi, J., Qi, X., Wang, X., Jia, J.: Pyramid scene parsing network. In: Proceedings of the IEEE Conference on Computer Vision and Pattern Recognition. pp. 2881–2890 (2017)

Author Index

J. Guiochet et al. (Eds.): SAFECOMP 2023, LNCS 14181, pp. 283–284, 2023.
https://doi.org/10.1007/978-3-031-40923-3

Printed in the United States
by Baker & Taylor Publisher Services